The Chemical Theatre

The Chemical Theatre

Charles Nicholl

· Routledge & Kegan Paul ·
LONDON, BOSTON
AND HENLEY

First published in 1980
by Routledge & Kegan Paul Ltd
39 Store Street,
London WC1E 7DD,
9 Park Street,
Boston, Mass. 02108, USA and
Broadway House, Newtown Road,
Henley-on-Thames, Oxon RG9 1EN

Set in 10/12 Linotron 202 Sabon by
Input Typesetting Ltd,
London
and printed in Great Britain by
Unwin Brothers Ltd,
Old Woking, Surrey

© Charles Nicholl 1980
No part of this book may be reproduced in
any form without permission from the
publisher, except for the quotation of brief
passages in criticism

British Library Cataloguing in Publication Data

Nicholl, Charles
   The chemical theatre.
   1. Alchemy in literature
   2. English drama – Early
   modern and Elizabethan, 1500–1600 –
   History and criticism
   3. English drama – 17th century –
   History and criticism
   I. Title
   822'.3'093   PR658.A/   80–40712

ISBN 0 7100 0515 6

Contents

1 The Hunting of the Alchemist   *page* 1

    The modern image · 1
    The satirical image · 7
    'A multis amatur Alchymia' · 14

2 Three English Alchemical Texts   23

    The Mirror of Alchimy · 23
    The Compound of Alchymy · 32
    The Hieroglyphicall Unit · 41

3 The Rise of 'Chymicall Physick'   55

    Paracelsus · 55
    Metal-brewing Paracelsians · 62
    Some literary reactions · 69

4 A New Chemical Light: Sources and Reflections   81

    An alchemical renaissance · 81
    The cloudy voice · 90
    'Of wit and of Alcumie' · 97

5 Alchemical Patterns in Jonson and Donne   107

    Eastward Hoe: a chemical morality · 107
    Aurum palpabile · 112
    Love's alchemy · 119

6 Alchemical Bearings on *King Lear*   136

    The *longissima via* · 136
    The Wheel of Fire · 144
    An alchemical reconstruction · 152

7 The Transmutation of King Lear — 154

The King · 155
The Dragon · 160
The daughter · 165
The mortification · 172
The Fool · 176
The fire · 185
The blackness · 191
The dew · 201
The Stone · 210
Coda: Thomas Tymme's chymicall pathology · 218

8 Shakespeare's Chemical Theatre — 225

The Secret Cave · 225
Magister ludi · 236

Notes — 240

Bibliography — 278

Index — 285

Illustrations

1. Lady Alchymia. From Conrad Gesner, *The Practise of the New and Old Phisicke* (1599)   page 14
2. Title-page from Roger Bacon, *The Mirror of Alchimy* (1597)   24
3. Title-page from George Ripley, *The Compound of Alchymy* (1591)   33
4. From Thomas Tymme's unpublished introduction to the *Monas Hieroglyphica* (c. 1602)   42
5. Paracelsus at age 45. Engraving by Augustin Hirschvogel (1538)   55
6. Chymist at work. From Conrad Gesner, *The Practise of the New and Old Phisicke* (1599)   63
7. Landscape with *lumen chemicum*. From Heinrich Khunrath, *Amphitheatrum Sapientiae Aeternae* (1609)   82
8. From anon., *Rosarium Philosophorum* (1550)   120
9. Killing the king. From Daniel Stolcius, *Viridarium Chymicum* (1624)   137
10. From anon., *Speculum Veritatis* (MS, seventeenth century)   144
11. From Lambspringk, *De Lapide Philosophorum Emblemata* (1678)   154
12. From Petrus Bonus, *Pretiosa Margarita Novella* (1546)   159
13. From Lambspringk, *De Lapide Philosophorum Emblemata* (1678)   161
14. From Michael Maier, *Atalanta Fugiens* (1618)   166
15. From Michael Maier, *Atalanta Fugiens* (1618)   173
16. From Daniel Stolcius, *Viridarium Chymicum* (1624)   176
17. From Michael Maier, *Atalanta Fugiens* (1618)   186
18. From Daniel Stolcius, *Viridarium Chymicum* (1624)   192
19. From Robert Fludd, *Mosaicall Philosophy* (1659)   196
20. From Leonhardt Thurneisser, *Quinta Essentia* (1574)   202
21. From anon., *Turba Philosophorum* (MS, sixteenth century)   202
22. From anon., *De Alchimia* (MS, sixteenth century)   208
23. From anon., *Rosarium Philosophorum* (1550)   210
24. From Daniel Stolcius, *Viridarium Chymicum* (1624)   213
25. Creede's device. From the title-page of Thomas Tymme, *The Practise of Chymicall Physicke* (1605) [enlarged]   219

26 From anon., *Geheime Figuren der Rosenkreuzer* (1785) 226
27 From Stefan Michelspacher, *Cabala* (1616) 229
28 Entrance to the amphitheatre. From Heinrich Khunrath, *Amphitheatrum Sapientiae Aeternae* (1609) 231
29 From Michael Maier, *Atalanta Fugiens* (1618) 234
30 From Johann Daniel Mylius, *Philosophia Reformata* (1622) 234
31 From Georgius Anrach, *Pretiosissimum Donum Dei* (MS, seventeenth century) 235
32 'Truly a magician.' Engraving of Shakespeare by Charles Turner (1824) 237

Acknowledgments

Illustrations are reproduced by kind permission of the following institutions: the Trustees of the British Museum, nos 1, 2, 3, 6, 7, 9(a) and (b), 11, 12(a), (b) and (c), 13, 14, 15, 16, 17(a) and (b), 18, 19, 20, 24, 25, 27, 28, 29, 30(a) and (b); the Bodleian Library, Oxford, no. 4; the Jung Collection, Küsnacht, by permission of Mr Hans Jung, nos 8, 23; the Biblioteca Vaticana, no. 10; the Bibliothèque Nationale, Paris, no. 21; the Bibliotheek de Rijksuniversiteit, Leiden, nos 22(a) and (b); the Bibliothèque d'Arsenal, Paris, no. 31(a) and (b).

· 1 ·

The Hunting of the Alchemist

The modern image

It has been said that the past is a foreign country, with its own peculiar language and customs. If so, the alchemist must be reckoned one of its oddest inhabitants. Today we understand alchemy chiefly as the pursuit of an impossible mirage: the transmutation of base metals into gold. Faced with the enigmas of the alchemist, we rest easy with his apparent failures. Few would take issue with the sceptical and 'modern' words of Francis Bacon, writing in 1605: 'Whosoever shall entertaine high and vapourous imaginations, insteede of a laborious and sober enquirie of truth, shall beget hopes and beliefes of strange and impossible shapes.'[1]

The modern image of alchemy tends in two directions: one scientific, the other magical. The first defines alchemy simply and chronologically as early chemistry. It is seen as a murky medieval half-light, out of which modern chemistry gradually began to emerge during the seventeenth century, notably with the publication of Robert Boyle's *Sceptical Chymist* in 1661. A trifle grudgingly, the alchemists are acknowledged as the 'fathers of laboratory technique',[2] the discoverers of certain basic chemical data, stumbled upon as if by accident. Bacon's *Advancement of Learning* (1605) again provides a prototype for this reading of the alchemist as ancestral and 'accidental' scientist:

> Surely to Alcumy this right is due, that it may be compared to the Husbandman whereof Aesope makes the Fable; that, when he died, told his sonnes that he had left unto them gold buried under ground in his Vineyard; and they digged over all the ground, and gold found they none; but by reason of their stirring and digging the mold about the rootes of their vines, they had a great Vintage the yeare following: so assuredly the search and stirre to make gold hath brought to light a great number of good and fruitfull inventions and experiments, as well for the disclosing of Nature as for the use of mans life.[3]

At about the same time, John Donne expressed a similar idea in his poem, 'Loves Alchymie':

> no chymique yet th'Elixar got,
> But glorifies his pregnant pot,
> If by the way to him befall
> Some odiferous thing, or medicinall.[4]

The distinction drawn by Bacon and Donne, between the alchemist's illusory quest and its real and fruitful by-products, has become standard in any evaluation of alchemy as part of a continuous scientific tradition. The mainstream of alchemical activity is almost always derided. Ferdinand Hoefer, for instance, a leading nineteenth-century historian of chemistry, describes the alchemical *Testamentum* of Raymond Lully[5] as 'a tissue of generalities' and 'devoid of sense' and lacking in 'any clear and positive experiments'.[6] In similar vein, Maxson Stillman speaks of the seventeenth century in terms of the 'growing tendency towards real and practical aims in chemistry' and the 'development of chemistry towards a real science'.[7] Generalizing, obscure, impractical; above all, unreal – as a science, it seems, alchemy is very unscientific.

This is where the other half of our modern image of alchemy takes over. For alchemy is popularly defined as one of the 'occult arts', and the popular stereotype of the alchemist is scarcely distinguishable from the pantomime magician with conical hat and bubbling cauldron. To us, the alchemist's avowed quest for certain miraculous substances – the Philosophers' Stone which converts all to gold, the Elixir Vitae which confers immortality – belongs to the realm of magic rather than science. About magic we have a choice. As with gods and ghosts – but not, for instance, with Newton's Law of Gravitation – we are invited to believe or disbelieve. Bacon provides once more the voice of the sceptic, castigating the 'superstitious conceits' and 'frivolous experiments' which typify 'this degenerate Naturall Magick, Alchimie, Astrologie and the like'.[8]

These are the twin parts of our conventional notion of alchemy: part primitive science, part impossible magic. We see how neat and flush the two fit together in our minds. Where science leaves off, magic (an 'art') begins. What is not possible in the real world of science is consigned to the dream world – the pantomime – of magic. There is a clear frontier, and if the alchemist is something of a riddle, it is perhaps because he crosses and recrosses this frontier as if it did not exist.

It is no accident that Bacon's *Advancement of Learning* provides a prototype for modern attitudes to alchemy. For Bacon – *bucinator novi temporis*, the trumpeter of a new age – was an early pioneer of modern 'scientific method'. His intent was purgative, to disperse these '*idola*' and 'phantoms of the mind' which, in his diagnosis, prevented a true understanding of Nature. He typified that process of separation central to the scientific development of the seventeenth century – a disentangling of physical properties from metaphysical speculation; of empirical experiment from pre-conceived cosmological patterns; of, in Bacon's own words, 'laborious and sober enquirie' from 'high and vaporous imaginations'. In this separation originate our own distinctions, that implicit frontier between science and magic. Our divided definition of alchemy has, in this sense, an historical origin. If we are to rediscover the strange world of the alchemist, to find in it something more than mere folly and fumbling, we must get back behind that Baconian partition to a time when 'science' and 'magic' were one and the same enterprise.

*   *   *

We call the dawning science of the seventeenth century 'mechanistic' not primarily because of the machines it invented, but because of its entire perception of Nature as a machine – inanimate, self-running, morally neutral. In his *Principia Philosophiae* (1644), Descartes wrote:

> The only difference I can see between machines and natural objects is that the workings of machines are mostly carried out by apparatus large enough to be readily perceptible by the senses . . . whereas natural processes almost always depend on parts so small that they utterly elude our senses.[9]

This is Descartes's automatic Nature. The role of the scientist is to discern that invisible 'apparatus', the moving parts of the universal machine. The Cartesian God is all but redundant: in Pascal's sarcastic summary, He has nothing to do but *'donner une chiquenaude pour mettre le monde en mouvement'* ('give a flick of the thumb to set the world in motion').[10]

We recognize in this the basis of our own 'scientific' view of Nature, a Nature operating by mechanical laws rather than divine inspiration. But this inanimate world is quite opposite and alien to the world of the alchemist, who saw the *anima mundi* – the world-soul – infused into everything. Nature, wrote Paracelsus,

> is not visible, though it operates visibly; for it is simply a volatile spirit, fulfilling its office in bodies, and animated by the universal spirit – the divine breath, the central and universal fire, which vivifies all things that exist.[11]

Among the names the alchemist gave to this spirit, one of the most common is 'quintessence'. Hidden and cradled in the physical world of the four elements (earth, water, fire and air) was the spiritual fifth element – the *Quinta Essentia*, or quintessence. The 'privy quintessence', wrote the English alchemist William Bloomfield in 1557,

> is not in sight, but resteth invisible;
> Till it be forced out of Chaos darke,
> Where he remaineth ever indivisible....
> This Chaos darke the Mettalls I do call,
> Because as in Prison it resteth them within;
> The seacret of Nature they keepe in thrall,
> Which by a meane we do warily out-twyne.[12]

When vapours rose from a heated liquid (distillation) or corroded solid (sublimation), the alchemist saw a spirit arising liberated from the embrace, the dark prison, of matter. The volatile vapours escaped upwards: this was the spirit's winged ascent (Latin *volatilis*=winged or flying) towards the 'heaven' of the circular microcosmic vessel. Within the animistic world of the alchemist, chemical procedure was a method of revelation, the 'meane' to coax out a hidden and spiritual 'seacret of Nature'. Scientific means, magical ends.

Many of the alchemist's operations were concerned with dissolving and disintegrating, breaking down the physique of substance to free the 'divine breath' or quintessential spirit within it. To him, this was a work of redemption: he was healing the corruption and chaos of matter. Already we glimpse something of the wider metaphorical meaning of 'transmuting base metals into gold'. This urge to penetrate matter also suggests something of the alchemist's relationship to the substances he studied. He was not the detached experimental scientist, the quantifying observer:[13] he wished to enter in and participate. Thus the title of a famous seventeenth-century text – *Introitus Apertus ad Occlusum Regis Palatium* [14] – invites us to enter 'the shut palace of the king', the 'king' being gold. Similarly, Francis Anthonie, writing in 1616, calls gold a 'most princely Pallace, drawing the beholders of his naturall luster to search what excellent Iewels bee contained within that outward wall of so rich promises'.[15] Matter beckoned the alchemist into its depths, promising marvellous relevations. '*Visita Interiora Terrae Rectificando Invenies Occultum Lapidem.*'[16] This not only tells us, 'visit the interior of the earth and by rectifying (i.e., purifying) you will find the hidden stone'. It is also an acrostic code which reads 'Vitriol'. Green vitriol (ferrous sulphate) was the substance from which the alchemist prepared his oil of vitriol (sulphuric acid), whose caustic corrosive properties were a means to dissolve the 'earthy' solidity of substance. Vitriol was a chemical vehicle for this 'visit' to the interior of the earth. Again, a scientific process harnessed to an occult intent.

To penetrate the 'shut palace' of matter, much was required of the alchemist. He must himself be cleansed and purified: 'Know that thou canst not have this science unless thou shalt ... wipe away all corruption from thy heart.'[17] Those that did not, warned the mysterious Basil

Valentine in his *Triumphant Chariot of Antimony* (1604), 'God hath set a cloud before their eyes, and such being blinded, may not know those secrets that lie hid under a Metallick Form.'[18] The alchemist approached his experiments with ritualistic reverence. Such was his involvement, that in some senses his journey into the mysterious interior of matter was also a venture into his own self. Gerhard Dorn, the late sixteenth-century Belgian alchemist, asserted: 'Thou wilt never make from other things the one that thou seekest, except there first be made one thing of thyself.'[19] The alchemist's probing, purifying, transforming operations became a kind of mirror for psychic discoveries and changes. The 'divine breath' and 'hidden stone' were to be found in man as well as matter. So Dorn commands: 'Transmute yourselves from dead stones into living philosophical stones.'[20]

This *participation mystique*, so alien to modern science, has made alchemy a fertile area for the psychological and religious interpretations of Jung. For him, alchemy was 'in essence, chemical research work into which there entered, by way of projection, an admixture of unconscious psychic material'.[21] He also found in it something of a 'missing link', seeing its wealth of exotic symbolism as providing a vehicle for psychic and religious archetypes,[22] and thus fulfilling a role which the post-medieval Church increasingly deserted:

> Alchemy is rather like an undercurrent to the Christianity that ruled on the surface. It is to this surface as the dream is to consciousness.... Whereas in the Church the increasing differentiation of ritual and dogma alienated consciousness from its natural roots in the unconscious, alchemy and astrology were ceaselessly engaged in preserving the bridge to nature, i.e. to the unconscious psyche, from decay.[23]

This symbolic 'bridge to nature' did not only lead inwards, linking the psychic fortunes of the alchemist with the chemical changes in his 'Hermetic vessel'. It also extended outwards, linking specific substance with the entirety of physical Nature. The alchemist's matter, his Raw Stuff, was a 'piece of earth'. The quintessence within it was a manifestation of the universal *anima mundi*. In the extreme of this analogy, the alchemist's operations reiterated the Creation itself:

> Moses, that ancient Theologue, describing and expressing the most wonderful Architecture of this great world, tels us that the Spirit of God moved upon the water; which was an indigested Chaos or mass created before by God, with confused Earth in mixture: yet by his Halchymicall Extraction, Separation, Sublimation, and Coniunction, so ordered and conioyned again.[24]

The Creation was thus a piece of 'Divine Halchymie' and alchemy a piece of divine creation. There is more than a touch of Faustian danger in the alchemist's larger aspirations as creator and redeemer of matter.

*   *   *

The revelations of spirit, the journey into matter, the capacity of chemical experiment to mirror both psychic and cosmic meanings – I have simplified and run together certain strands of alchemical philosophy and will return to them in greater detail in the following chapters. The point here is how these strange magico-mystical conceptions are all contained and encoded in chemical procedures, recognizable in themselves as 'scientific'. Our divided image of the alchemist is wrong to separate the groping scientist from the grandiose magician. For the alchemist, scientific procedure was a mounting of symbolic chemical events, and his vessel an arena in which invisible and magical potentialities were visibly enacted. It is in this sense that I shall come to define alchemy as 'chemical theatre'.

Our hunting of the alchemist – our effort to recover the essential unity and meaning of his endeavours – takes us back to the years around 1600. All the authors I have briefly cited – Paracelsus, Bloomfield, Anthonie, Valentine, Dorn – were working and writing in the sixteenth and early seventeenth centuries. The words about the 'Divine Halchymie' of the Creation were written by an obscure Suffolk parson called Thomas Tymme. His work, *The Practise of Chymicall and Hermeticall Physicke*, was published in London in 1605, the same year as Bacon's *Advancement*. It was precisely at the time when modern science began to formulate its priorities – secular, empirical, mechanistic – that alchemical philosophy, which is none of these things, was at its most influential and sophisticated. Certainly alchemy was doomed, and we shall see how the portmanteau term 'alchemist' had begun by this time to be divided into practical 'chymist' and esoteric 'philosopher'. But for a time, the very threat of the 'new Philosophy' that 'calls all in doubt'[25] urged on the alchemist a special self-awareness and ingenuity. As Jung notes,

> The process of fission which separated the φυσικά (physika) from the μυστικά (mystika) set in at the end of the sixteenth century and produced a quite fantastic species of literature whose authors were, at least to some extent, conscious of the psychic nature of their 'alchemystical' transmutations.[26]

It is with this 'fantastic species of literature' and its sphere of influence that this book is concerned. 1605 was not only the year of Bacon's *Advancement* and Tymme's *Chymicall Physicke*. It was also the year that Shakespeare was writing *King Lear* and Ben Jonson's *Eastward Hoe* was on stage. Alchemy was part of the air they breathed, and like everything else they breathed, it found its way into the great and inclusive world of their drama. To show this is the purpose of my book.

It would be pointless and sentimental to regret the scientific revolution interposed between us and alchemy. Yet perhaps, after so much 'laborious and sober enquirie', we can return to the 'high and vapourous imaginations' of the alchemist and find in them something substantial, even something precious. Bacon spoke of putting Nature 'on the rack' and forcing it to reveal its secrets. By contrast, Elias Ashmole offered his great collection of alchemical poetry with the words: 'There you may meet with the Genii of our Hermetique Philosophers, learne the language in which they woo'd and courted Dame Nature.'[27] We might ask how much more of our torture Nature can take, and look on the alchemist who courted her, who ceaselessly preserved the 'bridge to nature', as something more than a quaint remnant from an alien past. Besides, the past may be a foreign country, but it is also a homeland from which we are all emigrés.

The satirical image

We have begun to rescue the alchemist from that murk of delusion and unreality to which a cursory modern attitude has consigned him, and at least to suggest that his grandiose claims might make some sense if one sees them as part of a now-forgotten philosophy of Nature. The influence of this philosophy on the Elizabethan and Jacobean mind has been greatly underrated. But to assess this influence, we must also recognize what opposed it: a tremendous resistance and hostility towards alchemy. Many Elizabethans and Jacobeans were quite as sceptical as we are about the possibility of transmuting base metals into gold. 'Such a beggerly science it is,' says the alchemist's boy in John Lyly's *Gallathea* (1585), 'that the ende is to have neyther gold, wit, nor honestie.'[28] The satirical image of the alchemist as deceiver and self-deceiver was a staple of the literature, epitomized by Dr Subtle, the 'chymicall cousoner' of Ben Jonson's *The Alchemist* (1610).

The paraphernalia of 'gold-making' – though in itself a distortion of the true alchemist's endeavour – made alchemy a fertile area for fraud. The smoke-seller, the puffer, the multiplier, the drone, the cook (especially 'Geber's cook'), the collier: these are some of the names given to the alchemical confidence-trickster. This was the 'wandering and roguing Alchymist',[29] the 'unskilfuller cousning kind of Alchumist' whose sole art was 'to picke mens purses',[30] the 'wandering Alchymist that promiseth golden mountains'.[31] 'Smoke' is a word that comes up again and again in describing him. The smoke of the furnace that clings to his beard and clothes; the smoke of chemical mishap when his retorts explode and all is 'flowne *in fumo*';[32] the smoke of the promised end, when the dream of endless gold vapours into nothing.

In framing their warnings and satires against the chemical con-man, Elizabethan and Jacobean writers drew on a common stereotype. Four earlier works were staple sources: Petrarch's *Dialogue de Alchimia* (1366), Chaucer's *Chanouns Yemannes Tale* (c. 1390), Erasmus' *Colloquium Alcumista* (1519) and the nineteenth chapter of Cornelius Agrippa's *De Vanitate et Incertitudine Scientiarum* (1530). The visual image of the rogue alchemist is remarkably consistent. He is always a ragged man. Chaucer's Host is much offended by the alchemical Canon's 'sluttissh' appearance, his 'baudy' and 'totore' gown.[33] In the *Ordinall of Alchimy* (1477), Thomas Norton warns against this same tattered itinerant, the 'fals man' who

> walketh from Towne to Towne,
> For the most parte in a threed-bare Gowne,
> Ever searching with diligent awaite
> To winn his praye with some fals deceit.[34]

He could also be recognized, according to Canon's yeoman, by his 'wan and ... leden hewe' and 'blered' eyes from working at the furnace, and from the 'smel of brymstoon' clinging to his clothes.[35] The features of this medieval portrait are still recognizable in Thomas Lodge's grisly account in *The Anatomie of Alchymie* (1595). His beggarly alchemists are

> like a crased clocke, that cannot chime:
> Olde, clothles, meatles, smelling brimstone still,
> Besmeer'd with cole-dust, from their furnace brought,
> Plagu'd with the palsie, (letchers common ill)
> By temp'ring of quicksilver quickly cought:
> Their riches are the droppings of their nose,
> Where els beside, the slaves are brought so low,
> That for three farthings, they will beg and glose,
> And sel their soules, & teach what ere they know.[36]

This is clearly an invective stereotype, but there was no doubt some reality behind it. Reginald Scot, for instance, quotes Chaucer's description of the ragged smoke-seller in his *Discoverie of Witchcraft* (1584), then adds that 'experience verifieth his assertion, that they looke ill-favouredlie, & are alwaies beggerlie attired'.[37] Scot devotes the fourteenth book of the *Discoverie* to the wiles of 'Alcumistrye' on the premise that 'in the bowels hereof dooth both witchcraft and coniuration lie hidden'.[38] Certainly by Shakespearean times, the rogue alchemist was part of a whole back-alley fraternity — a 'magical underworld' of conjurors, charmers, blessers, figure-casters, almanac-men, mountebanks, quacksalvers, necromancers and cunning-men. Their real existence is well testified, both in catch-penny pamphlets describing the 'subtill practise' of Judith Philips (1595), the 'lewd cousnages' of John and Alice West (1613) and the 'notorious life' of John Lambe (1628), as well as in more general exposés, as Scot's *Discoverie*, Samuel Harsnett's *Declaration of Egregious Popishe Impostures* (1603) and John Melton's *Astrologaster* (1620). Some arose from this underworld to attain respect and prosperity: the astrologer-physician Simon Forman was worth £1,200 when he died in 1611.[39] Others found powerful patrons, as John Lambe did in the Duke of Buckingham, but this was not always enough: leaving the Fortune theatre one summer afternoon in 1628, Lambe was beaten to death by a mob of apprentices denouncing him as the 'Duke's devil'.[40] Such was the status – half-feared, half-derided – of the shady magician. This was a brush that tarred many genuine exponents of the 'magico-scientific' tradition – brilliant men like John Dee and Thomas Hariot, both of them publicly disparaged as 'conjurors'.[41] So, too, the genuine alchemist was at pains to dissociate himself from 'that wicked swarm of smoke-sellers, whose delight is to cheat'.[42] Two seventeenth-century alchemists devoted whole tracts to castigating them: Michael Maier's *Examen Fucorum Pseudo-Chymicorum* (1616) and Gabriel Plattes's *Caveat for Alchymists* (1655).

*   *   *

As Dr Subtle impressively intones that 'bright Sol is in his robe', and 'we have a med'cine of the triple soule', and the work has 'past the Philosophers Wheele, in the lent heat of Athanor, and's become Sulphur o' Nature', the sceptical Surly mutters aside: 'What a brave language here is! next to canting!'[43] A tantalizing alchemical rhetoric – the 'woords of art and devises to bleare mens eies'[44] that Scot warns against – was the smoke-seller's first weapon. And if he was an adept, it was not in the mysterious art of transmutation, but in the conjurer's art of sleight-of-hand. Hugh Plat, in *The Jewel House of Art and Nature* (1594), warns:

> let every man that is besotted in this art . . . take heed also of all false and double bottoms in crusibles, of all hollow wands or rods of iron, wherewith some of these varlets do use to stir the mettal and the medicine together: of all Amalgames and powders, wherein any gold or silver shal be craftily conveyed; of Sol and Luna [i.e., gold and silver] first rubified, and then projection made on it, as if it were on Venus [i.e., copper] herself: but specially of a false back to the Chimney or furnace, having a loose brick or stone closely joynted, that may be taken away in another room, by a

false Simon that attendeth on the Alchymists hem, or some other like watch-word, who after the medicine and the Mercury put together in the Crusible, entertaineth Balbinus with a walk, and with the volubility of his tongue, until his confederate might have leisure to convey some gold or silver into the melting pot.[45]

This is a fair compendium of the ploys for replacing the base metal with the desired precious one. Dr Subtle has the trick of 'coz'ning with a hollow coal', a technique earlier perfected by Chaucer's rogue Canon.[46] The fire is heated, the crucible filled with quicksilver and the tincture that 'shal make al good' (redeem all metals from baseness) projected onto it. Then, while stoking the fire, the Canon places over the crucible his doctored piece of coal,

> In which ful subtilly was maad an hole,
> And therinne put was of silver lemaille
> An ounce, and stopped was, withouten faille,
> This hole with wex.[47]

The coal burns, the wax melts, the silver filings flow into the crucible and – eureka – the mercury is transmuted into silver. After a couple more tricks, the apparently miraculous powder is sold for £40 and the Canon makes himself scarce. 'Lo, thus byjaped and bigiled' was the gullible victim.

The smoke-seller's simulated transmutation was not the only sharp practice associated with alchemy. Counterfeiting, adulterating and coining were also popular and profitable forms of crooked chemistry. Frank Quicksilver, the 'runagate' goldsmith's apprentice in *Eastward Hoe* (1605), promises:

> I will blanche Copper so cunningly, that it shall endure all proofes but the Test: it shall endure malleation, it shall have the ponderositie of Luna, and the tenacitie of Luna, by no meanes friable.[48]

He means that he will whiten copper so that it counterfeits silver – Luna (moon) is the planetary name for silver, as Sol (sun) is for gold. His false silver will, he claims, pass all the assayer's examinations (hammering, weighing, etc.) except for cupellation, 'the Test'.[49] He goes on to explain the process for 'blanching' copper:

> Take Arsnick, otherwise called Realga (which indeede is plaine ratsbane); sublime 'hem three or foure times, then take the sublimate of this Realga and put 'hem into a glasse, into *chymia*, and let 'hem have convenient decoction naturall, foure and twentie houres, and he will become perfectly fixt; then take this fixed powder, and proiect him upon wel-purged Copper, *et habebis magisterium*.

Quicksilver's argentiferous agent is arsenic disulphide ('realgar'), purified by sublimation, by *chymia* (described in Dorn's *Dictionarium Paracelsi* (1583) as 'the art of separating the pure from the impure, and making essences'[50]) and decoction (cooking or boiling down, again to extract essences). He rounds off his recipe with an alchemical flourish – 'and thus you have the master-work' – an ironic echo suggesting that the alchemist's vaunted transmutations were no more than cunning counterfeits. *Eastward Hoe* was a collaboration between Jonson, Chapman and Marston (and earned the first two a spell in prison), but this passage is certainly Jonson's, a foretaste of his virtuoso chemical display in *The Alchemist*. I suspect that Jonson's source was Paracelsus' *De Natura Rerum* (1573), which describes how to 'sublimate copper and reduce it by fixed arsenic to a white substance, as white as Luna'.[51]

Frank Quicksilver has another trick:

> *Quicksilver*: Ile take you off twelve pence from every Angell, with a kind of *aqua fortis*, and never deface any part of the image.
> *Petronel*: But then it will want weight?
> *Quicksilver*: You shall restore that thus: take your *sal achyme*, prepar'd, and your distild urine, and let your Angels lie in it but foure and twenty howres, and they shall have their perfect weight againe.[52]

An angel was a gold coin worth 10 shillings, so called for its device of the archangel Michael slaying the dragon. Quicksilver offers a technique for creaming off some of the gold with a caustic solvent (*aqua fortis*, strong water, is usually nitric acid) and then restoring the coin to its original weight with some unnamed salt (*achyme* simply means purified, literally 'without juice'). This was a popular practice. Dr Subtle is threatened with the hangman for 'laundring gold and barbing it'.[53] 'Laundering' refers to the same caustic bath that Quicksilver recommends, 'barbing' to the clipping or shaving of a coin's edge. It was not until the Restoration that 'mill coinage' was introduced, coins thick enough to have some legend or grain round the edge, thus protecting them from the gold-barber. Having creamed or shaved off your percentage, the adulterated coins, or 'light gold', were easily disposed of, as Quicksilver explains:

> The gallants fall to play; I carry light golde with me; the gallants call, 'Coozen Francke, some golde for silver!' I change, gaine by it; the gallants loose the gold, and then call, 'Coozen Francke, lend me some silver!'[54]

Blanched copper, laundered angels, light gold – this was a time when coinage had a real rather than symbolic value, and when gold and silver from the New World was glutting Europe: the stock of precious metals in Europe is estimated to have trebled during the sixteenth century, and perhaps as much as £3 million worth of bullion entered England.[55] It was a coiner's paradise, and the State Papers of the time are full of shadowy figures accused of the practices Quicksilver is so expert in. In 1570, for instance, an itinerant bookseller named William Bedo was in trouble at the Tower for having 'lightened' and 'inbased' £10 worth of silver coins.[56] He had acquired 'the art how to demynysshe any silver coigne in the wayght' from one John Buckley, an Oxford student and reputed 'cunning-man'. He in turn had taken it, we learn, from 'a booke made by John Baptista Neappolitanus, who wreyteth of naturall magyke'. This would be G. B. della Porta's *Magia Naturalis* (1558), a popular compendium of sciences practical and esoteric. Its fifth book, treating of alchemy and 'how metals may be altered and transformed, one into another',[57] contained a wealth of information for the would-be coiner. Knowledge was a saleable commodity, then as now, and for the price of a 'tablett of gold', Buckley had translated della Porta's recipes from the Latin for Bedo.

Buckley the cunning-man, scratching a living with 'cristall stone' and esoteric tomes on 'estromancy, gematry and alcamistrye';[58] Bedo the coiner, in his lodgings 'by the waye', creaming off silver with his magic recipe and 'a little fyre in a fyre shovell'; these are the shadowy proto-types behind Dr Subtle and Frank Quicksilver, the real low-life scenes of Elizabethan alchemy, an odd, secretive, needy world. Another such figure was John Poole, imprisoned at Newgate 'upon suspicon of coiginge', and getting into deeper trouble while there for certain 'trai-torous speeches' reported to the authorities.[59] This blend of crooked chemistry and subversive sentiment obviously appealed to the man Poole met in Newgate in September 1589, Christopher Marlowe. Marlowe spent a fortnight in 'the Stink', after a street fray in which an inn-keeper's son was killed. And when, four years later, the Privy Council received evidence about his various unorthodoxies – religious, political, sexual – among them was the claim

> that he had as good Right to coine as the Queene of England, and that he was acquainted with one Poole, a prisoner in Newgate, who hath greate skill in mixture of mettals, and having learned some thinges of him, he ment through help of a cunninge stamp maker to coin ffrench crownes, pistoletes and English shillinges.[60]

The conversations of Poole and Marlowe in Newgate prison are sadly unrecorded. They are a kind of lost text, a scurrilous counterpart to the moments of alchemical poetry in Marlowe – the Elixir we hear of in *Tamburlaine* (1587), for instance,

> which a cunning alchemist
> Distilled from the purest balsamum

> And simplest extracts of all minerals....
> The essential form of marble stone,
> Temper'd by science metaphysical
> And spells of magic.[61]

* * *

To the serious alchemist, these various smoke-sellers, counterfeiters and coiners were simply those 'lewd persons' through whom 'the worthy science of Alchymie is come in such disdain'.[62] To the sceptic, however, all alchemy was imposture and coining was its natural nadir. Thus Thomas Lodge:

> In briefe, when other subtill shifts doe faile,
> They fall to coyning, & from thence by course
> Through hempen windowes learne to shake their taile,
> And love to die so, lest they live farre worse.[63]

The 'hempen window' is, of course, the hangman's noose. For nearly three centuries – from 1404 to 1689 – alchemy was technically illegal, a felony punishable by death and forfeiture of goods.[64] This was aimed against profiteering and forgery, and the genuine 'philosopher' was seldom molested by the law. But the atmosphere of post-Reformation England fostered an intense suspicion of all forms of magic as popish and diabolical, and the line between genuine and criminal alchemy was perilously vague. All this contributed to the air of secrecy and heterodoxy which surrounds the alchemist.

Even at this early stage of our rediscovery of the alchemist, opposed and contradictory readings emerge. On the one hand, the strange but compelling mysticism of his chemical rituals, the quest for hidden spirits and magical stones, the psychic and cosmic themes woven into his philosophy of transformations. On the other hand, the shady and predatory *demi-monde* of the smoke-seller and coiner, and the recurrent image of the alchemist as either a menace or a fool. Fascination and hostility are the twin poles of the attitude to alchemy in Shakespearean England. 'It is a wonder to see', wrote Hugh Plat in 1594,

> how every Art hath gotten his counterfeit in these days. How Logick is turned into Sophistry, Rhetorick into flattery, Astronomy into vain and presumptuous Astrology, that ancient and divine science of Alchimy into Cementations, Blanchers and Citrinations, ending commonly either in cosenage, coinage or *in capistro*.[65]

Alchemy as an 'ancient and divine science'; alchemy as a short-cut to the gallows (*capistrum* = halter): the mind of the time fed on such paradoxes. The presence of Shakespeare's Prospero and Jonson's Dr Subtle on the London stage within a year of one another is a version of this dialectic.

'A Multis Amatur Alchymia'

1 Lady Alchymia. From Conrad Gesner, *The Practise of the New and Old Phisicke* (1599)

It is said that 'Alchymia is loved by many and yet she is a virgin'.[66] Whether conceived of as a magic, a science, a folly or a crime, it cannot be denied that alchemy was a forceful presence in Shakespearean England. Its many 'lovers' had many reasons for their passion: the word 'enrichment' perhaps encompasses them all, but an enrichment whose exact nature was variously interpreted. It was a phenomenon of the age, pursued not only by the doughty alchemist *per se*, but by a host of enthusiastic amateurs that included some of the most famous names of the day. Of the Englishmen contemporary with Shakespeare who might actually be called 'alchemists', few are nowadays heard of outside specialist circles. A selective list might include: Thomas Charnock (1524–81), Miles Bloomfield (1525–1603), John Dee (1527–1608), Samuel Norton (1554–1604), Edward Kelley (1555–95), Alexander Seton (d. 1604), Thomas Tymme (1574–1620), Robert Fludd (1574–1637) and William Backhouse (1593–1662).[67] Even here a tremendous variety of approaches to alchemy is contained, as we shall see, but all these men were involved in some form of esoteric or symbolic chemistry, and all would have asserted the attainability of some miraculously transforming substance called, among many other names, the Philosophers' Stone. To them should certainly be added those who embraced alchemy as a means of preparing medicines and who practised the controversial 'chymicall physick' propagated by Paracelsus: John Hester (d. 1593), Francis Anthonie (1550–1623), Simon Forman (1552–1611), Thomas Moffett (1553–1604), James Forester (d. 1613) and others. Many of these men we shall meet with in the following chapters.

Even more indicative of the ubiquity of alchemy in Shakespeare's day is the range of famous dabblers. The poets Philip Sidney and Edward Dyer, for instance, were closely associated with Dr Dee. In his contemporary biography of Sidney, *Nobilis*, Thomas Moffett says of him:

> Not satisfied with the judgement and reach of common sense, with his eye passing to and fro through all nature, he pressed into the innermost penetralia of causes; and by that token, led by God, with Dee as teacher, and with Dyer as companion, he learned chemistry, that starry science, rival to nature.[68]

Dee had long been associated with Sidney's family, and probably began tutoring Sidney in alchemy and related magical pursuits in the early 1570s.[69] Thereafter Sidney regularly visited Dee's house at Mortlake, with its enormous library of occult and scientific works: so too did Sidney's uncle, the Earl of Leicester, another of Dee's former pupils.[70] Sidney's sister Mary Herbert, Countess of Pembroke, patroness of many leading writers, was also 'a great Chymist, and spent yearly a great deal in that study'.[71] She employed Walter Raleigh's half-brother Adrian Gilbert, another close associate of Dee's, as her 'laborator' at Wilton House. The member of the Sidney circle most devoted to Dee was the poet and diplomat Edward Dyer, now chiefly remembered for his elegy on Sidney ('Salute the stones, that keepe the lims, that held so goode a minde') and the poem, 'My Mind to Me a Kingdom Is'. His friendship with Dee went back at least to 1566, and he stood godfather for Dee's son Arthur in 1579. In Aubrey's words, Dyer 'laboured much in chymistry, was esteemed by some a Rosie Crucian'.[72] He fell particularly under the spell of Dee's bizarre associate, Edward Kelley, and spent several months in the winter of 1590–1 as Kelley's pupil in the secret art. Kelley later wrote to him, recalling

> what delight we tooke together, when from the Metall simply calcined into powder after the usuall manner, distilling the Liquor so prepared with the same, we converted appropriat bodies (as our Astronomie inferiour teacheth) into Mercury, their first matter.[73]

Such were the obscure delights of alchemy – delights, however, pursued by minds that were anything but obscure.

Another well-known alchemical enthusiast was Sir Walter Raleigh. Aubrey notes: 'He was a great Chymist, and amongst some MSS receipts I have seen some secrets from him. He studied most in his Sea-Voyages, where he carried always a Trunke of Bookes along with him.'[74] Raleigh's long association with men like Thomas Hariot, the mathematician, and the Earl of Northumberland, popularly known as the 'Wizard Earl', had earned him a general reputation for dabbling in magic. This reached a peak in the early 1590s, when a pamphlet by the Jesuit Robert Parsons denounced 'Sir Walter Rawley's school of Atheism', where 'the olde and the new Testaments are jested at, and the schollers taughte . . . to spell God backwarde'.[75] The commission convened to investigate these allegations found little to get its teeth into, and in reality the 'atheism' was probably inquisitive discussion and scientific experiment, mathematical, astronomical and chemical. Raleigh's particular alchemical interest was in medicinal distillation. In his *History of the World* (1614) he writes how

> a skilful and learned Chymist can as well by separation of visible elements draw helpful medicines out of poison, as poison out of the most helpful herbs and plants (all things having in themselves both life and death).[76]

His fame as a distiller was considerable: when, in November 1612, James I's son Prince Henry lay dying, the Queen herself persuaded Raleigh to compound his 'Great Cordial' for him. The legend of this cordial persisted through the century: Robert Boyle, the 'sceptical chymist' himself, had the recipe and, according to Aubrey, 'does great cures by it'.[77] In 1662 the diarist John Evelyn recorded accompanying Charles II to 'monsieur Febure, his chymist' to see his 'accurate preparation' of Sir Walter's cordial.[78] Lefevre's version was a cocktail of distilled herbs, mixed with powders of pearl, red coral, deer's horn, ambergris, sugar, musk and antimony, but it has been plausibly suggested that Raleigh's secret ingredient was some form of quinine from his South American expeditions.[79]

Most of Raleigh's last years – December 1603 to March 1616 – were spent imprisoned in the Tower, and there he devoted himself to alchemical experiment. As the new Lieutenant reported in 1605, 'in the Garden he hath converted a little Hen-house to a still house, where he doth spend his time all the day in his distillations.'[80] Later that year his close friend Henry Percy, ninth Earl of Northumberland, joined him in the Tower for implication in the Gunpowder Plot. The 'Wizard Earl', another alchemical enthusiast, immediately set about equipping his own laboratory. From his accounts[81] we learn that 'bricks tyles and necessaries for makinge the furnace' cost 18s 6d, two lead cisterns 15s 10d, two stills 8s 6d, and various glassware 31s 6d. It appears that the Earl's alchemical interests tended toward the gastronomic. His distiller Roger Cook was set to work brewing a 'distillacone called *Spiritus dulcis*', whose ingredients were 8 gallons of Sack (dry sherry wine), 4 lb of sugar candy and some 'spirits of roses'. The wide range of alchemical texts still preserved in the Earl's library at Alnwick Castle shows, however, that his interest in alchemy extended further than the preparation of exotic liqueurs.[82] Northumberland is typical of the scholar nobleman with occult leanings: another was the Earl of Derby, the aptly named Lord Strange, whose mysterious death in 1594 was popularly attributed to witchcraft. Poets, courtiers, countesses and earls: all lovers of Lady Alchymia.

* * *

In 1599 Sir John Davies rounded off an acrostic 'hymn' to Queen Elizabeth with this stanza:

> R udenesse itselfe she doth refine
> E ven like an Alchymist divine,
> G rosse times of iron turning
> I nto the purest forme of gold
> N ot to corrupt till heaven waxe old,
> A nd be refin'd with burning.[83]

The comparison of the Queen to a successful alchemist is only a poet's simile, but presumably one acceptable to her. Throughout her long reign, Elizabeth maintained a definite – if somewhat covert – contact with alchemy. The Somerset alchemist Thomas Charnock, author of the *Breviary of Philosophy* (1557), tells us:

> I dyd dedicate a booke off philosophie to Queene Elizabeth and delyvered him to hir . . . secretarye Sicyll [i.e. William Cecil]: but be cause the Quene and hir counsell had set goone a work in Somerset place in London before I came and had wrought there by the space of one yere, therefore my booke was layde a syde ffor a tyme.[84]

Charnock made his journey to London in 1566: the 'work' – that is, the alchemical *opus* – he mentions in progress at Somerset House would have been that of Cornelius de Lannoy, a shadowy figure who was, according to Edward Dyer, the Queen's private alchemist at this time.[85] Elizabeth certainly figured in many alchemists' dreams as the ultimate patron. William Bloomfield, for instance, dedicated his *Quintaessens, or the Regiment of Life* (c. 1574) to her,

> Havyng no thing that I may better pleasure your highnes withall, then this hid treasure that God of his mercy hath gyvyn unto me, myne intent is to revele the same. That to such a singular princess a most preciose & singular Iowel or preciose perle.[86]

About the same time, one John Peterson of Lübeck delivered to the Queen three phials full of mysterious liquids, together with a letter containing an 'Apollogie to the noble science of Alcumey' and a 'confydent affirmacion of the wonderfull riches (excedynge al comparison) to be by them attayned'.[87] And the words of Samuel Norton in his *Key of Alchimie* (1577) suggest something of the symbolic status which the 'Virgin Queen' so assiduously cultivated. Referring to an old text, apparently by his great-grandfather Thomas Norton, Samuel writes:

> That which I most of all desire to come to pass, is that which hee intimates in his 6th Chapter, where speaking of the Stone to be revealed to the kings of this Land, it shall be found he saith:
>
> > By the fortune & by the grace
> > Of a woman faire of face
>
> And what know I, Oh Queene! whether it be yr. selfe or noe?[88]

In Norton's hopes, Queen Elizabeth presides over the long-sought-for revelation of the Stone: a dream not unlike Shakespeare's famous vision of England as 'This earth of maiesty.... This precious stone, set in the silver sea'.[89]

The Queen's long association with John Dee, though based on much more than his alchemical interests, would certainly have touched on them. Ashmole records how, along with gifts of money, she 'sent him word ... to doe what he would in Alchymie and Philosophy, and none should controule or molest him'.[90] Dee himself recalls how 'her Majestie very graciously vouchsafed to account herself my schollar in my booke' – his *Monas Hieroglyphica* (1564) – and how she said, 'if I would disclose unto her the secretes of that booke, she would *et discere et facere*'.[91] Dee's *Monas* encompasses a whole complex of magical systems, among which alchemy is prominent. The Queen's promise to 'both learn and do' suggests a half-serious touch of alchemical ambition. Hume, on the authority of Camden, quotes her words to the Earl of Essex, in about 1600. Receiving an apology from him, she replied

> that he had tried her patience a long time, ... that her father would never have pardoned so much obstinacy; but that, if the furnace of afflication produced such good effects, she should ever after have the better opinion of her chémistry.[92]

The tone is jesting, but 'her chemistry' might again suggest that the Queen herself dabbled with furnace and alembic. Whether as patron or experimenter, Queen Elizabeth shows us that the fascination for alchemy percolated up to the highest level of English society.

*   *   *

No account of the alchemical world of Shakespeare's day would be complete without briefly charting 'The Rise and Fall of Edward Kelley'. No one expresses more clearly the power and appeal, the ambiguous charisma, of the alchemist at this time. Nowadays Kelley is universally condemned as a charlatan – 'the grand Imposter of the world', 'a terrible zombie-like figure', 'an egregious scoundrel',[93] to take a cross-section of slanders over the centuries – but in the early 1590s his reputation was at its height, as the one-time apothecary's apprentice lorded it over the Imperial Court of Rudolf II at Prague, brushing off the repeated entreaties of Elizabeth's chief adviser Lord Burghley for him to return to England. 'I wolde be contented to obey your friendly coniuration,' he wrote haughtily from Prague, but being

> seased in lands of inheritance yielding £1500 yerely, incorporated to the kingdom in the second order, of some expectation and use more than vulgar, of his Majesties pryvy cownsell, ... Chief Regent in and over all the lands and affayres of the Prince Rosenberg: I can not see how I might easily or honestly departe, much les so steale away.[94]

The shadow Kelley cast on the contemporary imagination can be gauged by his frequent appearance in the writing of the time. 'O humanity, my Lullius, or O divinitie, my Paracelsus' wrote Gabriel Harvey in exalted mood in *Pierces Supererogation* (1593): back came the reply from his enemy, Tom Nashe, 'Let him call uppon Kelly, who is better than them both.'[95] In a letter from Plymouth in 1597, John Donne describes the poverty there: 'I do not think that 77 Kelleys could distill £10 out of all the towne.'[96] In 1610 Johnson's Dr Subtle is eulogised as 'an excellent artist ... a divine instructer ... A man the Emp'rour has courted above Kelley.'[97] And still in 1626, thirty years after Kelley's mysterious death, John Fletcher epitomizes an alchemist's skill with the comment: 'He learnt it of Kelley in Germany. There's not a Chimist in Christendome can goe beyond him for multiplying.'[98] All these references are ironic: but the name of Kelley sticks.

Kelley's early life is a tissue of vague allegations and shady circumstances. Fraud, necromancy and coining all appear on the record: for the latter, in Lancaster about 1580, he had his ears cropped. He emerged from the underworld one Saturday morning in March 1582 when he presented himself at Dr Dee's house at Mortlake and offered his services as a 'skryer', or spirit medium. Among Dee's many magical pursuits – mathematical, cabbalistic, alchemical, astrological – was a particular penchant for invocatory 'angel-magic'. To Kelley's enquiries, Dee

> confessed my self long time to have byn desirous to have help in my philosophical studies throwgh the Cumpany and information of the blessed Angels of God. And there uppon, I brought forth to him my stone in the frame.[99]

This was Dee's 'shew stone', or crystal ball, a small globe of smoky polished quartz the colour of moleskin: it is now at the British Museum, along with his 'magic mirror', a disc of polished obsidian apparently of Aztec origin. To these, in Dee's words, 'were answerable *aliqui angeli boni*' – good angels, rather than daemonic 'familiar spirits' of the Faustian type, an essential distinction in Renaissance magic. The stone was set up, prayers were made 'to God and his good Creatures for the furdering of this Action', and within a few minutes Kelley 'had sight' of an angel in the stone. The angel identified himself as Uriel, the seance proceeded smoothly, and the successful skryer was engaged at £50 *per annum*. Edward Kelley had arrived.

For the next seven years, Kelley was Dee's 'intimate friend, and long companion in philosophicall studies and chemicall experiments.'[100] For Kelley soon revealed an alchemical prowess to match his skill as a spirit medium, and Dee's diaries make it clear that he considered Kelley his alchemical 'master' who 'did open the Great Secret'[101] to him. In September 1583 the pair set out with wives and children for Bohemia, at the invitation of Count Laski, Palatine of Sieradz, a devotee of alchemy and the occult. 'Alchemy was the greatest passion of the age in Central Europe,' writes Evans,[102] and none was more passionate than the Holy Roman Emperor, Rudolf II. At his court at Prague – then 'the metropolis of alchemy, and the headquarters of adepts and adeptship'[103] – Dee and Kelley presented themselves in the summer of 1584. They remained at Prague for two years, performing their 'angelic conferences' and alchemical transmutations, until Rudolf's Catholic advisers succeeded in getting them expelled as conjurors and spreaders of '*una nuova superstitione*'. It is interesting that the Papal Nuncio speaks of '*Giovanni Dee e il Zoppo suo compagno*': this suggests that, as well as having lost his ears in the pillory, Kelley was a cripple, '*zoppo*' meaning 'lame'.[104] They were sheltered at Trebon, in southern Bohemia, by the powerful Count Rosenberg (Vilem Vok of Rozmberk), but by 1588 Dee and Kelley had quarrelled irrevocably. Dee embarked homewards early in 1589, but Kelley returned to Prague to dazzle Rudolf with alchemy once more – to such effect that he was 'feted by the Emperor, accepted as a citizen of Bohemia . . . and granted a patent of Imperial nobility'.[105] Reports of his success filtered back to England, and Lord Burghley grew increasingly desperate to tempt Sir Edward, as he now was, back home, 'to honour her majesty as a loyal natural subject, with the fruits of such knowledge as God hath given him'. If Kelley himself could not be persuaded back, perhaps he might deign to send the Queen some 'small portion' of his fabled transmuting tincture, just enough 'as might be to her a sum reasonable to defer her charges for this summer for her navy, which we are now preparing to the sea, to withstand the strong navy of Spain.[106] This is the measure of Kelley's power: the Queen's Secretary of State begging a portion of his magic tincture to finance a campaign against the 'Old Enemy'.

His glory was, however, brief. He fell from the obsessive Emperor's favour as spectacularly as he had risen. In 1591 he was imprisoned. Thereafter the story grows confused: torture, release, reincarceration. It is generally said that in 1595 Kelley fell from a window while trying to escape and died from his injuries: Dee's diary recorded brusquely 'the newes that Sir Edward Kelley was slayne'.[107] But Evans has evidence[108] that Kelley was alive in 1597, at the castle of Most. The famous Paracelsist physician, Oswald Croll, appears to have visited him there. Kelley's life recedes back into that smoky obscurity whence it emerged. A few of his alchemical writings remain – two poems attributed to him in Ashmole's *Theatrum Chemicum Britannicum* (1652); the *Fragmenta*, edited by Ludwig Combach, published at Geismar in 1647; the two tracts *De Lapide Philosophorum* (1676) – but perhaps his self-generated, semi-daemonic legend was his true masterpiece.

An incredible series of attested transmutations remain from his years at Prague and Trebon. The Frenchman Nicolas Barnaud, whose *Triga Chemica* (1599) will be of interest to us later, witnessed a drop of Kelley's red tincture transmute a pound of mercury into gold at the house of Tadeas Hajek, Rudolf's 'senior alchemical adviser'.[109] At Trebon in December 1586, Dee recorded that upon an ounce and a quarter of crude mercury 'E.K. made projection with his powder in the proportion of one minim, and produced nearly an ounce of best gold'.[110] This was also witnessed by the English gentlemen Edward and Francis Garland. In 1588, Edward Dyer was present at a transmutation in Prague, and later asserted to John Whitgift, Archbishop of Canterbury:

> I am an eyewitness thereof, and if I had not seen it, I should not have believed it. I saw Master Kelley put of the base metal into the crucible, and after it was set a little upon the fire, and a very small quantity of the medicine put in, and stirred with a stick of wood, it came forth in great proportion perfect gold, to the touch, to the hammer, to the test.[111]

Many alchemical writers of the day cited Kelley's transmutations – influential works like Libavius' *Alchymia* (1597) and Hoghelande's *Historiae aliquot Transmutationis* (1604), as well as contemporary Bohemian manuscripts which 'typically bear a note like: "this was used by Edoardus Gelleus, the English alchemist" '.[112] In the notes of his *Theatrum*, Ashmole further dilates the testimony of Kelley's prowess:

> For neerer and later Testimony, I have received it from a credible person, that one Broomfield and Alexander Roberts told him they had often seen Sir Ed: Kelly make projection, and in particular upon a piece of Metall cut out of a Warming pan, and without Sir Edwards touching or handling it, or melting the Metall (onely warming it in the fire) the Elixir being put thereon, it was transmuted into pure Silver.[113]

These are credible witnesses of an incredible performance. What matters is not the documentary truth of these reports, nor the dutiful scepticism of our reactions, but what they tell us about the standing of Kelley, and of the alchemist in general, in the last years of the sixteenth century. That is their 'truth' – the truth, perhaps, of theatre, where events happen because actor and audience will them to happen. Kelley's Bohemian alchemy, whether or not it turned crude mercury into philosophical gold, was a series of performances, with some of the most powerful figures in Europe, political and intellectual, as audience. A few years later, Shakespeare's Theseus says of the players in *A Midsummer Night's Dream* (c. 1595): 'The best of this kind are but shadowes: and the worst are no worse, if imagination amend them.'[114] Drama is a shadow-world; transmutation is a fantasia. But their effects on the mind of the time were true, real and profound.

· 2 ·

Three English Alchemical Texts

The Mirror of Alchimy

To recognize the insistent presence and popularity of alchemy in Shakespeare's England is the first step towards assessing, perhaps re-assessing, its influence on the time. The next and more vital step is to define the theoretical principles of alchemy, that strange philosophy to which all alchemical endeavour, from the criminal to the mystical, was somehow anchored. The dubious promises and smoky deceits of alchemy certainly found their way into the literature of Shakespeare and his contemporaries. But alchemy has another, hitherto neglected, presence in their work: a matter of themes and patterns, symbols and terminologies, the whole exotic paraphernalia of alchemy as a philosophy of transformations. Even an out-and-out sceptic like Ben Jonson made deep inroads into this philosophy, if only to give added sting to his critique of it.

It is the many alchemical publications of the day which offer the significant clues to this presence. Elias Ashmole said of the alchemical philosophers: 'they have written more than they would speake, and left their lines so rich, as if they had dissolved Gold in their Inke, and clad their Words with the Soveraigne Moysture.'[1] This might be overstating the matter, but certainly they were weird and wonderful writers, and their texts give a much richer picture of what alchemy meant to anyone who penetrated beyond its most superficial manifestations. Though fraught with vagaries and obscurities, these writings also offer a precise chronological index: I have spoken rather vaguely of 'Shakespeare's England', and from now on will concentrate more exclusively on alchemical texts published between 1590 and 1612, a span that encompasses the creative career of Shakespeare, and the works of Jonson and Donne significant to this book. The first of the texts to which we turn is *The Mirror of Alchimy*, published in London in 1597.

The *Mirror of Alchimy* consists of four tracts: the title-tract, attributed, probably falsely, to 'the thrice-famous and learned Fryer, Roger Bachon'; the *Smaragdine Table of Hermes Trismegistus*, together with 'a briefe commentary of Hortulanus the philosopher' upon it; the *Booke of the Secrets of Alchimie*, by 'Galid, sonne of Iazich'; and the *Discourse*

THE
Mirror of Alchimy,

Composed by the thrice-famous and learned
Fryer, *Roger Bachon*, sometimes fellow of
Martin Colledge: and afterwards of
Brasen-nose Colledge in
Oxenforde.

Also a most excellent and learned discourse of
the admirable force and efficacie of Art *and* Nature,
written by the same Author.

With certaine other worthie Treatises of
the like Argument.

Vino vendibili non opus est hedera.

LONDON
Printed for Richard Oliue.
1597.

2 Title-page from Roger Bacon, *The Mirror of Alchimy* (1597)

of the Admirable Force and Efficacie of Art and Nature, again attributed to Roger Bacon. The immediate history of the book is clear enough: the two Bacon tracts were translated from French versions – *Le Miroir d'Alquimie* and *L'Admirable Pouvoir et Puissance de l'Art et de la Nature*, both published at Lyons in 1557 – while the other two tracts were based on Latin versions in the miscellany *De Alchemia* (Nuremberg, 1541).[2] Any earlier history, however, is more nebulous. The thirteenth-century *'doctor mirabilis'* Roger Bacon certainly included alchemy among his magical and scholarly pursuits, but it is probable that the *Mirror* is a later text 'fathered' on to him by its anonymous author.[3] This is a recurrent habit of alchemical literature: none of the famous texts of 'Raymond Lully', for instance, were actually written by the Spanish mystic Ramon Lull. The second tract in the *Mirror*, the Hermetic *Tabula Smaragdina*, or *Emerald Table*, is one of the ancient classics of alchemy: the earliest-known version is in an eighth-century Arabic text,[4] but it probably goes back to earlier Greek sources. Its commentator, Hortulanus, is a shadowy medieval figure: Ashmole identifies him with John Garland (*fl.* 1230),[5] and he was certainly known to John Gower, who mentions 'Ortolane' in his *Confessio Amantis* (*c.* 1383).[6] Finally, 'Galid sonne of Iazich' takes us back to the seventh century, the Umayyad prince Khalid ibn-Yazid being reputedly the first of the great Moslem alchemists.[7] I mention all this because it gives some idea of the continuity of alchemical tradition. Ancient texts and sayings – Egyptian, Greek, Islamic, European – were constantly reiterated, and alchemy drew much dignity from its sense of contact with extreme antiquity. For our purposes, however, the *Mirror of Alchimy* is an Elizabethan digest of basic alchemical theories. As such it is invaluable.

* * *

The book begins with some simple definitions of alchemy. For instance, 'Alchimy is a science, teaching howe to transforme any kind of metall into another: and that by a proper medicine, as it appeareth by many Philosophers Bookes.'[8] This is the basic assumption and aspiration of alchemy. Essentially it refers back to the conception of matter formulated by Aristotle in the fourth century BC. This postulates a duality of 'matter' and 'form', suggesting a basic *prima materia* – pre-existent, formless, primaeval matter – on to which various gradations of form and identity are impressed. The most simple distinguishing forms are the four 'qualities', cold, hot, moist and dry. Out of paired combinations of these qualities arise the four elements: earth (cold and dry), water (cold and wet), fire (hot and dry) and air (hot and wet). And out of different permutations of the elements are formed all the specific substances of the material world. The elements are clearly inter-convertible. Water heated turns to steam: by replacing a cold quality with a hot one, the element water is converted into air. Similarly, a solid giving off vapour is exchanging its inherent earth for air. The phenomenon of chemical change was a mobility among forms: trans-form-ation. On the basis of this theory, it was thought possible to transmute one metal into another, to remove the form of lead (or copper, or mercury) and impose the form of gold. Within the prevailing animistic conception of matter, this was seen as the 'death' or 'corruption' of lead, followed by its 'rebirth' or 'generation' in purer form. 'No generation without corruption' is a favourite alchemical catch-phrase. Having formulated this possibility, the alchemists set about their search for the chemical agent to make it happen. Thus another definition in the *Mirror of Alchimy* reads: 'Alchimy therefore is a science teaching how to make and compound a certaine medicine, which is called Elixir, the which when it is cast upon mettals or imperfect bodies, doth fully perfect them in the very proiection.'[9] This, the 'Universal Medicine' or Elixir (Arab *al-iksir*, 'the powder'), was also known as the Tincture, the Philosophers' Stone[10] and a host of other names. 'The earthly Philosophers' Stone and its substance have a thousand names,' asserts Johann Siebmacher in the *Hydrolithus Sophicus* (1619), the title of his treatise – 'the water-stone of the wise' – being one.[11] A book published in 1652[12] was entirely devoted to synonyms for the Stone, listing some 170 – 'the shaddow of the Sun', 'the spittle of Lune', 'the lesse worlde', etc. – and Dom Pernety's *Dictionnaire Mytho-Hermetique* (1787) lists over 600, ranged from Absemir to Zumelazuli![13] We will generally content ourselves with the most familiar name: the Stone.

The *Mirror* says of this transmuting elixir that it 'perfects' imperfect metals. This introduces the whole idea of gold, the metal universally desired as the end-product of transmutation. Gold, we learn from the *Mirror*, is 'a perfect masculine bodie, without any superfluitie or diminution'.[14] It was a quality of perfection and harmony which the alchemist sought to impose on baser matter. In this lies perhaps the crux of alchemy's ambiguity. The traditional promise of the alchemist to 'make gold' was frequently interpreted as a passport to endless financial wealth: the success of Edward Kelley reminds us of the sway this promise held over the imagination of the time. But the more elevated alchemist insisted that what he created was not 'common gold' at all. It was 'our gold' or 'philosophical gold': '*aurum nostrum non est aurum vulgi*.'[15] So it is asserted in the *Mirror* that 'the gold ingendered by this Art, excelleth all naturall gold in all properties, both medicinall and others'.[16] This is essentially an emphasis on gold as a condition rather than a mere substance. In making philosophical gold, the alchemist was creating in matter a condition of harmony and incorruptibility. Philosophical gold was matter redeemed from baseness, dividedness and corruption. As such, it had sympathetic curative powers: the 'medicinall' associations of alchemical gold are its harmony in action; the elixir which created it merges into the 'elixir of life', promising human incorruptibility.[17]

In his work of perfecting metals, the alchemist insisted that he was pursuing the intentions of Nature. As the *Mirror* says, 'Nature alwaies intendeth and striveth to the perfection of Gold, but many accidents comming between change the mettalls.'[18] Metals, it was supposed, 'grew' within the mines and mountains. The alchemist harnessed and accelerated their natural evolution towards perfection. The alchemist's purpose, says Arnald of Villanova, is 'that what remains uncompleted by Nature may be completed by Art',[19] and this is a claim frequently echoed in alchemical literature. Also in Jonson's *Alchemist*, where Subtle says that lead and other metals 'would be gold if they had time',

> for 'twere absurd
> To thinke that nature in the earth bred gold
> Perfect i' th'instant.

In this way 'our Art doth furder' Nature's innate tendency towards gold.[20] In Shakespeare's *Winter's Tale* (1610), the King of Bohemia says:

> This is an Art
> Which do's mend Nature: change it rather, but
> The Art it selfe is Nature.[21]

In their context, these are also words resonant of alchemy.

These are some of the basic ingredients in the alchemical idea of transmutation: a belief in metallic identity as alterable form; a search for an Elixir or Stone to achieve the alteration; an emphasis on gold-making as the creation of material perfection rather than mere lucre; an insistence that alchemical art followed the footsteps of nature, pursuing her own 'entelechy' or perfecting intention. It is essential to see at this point that the alchemist's search for 'philosophical gold' and his desire to prepare the transmuting Stone are one and the same quest for perfection. Philalethes writes:

> Whoever wishes to possess this secret Golden Fleece, which has virtue to transmute metals into gold, should know that our Stone is nothing but gold digested to the highest degree of purity and subtle fixation to which it can be brought by Nature and the highest effort of our Art; and this gold thus perfected is called 'our gold', no longer vulgar, and is the ultimate goal of Nature.[22]

Not all writers make this identification so explicit, but 'our gold' and the Stone have always the same properties. If gold is praised for proportion and unity, so is the Stone:

> Compound ye our Stone
> Equall, that in him repugnance be none,
> Neither division, as ye proceede.[23]

If gold is incorruptible, so too is the Stone, of which Michael Sendivogious writes in the *Novum Lumen Chemicum* (1604):

> The Sages call it the Phoenix and Salamander. Its generation is a resurrection rather than a birth, and for this reason it is immortal and indestructible. Now, whatever is conceived of two bodies is subject to the law of death, but the life of this fruit is a separation from all that is corruptible about it.[24]

The constant in alchemy is an idea of transformation. From Raw Stuff to Philosophers' Stone, from common gold to philosophical gold, from base metal to precious metal: these are different versions of the same redemptive perfecting process, in which a corrupt 'form' is removed and a pure one instated. The creation of the Stone was undoubtedly the most important of these transformations. The alchemist's *magnum opus* – the arduous 'great work' whose end-product was the Stone – was a kind of blueprint for all subsequent transmutations performed by the Stone. The Stone was capable of perfecting matter precisely because it was itself matter perfected.

*   *   *

Loss and restoration of form: this is the basic rhythm of alchemical transformation. It is expressed in the formula, '*solve et coagula*': an injunction to dissolve and congeal. Chemically this is typified in the process of sublimation, reducing a solid to vapour (*solve*) and then condensing the vapour to purified solidity (*coagula*). Another formula reads: '*fac fixum volatile et volatile fixum*', make the fixed volatile and the volatile fixed. This appears in Trismosin's *Aureum Vellus* (1598) as:

> *Si fixum solvas faciasque volatile*
> *Et volucrem figas, faciet te vivere tutum.*

William Backhouse translates:

> If thou dissolve the fixt & make it fly,
> And fix the bird, thou shalt live happily.[25]

Backhouse expressed the same idea more enigmatically in his own poem, 'The Magistery' (1633):

> The Eagle which aloft doth fly
> See that thou bring to ground;

> And give unto the Snake some wings,
> Which in the Earth is found.[26]

This interplay of fixed and volatile, solid and vaporous, is central to alchemy, for the alchemist saw it as an interplay between the bodily and spiritual aspects within matter. The *Mirror* makes this clear in its description of *solve et coagula*:

> Solution and congelation shalbe in one operation, and shall make but one worke.... And this solution and congelation which wee have spoken of, are the solution of the bodie and the congelation of the spirite, and they are two, yet have but one operation. For the spirits are not congealed except the bodies be dissolved, as likewise the bodies is not dissolved unlesse the spirit be congealed.

Here we see that the process of removing or breaking down metallic form is essentially a resolving of bodily matter into spirit. The keynote of the alchemist's perfecting intention is spiritualization, and dissolution is the body's route to spirit:

> the spirit wil not dwel with the body, nor be in it, nor by any means abide with it, untill the body be made subtil and thin as the spirit is. But when it is attenuate and subtill, and hath forsaken his grosnesse and corpority, and is become spirituall, then shall he be mingled with the subtill spirits, & imbibed in them, so that both shall become one and the same, & they shall not be severed, like as water put to water cannot be divided.[27]

Inextricably linked with this purifying loss of form is the *coagula* which congeals the spirit back into material form. Thus, says the *Mirror*, 'this work or masterie is a coniunction or marriage of the congealed spirit with the dissolved bodie'.[28] The process is a transforming circle, the substance returning once more to solidity, but now divested of 'grosnesse and corpority'. 'Corporeall things in this regiment are made incorporeall, & contrariwise things incorporeal corporeall, and in the shutting up of the worke, the whole body is made a spirituall fixt thing.'[29] This 'spirituall fixt thing' is one definition of the Stone: matter suffused with spirit. Another term often used is '*corpus subtile*', or subtle body. Philalethes expresses this paradoxical condition:

> It is called a stone, not because it is like a stone, but only because by virtue of its fixed nature, it resists the action of fire as successfully as any stone.... If we say that its nature is spiritual, it would be no more than the truth; if we described it as corporeal, the expression would be equally correct; for it is subtle, penetrative, glorified, spiritual gold. It is the noblest of all created things after the rational soul, and has virtue to repair all defects both in animal

and metallic bodies, by restoring them to the most exact and perfect temper: wherefore it is a spirit of quintessence.[30]

It is in this work of dissolving and subliming, of reducing physical substance to spirituality, that the alchemist's fire looms so large. The furnace – or, more grandly, the 'Philosophers' Athanor' (Arab *al-tannur*, 'furnace') – was the centrepiece of the laboratory, and much ingenuity went into regulating the degrees of heat. Norton distinguishes thirteen heats in the *Ordinall*, ranging from that 'wherewith Pigg or Goose is scalded' to the 'Heat of Projection'.[31] The *Mirror* asserts:

> To immitate Nature, we must needes have such a furnace like unto the mountaines, not in greatnesse, but in continuall heate, so that the fire put in, when it ascendeth, may finde no vent, but that the heat may beat upon the vessell.[32]

Fire was the great agent of transformation, the purifying destroyer: 'his [i.e. the Stone's] perfection and proceeding consisteth in the fire, which is the cause of his life and death'.[33] Equally important was the vessel, the place of transformation: a variety of chemical ware – retorts, alembics, cucurbites, etc. – being compressed into a single prototype vessel known as 'Our Glass' or the 'Philosophers' Egg' or the 'vase of Hermes'. The *Mirror* says that the vessel must be round and made of glass, and these specifications are invariable throughout alchemical texts.

Enclosed in the vessel and consigned to the furnace, solid turned to vapour and vapour to new solid. 'Body' and 'spirit', opposed and contradictory in base substance, were harmonized in the *corpus subtile* of the Stone. The alchemist conceived of his work as an elimination of the discords and divisions inherent in physical nature. The idea of the duality of matter, and the necessity of overcoming it, is vital to alchemy. The opposites were expressed in many ways. Hortulanus writes in the *Mirror* that the matter

> is divided into two principle parts by Art: into the superiour part that ascendeth up, and into the inferiour part which remaineth beneath fixe and cleare.... The inferiour part is the Earth, which is called the nurse and ferment; and the superior part is the Soule, which quickneth the whole Stone and raiseth it up.[34]

This idea of the 'soul' or *anima* in matter is clearly related to the quintessence, the spiritual fifth element hidden within the 'prison' of the four material elements[35]. Spirit and body, soul and earth, quintessence and element, volatile and fixed, eagle and serpent: all these are opposite tendencies within matter which the alchemist set about to reconcile. His *opus* of transformation turned each into its opposite, dissolving body and congealing spirit to make them 'one thing', the Stone: 'The Stone is one, the Medicine one, which, however, according to the philosophers, is called *Rebis* (two-thing), being composed of two things, namely a body and a spirit.'[36]

It is in this context that we make our first acquaintance with that complex and elusive alchemical figure, 'Our Mercury' or the 'Mercury of the Philosophers'. For among the many alchemical characterizations of the opposites within matter, a vital one codes them as 'mercury' and 'sulphur'. The *Mirror* tells us: 'The naturall principles in the mynes are Argent-vive and Sulphur. All mettalls and minerals, whereof there be sundry and divers kinds, are begotten of these two.'[37] Mercury ('argent-vive' means quick, literally 'living' silver) and sulphur are described as the 'principles' out of which all metals evolve or are 'begotten'. Gold, for instance, is 'engendred of Argent-vive pure, fixed, clear, red; and of Sulphur cleane, fixed, red, not burning', while lead derives from the same two in an 'impure, not fixed, earthy, drossy' condition.[38] This states the theory at its simplest, as an idea of metallurgical evolution: mercury and sulphur as the basic constituents of the 'first matter' out of which all metals grow. Sendivogius, who compares an 'Alchymist without Sulphur and Mercury' to a 'Prince without a people', describes the theory slightly differently:

> The first matter of metals is twofold, but the one cannot make a metal without the other. The first and principal is the humidity of the Air mixed with heat, and this the Philosophers called Mercury, which is governed by the Beams of the Sun and Moon in the Philosophical Sea. The second is the dry heat of the Earth, which they called Sulphur.[39]

Esoteric details apart, the operative word here is 'called'. We see that these should not be understood as the actual chemical substances, mercury and sulphur. Rather they are certain basic properties and principles for which those substances are useful chemical names. As an anonymous tract in the *Musaeum Hermeticum* (1625) explains, to take actual mercury and sulphur as the origin of all metals

> makes nonsense of the dictum of the Sages. For ordinary quicksilver is an imperfect metal, and itself derived from the original substance of all metals. . . . Nor can common sulphur be the first substance of the metals, for no metal contains so much impurity as common sulphur.[40]

These primal substances, or vital principles, of matter are not mere chemicals: they are 'Our Mercury' and 'Our Sulphur', and just as '*aurum nostrum non est aurum vulgi*', so 'our Mercury is not common mercury: for all common mercury is corporal, specific and dead, while our Mercury is spiritual, female, living and life-giving'.[41] Philosophical Mercury is too complex a figure to dwell on now, and anyway only appears in embryonic form in the *Mirror*. It – or she – will be of vital interest later. The important general point is this: that 'mercury' and 'sulphur' are descriptive of the binary principles, the *yin* and *yang*, of matter. Mercury is associated with the moist, vaporous, volatile, spiritual, female aspects of matter; Sulphur with the solid, combustible, fixed, bodily, masculine aspects. As such, 'Our Mercury' becomes closely identifiable with the quintessential spirit and quickening *anima* inside matter.

* * *

There we must leave the *Mirror of Alchimy*. We must remember that, though the text is late Elizabethan, the theories it expounds belong to the earliest phase of European alchemy and have none of the sophistications and symbolic richness of later works. Nevertheless it provides us, and no doubt provided the curious Elizabethan, with a basic programme of alchemical tenets. We see the purpose of the alchemist's transformations: to divest matter of its corrupt and base form, to eliminate the division and duality inherent in material nature. We see the preciousness towards which he aimed – Stone, elixir, philosophical gold – as matter invested with supernatural properties of harmony, incorruptibility and spirituality. We hear the terms which echo through alchemical literature of all types: art and nature, body and spirit, sulphur and mercury, dissolve and congeal, furnace and vessel. These, at any rate, are constant landmarks in the shifting sands of alchemical philosophy.

The Compound of Alchymy

The Compound of Alchymy, 'a most excellent, learned and worthy worke, written by Sir George Ripley, Chanon of Bridlington in Yorkshire',[42] was published in London in 1591. Though appearing six years earlier than the *Mirror*, it in fact represents a later and more advanced phase of alchemical tradition. The tracts in the *Mirror* are still firmly entrenched in alchemical conceptions received into Europe from Arabic sources. While encyclopedists and scholastics like Bartholomew the Englishman, Roger Bacon and Albertus Magnus all contributed to the instatement of alchemy as a vital branch of medieval natural philosophy, it was not until the early fourteenth century that a specifically European alchemical literature began to emerge – such seminal works as the *Testamentum* of Raymond Lully, the *Rosarium Philosophorum* of Arnald of Villanova, the *Summa Perfectionis* of Geberus and the *Pretiosa Margarita Novella* of Petrus Bonus. At this time too, the first stirrings of English alchemical activity were recorded: we hear of the 'art which

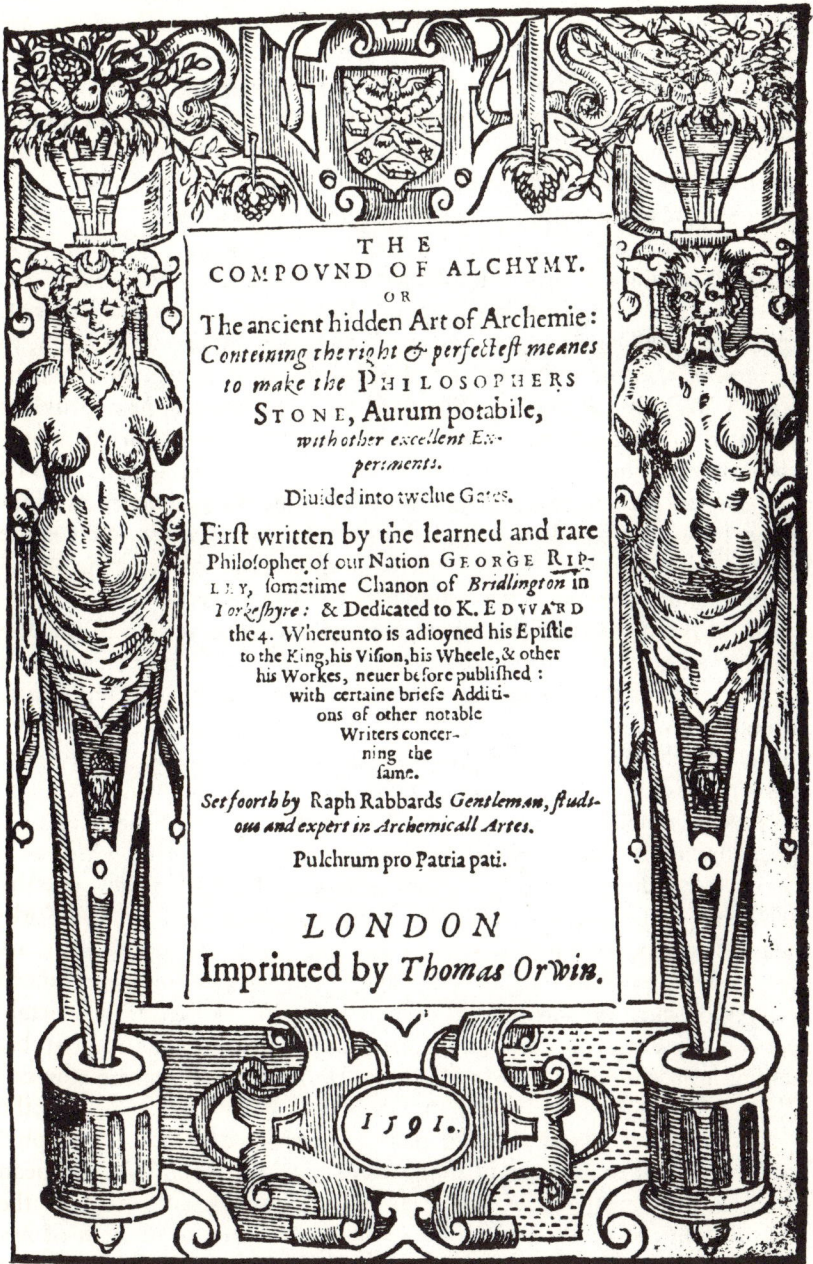

3 Title-page from George Ripley, *The Compound of Alchymy* (1591)

men call alchemy' in the St Albans *Gesta Abbatum* of 1319, and of *'alkymiste moderni temporis'* in Walter of Odington's *Icocedron* (c. 1330).[43] By the mid-fifteenth century, the great medieval tradition of English alchemy was in its heyday.

By common consent, the greatest of the medieval English adepts was Sir George Ripley. He was born in Ripley, Yorkshire, in about 1415, but tells us he travelled to 'farre countryes' to learn the secrets of alchemy.[44] He mentions his studies at Louvain and Rome, and alchemical lore associates him with the Knights of St John at Rhodes.[45] By 1471, he was ensconced as Canon Regular at the Augustinian priory of Bridlington, Yorkshire. It was in that year that he composed his *Compound of Alchymy*, a long alchemical poem in rhyming seven-line stanzas (ABABBCC), dedicated to Edward IV.[46] He died in about 1490. The *Compound* was the most celebrated of his works and, when published in 1591 by Ralph Rabbards, was the first to be printed. Many others circulated in manuscript. Of great interest to us later will be an Elizabethan manuscript translation[47] of his Latin poem, *Cantilena*, a curious and beautiful alchemical epithalamion celebrating the *'nuptiae Mercurii'*. Other important works of Ripley's include the *Medulla Alchemiae* (1476), the *Mystery of Alchymists*, the *Vision*, and the verses inscribed on the illuminated manuscript known as the Ripley Scrowle.[48]

This is the context of the *Compound of Alchymy*. It arrived on the Elizabethan bookstalls from the heart of the medieval English alchemical tradition. Unlike the *Mirror*, it is devoted to describing a single unfolding process: the *magnum opus* or great work of transformation, whose end product is the Philosophers' Stone. The work is divided into twelve chapters, described as the twelve 'gates' leading into the 'strong Castle of our wisdome'.[49] These gates are the twelve stages of the *magnum opus*. Already we find a crucial alchemical image: the *opus* as a journey, an expedition into a 'castle' of secret wisdom. The castle is also matter itself, that 'shut palace' which the alchemist sought to penetrate by chemical means, making his visit to the 'interior of the earth'. Each of the gates is an obstacle to be overcome, a chivalric ordeal summoning reserves of spiritual fortitude and chemical skill. The castle is further described by Ripley as 'the Philosophers Castle . . . round as any Bell'[50], making it an emblem of the circular *vas Hermetis*, or Hermetic vessel:

> In one glasse must be done all this thing,
> Like to an Egge in shape and closed weele.[51]

The substance to be worked on, the vessel which contains it, the wisdom which transforms it: all these are implied in Ripley's image of the castle. Into this stronghold of secrets, physical and metaphysical, the *Compound* promises to guide us,

> That this Gate shalbe to thee unlocked...
> And all the Castle thou holdest at thy will.[52]

The work's effectiveness as a guide is, perhaps, debatable. It is bafflingly obscure in parts, its rough doggerel verse tending to blot out rather than discreetly veil its meanings. Sherwood Taylor has attempted to relate the processes described by Ripley to modern chemical procedure and concluded that they are 'not susceptible of any physical explanation'.[53] Nevertheless, if approached on their own criteria – as successive stages of a journey inwards, a journey for which chemistry provides the vehicle but not the total meaning – they do form a pattern, a process no less real for being scientifically 'invalid'.

Ripley's twelve stages, or gates, are: calcination, solution, separation, conjunction, putrefaction, congelation, cibation, sublimation, fermentation, exaltation, multiplication and projection. Though there are various versions of the number and sequence of operations in the *magnum opus*, there is a case for taking Ripley's sequence as definitive of the *opus* as it was conceived in Shakespeare's day. For in the same year that the *Compound* was published, 1591, another alchemical work appeared in London: John Hester's translation of *De Ortu et Causis Metallorum* (1576) by the French alchemist and physician, Joseph Duchesne (Quercetanus). In this – published as *A Breefe Aunswere of Iosephus Quercetanus... concerning the Original and Causes of Mettalles* – we find an exposition of the 'operations of the philosophic stone' which exactly agrees with Ripley's twelve stages. The same sequence is also preserved intact in the *Philosophia Reformata* by Johann Daniel Mylius, a popular alchemical emblem-book published in 1622.[54] These authors may have had different perceptions of the wider implications of the *opus*, but the chemical coding of its process is the same. We look to the *Compound*, with a little help from Quercetanus, to map out the basic route of the alchemist's 'great work'.

*   *   *

We begin, in the Preface of the *Compound*, with a description of the Raw Stuff.[55] Not a chemical description, of course, but a comparison of it with the *prima materia* out of which God fashioned the world:

> In the beginning when Thou mad'st all of nought,
> A globous matter and darke under confusion
> By Him the beginning marveilously was wrought,
> Conteyning naturally all things withoute division,
> Of which in six dayes he made cleer distinction.

The Creation was a process of organization, an imposing of 'clear' (or as Ashmole renders it, 'dear') 'distinction' on the hotch-potch of the primal chaos. The alchemical *opus* will reiterate that process:

> So through his will and power, out of one Mas
> Confused was made each thing that being is . . .
> Ryght soe in our practize must it be.[56]

The Raw Stuff to be worked on is a *massa confusa*, a microcosm of original chaos, and the Stone will be a microcosm of the created world: 'In Philosophers bookes whoso list to see, Our Stone is called the lesse world.'[57] The chaotic and corrupt nature of the Raw Stuff is also described as monstrous and reptilian: 'our Toade of the earth',[58] for instance; also serpent and dragon, though these have a more complex significance. We also hear, in these prefatory remarks to the *Compound*, of the importance of the quintessence:

> We have our heaven incorruptible of the quintessence,
> Ornate with Signes, Elements and Starres bright,
> Which moysteth our earth by subtill influence.[59]

Just as God's demiurgic power worked through the spirit that 'moved upon the face of the waters',[60] so the alchemist's creation involves a manipulation of this heavenly quintessence. It is identified with 'Our Mercury' –

> Not the common called Quicksilver by name,
> But Mercurie without which nothing being is;
> All Philosophers record and truely saine the same:
> . . . Mercurie of mettalls essentiall
> Is the principle of our Stone materyall. . . .
> It is a Soule, a substance bright:
> Of Sunne and Moone a subtill influence,
> Whereby the earth receiveth resplendence.[61]

This is the basic character of the Raw Stuff: a corrupt and confused earth, and within it Our Mercury, a subtle quintessential soul, vector of heavenly influences. This Mercury is the 'principle of our Stone': earth will be transformed by the revealing and exalting of the spirit within it. So the *opus* begins. The first gate into the castle of matter is calcination. The laboratory meaning of calcination is the reduction of a solid to calx, or fine powder. It is here to be understood as a preparatory purifying operation: it is the 'purgation' of the stuff. Hester's Quercetanus translation describes this first stage more concisely than Ripley: 'This matter after they have founde it, first they thinke meete to be calcined, and purged from all his filths'.[62] In cleansing the matter of its grosser impurities, Ripley warns, care must be taken that 'of radicall humiditie it looseth none'.⁶³ This 'moisture at the root' (Latin *radix*, root) is once again Our Mercury, which must be preserved unharmed at all costs.

The second gate is solution, and here the crucial work begins. Solution

> sheweth out what erst was hid from sight,
> And maketh intenuate things that were thicke also;
> By vertue of our first menstrue cleare and bright,
> In which our bodies eclipsed been of light:
> And of their hard and drye compaction subtilate
> Into their owne first matter kindly retrogradate.⁶⁴

Here the solid matter is dissolved into liquid form, made 'intenuate' (i.e. attenuated, thinned out) by the action of a powerful solvent ('menstrue'). The body is 'eclipsed', all shape and substance becoming fluid. This is a return to its 'first matter', or *prima materia*:

> Behold how Ice to water doth relent,
> And so it must, for water it was before,
> Right so againe to water our earth is went...
> Each mettall was once water mynerall,
> Therefore with water they turne to water all.⁶⁵

Dissolution is thus a stripping away of metallic form, to return to a preexistent condition of liquidity. As Quercetanus describes it, 'making thin his grosnes into a certaine liquide substance, as into his first matter, which they call minerall water'.⁶⁶ The perfecting process is now under way: 'the bodie is from his first forme alterate.'⁶⁷ This solution will also, says Ripley, reveal something 'that erst was hid from sight': it will free the hidden spirit within matter. He expresses that all-important rhythm of *solve et coagula* with the words:

> Our Solution is cause of our Congelation;
> For the Dissolution on the one side corporall
> Causeth Congelation on the other side spirituall.⁶⁸

The dispersal of the body and the enrichment of the spirit are inextricable, a curious metaphysical version of Einstein's conservation of mass and energy. Yet in Ripley's chemical journey, it is not until the sixth gate that we reach congelation. In the limbo between *solve* and *coagula*, in the suspended state of formlessness between the loss and resumption of solidity, certain crucial events occur. The three gates which follow solution are a description of what happens *inside* that circle of *solve et coagula*.

What first of all happens is a separation, the third gate of the *Compound*. Separation (often called *Separatio Elementorum*) is a breaking apart and dispersal of the four elements –

> Makyng division of qualities elementall
> Into a fifth degree till they be turned all.

The 'fifth degree' refers once more to the quintessence, for this separation is the vital partition of the bodily and spiritual aspects of matter.

> Separation doth each part from other divide,
> The subtile from the gross, the thick from the thinn...
> Thy Matter dividing into partes two,
> So that the subtile from the gross thou take
> Till earth remaine beneathe in colours bloe,
> That earth is fixed to abide all woe:
> The other parte is spirituall and flying,
> But thou must turne them all into one thing.[69]

The breaking of the elements releases the 'spirituall and flying' *anima*, the Mercury of the Philosophers. As always, the stress is on fire as the agent of this violent but ultimately transforming extraction. In the *Compound*, as in many other texts, this does not just mean the elemental fire of the furnace: there is another secret fire, the *ignis innaturalis* or 'fire against nature':

> Fire against Nature must doe thy bodies woe;
> This is our Dragon as I thee tell,
> Fiercely burning as the fire of hell...
> Therefore make fire thy glasse within,
> Whych burneth the bodie much more than fire Elementall.[70]

This inner fire, the dragon, can be interpreted as a caustic solvent, just as the 'Lyon greene' is said to be oil of vitriol (sulphuric acid). Often, however, this fire is actually located inside the Raw Stuff, the matter's 'own proper heat', a self-dissolving principle activated by the alchemist. We shall return later to the paradoxes of the alchemical dragon.

Matter thus resolved into the component parts of its duality, the next step is to make of them 'one thing'. This is the fourth gate to be unlocked, conjunction:

> After the Chapiter of naturall Separation
> By which the elements of our Stone disservered be,
> The chapter here followeth of secret Coniunction,
> Which Natures repugnant ioyneth to perfect unitie,
> And so them knitteth that none from others may flie
> ... This Coniunction is nothing els
> But of dissevered qualities a copulation.[71]

This idea of conjoining the separated opposites in matter is often described as the 'Chymicall Wedding' and depicted in alchemical emblems as a coitus, often between a naked king and queen. It is a marriage 'betwixt the agent and the patient ... male and female, Mercury and Sulphure vive'.[72] In this coitus is 'the sperme conceived',[73] the seed of new form, eventually to be born as the 'Philosophers' Child' or *'filius macrocosmi'*, i.e. the Stone.

But the Chymicall Wedding is a strange deathly embrace, for it shades into 'clowdes and eclipses',[74] the blackness of putrefaction. This is the fifth gate, the nadir of matter's journey through formlessness. Putrefaction is the 'fleying' of the body, and 'the killed bodies into corruption foorth leading'. The matter undergoes 'paines by heat' and languishes 'in darknes of purgatorie withouten lights' until all becomes 'powder black as a crowes bill'.[75] This is the alchemical death, the final loss of form:

> The blacknes shewing shall tell thee when they dye,
> For they togeather like liquid pitch that tide,
> Shall swell and burble, settle and putrefie.[76]

This death is, of course, a prelude to new life: the resurrection of matter in purer form. Putrefaction is in fact the gestation of that seed of new form conceived in the *coniunctio*. Without putrefaction, Ripley says, 'no seed may multiply',[77] and Quercetanus asserts that 'there is nothing borne or groweth ... but first this putrefaction is made'.[78] This is a universal alchemical theme: no generation without corruption. And after the corruption:

> like as soules after paines transitorie
> Be brought into Paradice where ever is ioyfull life,
> So shall our Stone (after his darknes in Purgatorie)
> Be purged, and ioyned in Elements withouten strife,
> Rejoyce the whitenes and beautie of his wife.[79]

The blackness, or *nigredo*, of the alchemical death is the turning point of the *opus*. It is the 'perfect meane of profound alteration',[80] an alchemical valley of death. 'Thus by the gate of blacknes thou must come in To light of Paradice in whitenes.'[81]

A separation, a wedding and a death: all this takes place within the circle of *solve et coagula*. The sixth gate, congelation, is a triumphant return to altered regenerate substance. The flying spirit is recaptured. The alchemist has now attained the White Stone (the first of the Stone's two forms, the Red Stone being the final triumph). Congelation (or coagulation) is

> of soft things induration of colour white,
> And confixation of spirits which flying are ...
> For when thy matter is made perfectly white,

> Then will the spirit with the bodie congealed be,
> But of that time thou maist have long respite,
> Ere it congeale like pearles in sight to thee,
> Such congelation be thou glad to see . . .
> Then hast thou a Stone most precious of all Stones.[82]

Quercetanus writes, in similar vein, of this moment when

> all the matter be white like unto pearles; then is there a true fixing and congealing of the shining spirits with the bodies. The Chimick philosophers do call this the White Thorne, and white Sulphur incombustible which never departeth from the fier.[83]

The attainment of the White Stone, 'this worke named the milke white Dyademe',[84] marks the completion of the first and most important stage of the *opus*. The remaining gates describe the processes for further exalting the White into the Red Stone. They are briefer and less tractable to analysis: Ripley covers the second stage of the journey in only 61 stanzas, as opposed to the 151 of the first six gates. Briefly, the stages are as follows. Cibation, a process of fortification, a 'feeding of our matter'[85] (Latin *cibus*, food); sublimation, apparently a repetition of earlier processes, 'to make the bodie spirituall . . . that the spirite may corporall bee';[86] fermentation, exaltation and multiplication, all obscure processes which increase the power and 'the greater nobilitie of the matter';[87] and finally projection, which measures the perfecting strength of the now completed Stone: 'In Proiection it shal be proved if our practise be profitable . . . if thy Tincture be sure and not variable.' All being well, 'a little of thy medicine' will tinge all metals into gold 'all fires to abyde'.[88] The great work is completed, matter is transformed, and

> there is harvest, that is to say an end
> Of all this worke after thine owne desire:
> There shineth the Sunne up in his Hemisphere,
> After the Eclipses in rednes with glorie,
> As king to raigne upon all mettals.[89]

*   *   *

I might fear to have fallen into the exegete's trap of *ignotum per ignotius* – explaining the obscure by the more obscure – till a glance at the *Compound* reminds me that no one could be more obscure than Sir George himself. I have pruned, interpolated and generally cajoled his text into some sort of shape, but also I have tried to take his *opus* on its own terms: as an unfolding process in which chemical operations and metaphysical conceptions blur into a single aspiration towards 'perfect matter'. The difficult, dangerous and secret journey into the 'strong castle' of matter is a kind of rescue operation, to deliver from its stronghold 'the Red Man and his White Wife'[90] – or Red and White Queen – who comprise the Stone. Though the details of this journey are often impossible to understand, the basic route is clear enough. The alchemist dissolves, disintegrates and 'kills' his matter in order to regenerate it in altered form. In the course of this chemical assault, the opposites within matter are released, distinguished and then 'married' together. The blackness of putrefaction is followed by the whiteness of rebirth, when the fled spirit returns to quicken the stricken matter to new life. Though Ripley insists on a distinction between solution/congelation and separation/conjunction, this is not vital. Essentially, they describe the same rhythm of breakdown and reconstruction, severing and synthesis, which is the keynote of the whole *opus*. As Ripley puts it:

> Loosing and knitting thereof be principles two
> Of this hard Science, and poles most principall.[91]

The shape of the work is thus circular, a going-out and coming-back: the Mercurial spirit is released in order to return and redeem, matter is brought to nothing in order to become new matter. Ripley expresses this idea of circularity when he calls the *opus* a turning of 'the Wheele of our Philosophie'.[92] We shall return to Ripley's philosophical wheel in a later chapter, and find it has many features in common with that 'wheel of fire' which bears King Lear to death and life again.

The Hieroglyphicall Unit

The appearance of the *Compound* and the *Mirror* on the London bookstalls of the 1590s is part and parcel of the great alchemical enthusiasm of the time. They were offered in a spirit of revival: antique English texts rescued from the dim and dubious pre-Reformation past and now available for the delectation of the interested Elizabethan reader. He might find in them much the same as we can today: in the *Mirror* a guide to the theories, metallic and metaphysical, which contributed to the concept of alchemical transformation; in the *Compound* a sense of the pattern and process of the *magnum opus* that produced the fabled Stone out of chaotic matter.

In the century between the death of George Ripley in 1490 and the publication of his work in 1591, alchemy underwent profound and vital changes. But these were not, essentially, structural changes. The basic theories and patterns of alchemy remained constant, as they had done for a thousand years before.[93] The great change was rather one of status

Nature, descendeth from Unity by the Binary, into the Ternarie; & yet not before such tyme as it ariseth from the Ternarie, by order of Degrees into Simplicity.

Sacret & Celestiall is this Adepted Philosophy, wherein whosoever desireth to have true knowledge, he same must bee contemplative & solitary, free from comon tumult. The Spirit of God doth breath where it listeth, illumining where he will, & whome he *toucheth* & shadoweth wth his Divine grace he leadeth into all knowledge of truth. Let him therefore wch shall receive such knowledge, give thankes to the Lord God, & let him be answerable to yt his knowledge in the deeds of Charity & in Christian lyfe, yt God may be glorified in such Scyence, & the worker of good work, receive ye reward of his owne eternall felicity in the kingdome of heaven. Amen.

A Light in Darknes.

[alchemical cipher text]

Thomas Tymme of Hasketon.

4 From Thomas Tymme's unpublished introduction to the *Monas Hieroglyphica* (c. 1602)

and atmosphere. The story of sixteenth-century alchemy is one of re-evaluation: a new reading of the purpose and scope of alchemical enterprise. The old adepts like Bacon and Ripley continued to be celebrated as exponents of traditional alchemical wisdom, and names like Lully, Geber, Arnald of Villanova, Avicenna, Morienus, Hermes Trismegistus, Aristotle and Plato continued to be cited with reverence. But the alchemical philosophy which flourished in the years around 1600 was conscious of a new, sophisticated and vitally contemporary identity. This 'new alchemy' was a largely European phenomenon, but one Englishman in particular was a formative figure in its emergence: Dr John Dee.

The name of Dee has cropped up regularly through these pages, whether as confidant to Elizabeth and Leicester, teacher of Sidney and Dyer, or unleasher of the manic Edward Kelley. His name was widely associated with alchemy – John Aubrey went as far as to claim him as the prototype of Jonson's Dr Subtle: 'He used to distill Egge-shells, and 'twas from hence that Ben Johnson had his hint of the *Alkimist*, whom he meant.'[94] Dee's rambling house at Mortlake, where he lived from about 1570 until his death in 1608, undoubtedly housed much alchemical experiment. When the house was ransacked in 1583 by an antimagician mob, Dee estimated that some £200 worth of chemical equipment was destroyed.[95] Though his interest in practical transmutation probably reached a peak under the influence of Kelley, Dee's involvement in alchemy was life-long. His *Monas Hieroglyphica* (1564) and *Testamentum* to John Gwynn (1568) attest his early enthusiasm for alchemical philosophy.[96] Forty years on, we find him conducting and carefully recording a six-week alchemical experiment in the winter of 1607–8.[97] He was then eighty years old, poor and abandoned: still he courted Lady Alchymia.

The vital point about Dee's passion for alchemy is the mental context for it – the context of all Dee's other passions. He was, for instance, a brilliant mathematician – 'ye prince of Mathematicians of this age',[98] as one contemporary enthused – and his *Mathematicall Preface* to the English translation of Euclid (1570) was profoundly influential. He applied his skills practically, as geographer and mechanician,[99] but also believed in mathematics (or 'mathesis') as a potent magic, drawing on Pythagorean ideas of mystical numerology to claim that the 'Mathematicall minde' can 'by good meanes mount above the cloudes and sterres'.[100] Dee was also renowned as astronomer and astrologer, and was in effect 'astrologer royal' to Queen Elizabeth. In this connection he was one of the first in England to espouse the revolutionary theories of Copernicus,[101] though once again his emphasis was magical, finding in heliocentricity a new vindication of the ancient sacred (especially Hermetic) associations of the sun. There was also Dee's long involvement in 'angel magic', with its complex use of Cabbalistic sigils (tablets engraved with secret names and Hebraic formulae) and invocations to achieve communion with the *angeli boni*.[102] This versatility of magical pursuits – mathematical, astrological, cabbalistic and alchemical – is the keynote of Dee's career as 'Elizabethan England's great magus'.[103] For his true place – as Drs Yates and French have recently stressed – is in the mainstream of the great magical tradition of the Renaissance, alongside such figures as Marsilio Ficino and Pico della Mirandola, Trithemius and Cornelius Agrippa, Giordano Bruno and Tommaso Campanella. This is the context of Dee's alchemy: it is one aspect of an interlocking network of magical techniques, one method of many by which the magus sought to probe and control the 'hidden causes of things'. To recognize the characteristic tone of alchemical philosophy in Shakespeare's day, one must understand alchemy as integral to the whole repertoire of Renaissance magic.

John Dee's major alchemical work, the *Monas Hieroglyphica*, was first published (in Latin) at Antwerp in 1564. This makes it somewhat earlier than the immediate Shakespearean period, 1590–1612, but it was still flourishing then: it was reprinted in 1591, defended in print by Dee in 1595, and included in the alchemical anthology, *Theatrum Chemicum*, in 1602.[104] Unfortunately no copy survives of the earliest English translation of the *Monas*. The translator was Thomas Tymme, the Suffolk parson whose prolific output included translations of Calvin and Peter Ramus as well as alchemical and mystical texts.[105] Everything survives but the translation itself – Tymme's 'Epistle Dedicatorie' to his patron, Thomas Baker; his 'forespeech to ye Reader'; his brief introductory gloss entitled 'A Light in Darknes'.[106] In this introduction, Tymme refers to pages of 'the Latin Theatrum' – i.e. to the text of the *Monas* in the *Theatrum Chemicum* of 1602 – so his translation was written (assuming it was) in or after that year. Perhaps it was after doing the *Monas* that Tymme turned to the two tracts (1603 and 1604) of Quercetanus which he published as *The Practise of Chymicall and Hermeticall Physicke* in 1605.[107] The title at the head of this section, 'The Hieroglyphicall Unit', is a purely conjectural name for this lost text – 'hieroglyphicall' is predictable, but 'unit' indispensable, being the translation Dee himself gave for Greek μονας (*monas*) in his *Mathematicall Preface*: 'Note the worde, Unit, to express the Greke Monas, & not Unitie, as we have all commonly, till now, used.'[108] This was the first recorded use of the word 'unit' in English. The *Hieroglyphicall Unit* is, then, a seminal English alchemical text of *circa* 1602, which was never published and may not even have been written! Quotation being difficult under this circumstance, I refer to the admirable modern translation by Prof. C. H. Josten, and follow his preference for 'monad' over 'unit'. Dee explained a unit as 'that thing Mathematicall, Indivisible', and this is the idea to be conveyed by *'monas'* or 'monad': oneness, indivisibility, unity.

The *monas hieroglyphica* which gives the work its title is, first and foremost, a hieroglyph *of* the monad – an actual visual design symbolic of essential oneness. It appears on the title-page so:

The text of the *Monas* consists, in effect, of an extended gloss on this symbol, twenty four theorems in which the monad is 'mathematically, magically, cabbalistically and anagogically explained'.[109] These are preceded by a lengthy dedication to Emperor Maximilian II (father of Rudolf II), in which Dee describes the work as a 'magic parable'.[110] It is a 'parable' on the meanings hidden within his talismanic monad, meanings, he claims, which will place 'supracelestial virtues and metaphysical influences' within the discerning reader's grasp.[111]. The monad itself, as a design, is composed of conventional symbols of equal astrological and alchemical significance. Its constituents are the signs for Mercury (☿), Aries (♈), Sun (☉) and Moon (☽).[112] The central point of the solar circle also represents the earth: it is what Dee calls the 'terrestrial centre' of the monad. Clearly the monad works in some way as a cosmogram of astrological import. More important for our purposes, it works as a talismanic synopsis of the alchemical process. The central place is occupied by Our Mercury, primarily symbolic of the quintessential spirit within matter. The appended sign of Aries represents fire, Aries being the first of the zodiac's 'fire signs'. The solar and lunar symbols represent gold and silver, the matter from which Our Mercury is typically extracted (as in Ripley's Mercury 'essentiall of Sunne and Moone'). Spirit, fire and matter: the basic constituents of the *magnum opus*. Dee's monad, a hieroglyph of essential oneness, is in this sense a symbol of the Stone, the 'one thing' which the alchemist achieved out of division and plurality. At the centre of the monad is the dot representing earth, the Raw Stuff on which all this is performed. Finally, the oval enclosure which houses the monad suggests the Philosophers' Egg, the vessel which contains the matter and in which new matter is 'hatched'. This is only a brief personal summary of the alchemical relationships suggested by the monad. Thomas Tymme interpreted it somewhat differently in detail, but also concluded that Dee's design was symbolic of the alchemical process *per se*:

In this Hieroglyphicall Monas of ☿ he hath comprehended the whole science and practise of Alchimie, in which one figure is set before you the character of the 7 Planets, and therein also a misticall significacon of the 7 Mettalles, whereof two are perfect, & the others imperfect, yet able to be perfected by Art and Nature; the which worke of Art and Nature concurring together he hath also inserted & closely couched in yt Figure.[113]

The centrality of Our Mercury to the visual design of the monad is confirmed by the words on an ornamental scroll surrounding the monad on the title-page. This reads: 'Mercury becomes the parent and king of all planets when made perfect by a stable pointed hook.' Our Mercury – in combination with fire, the 'pointed hook' of the Aries symbol[114] – is thus defined as the *agent* of transformation, regenerating and 'reigning over' all metals. We begin to see a vital identification: of Our Mercury – the volatile spirit and divine breath hidden within matter – as the secret transforming substance itself. This was certainly how Tymme interpreted it. Dee's 'whole purpose and drift', he wrote, 'is to give unto ☿ [Mercury] the Mastery in alchemy, & the α and ω in the worke.'[115] Another motto on the title-page is biblical, from Genesis: '*De rore caeli et pinguedine terrae det tibi deus.*' This the Authorized Version (1611) renders: 'May God give thee of the dew of heaven and of the fat of the land.' The *ros caeli* or 'dew of heaven' once more suggests Our Mercury, variously described in alchemical writings as 'divine rain', 'silver rain', 'drop of wet dew', 'May dew', etc.[116] The 'fat of the earth' suggests the contrary aspect of matter, the fixed and bodily qualities codified as Our Sulphur. 'Of the fatnesse of the earth .... Sulphur is engendred,' says the *Mirror*.[117]

This stress on alchemical Mercury is typical of the kind of alchemy which Dee's *Monas* expresses, an alchemy intimately involved in the wider aims of Renaissance magic. It is essentially a stress on spirit and, like the alchemist, the magician was vitally interested in the manipulation of spirit. The purpose of *magia naturalis* was – as conceived by Marsilio Ficino, Florentine Platonist and formative figure in the Renaissance magical tradition – to 'draw down the life of heaven'.[118] The magician sought to attract and control invisible, supernatural forces emanating from 'above', particularly from the planets. His magic was a channelling of celestial influences (literally, 'in-flowings') into special objects, places or moments. 'The magus', wrote Paracelsus in his *Philosophia Sagax* (1537), 'can transport many meadows of heaven into a small pebble, which we call 'Gemaheu' or 'Imago' or 'Character'. For these are containers in which the magus keeps sidereal forces and virtues as in a box.'[119] To achieve this influx of heavenly powers, some agent or vehicle was required. This Ficino called the '*spiritus mundi*' – the spirit, or breath, of the world. The great German magician, Heinrich Cornelius Agrippa, described this same *spiritus* in his *De Occulta Philosophia* (1533): it is

> such a kind of spirit required to be, as it were, the medium whereby Celestiall Souls are joyned to gross bodies, and bestow upon them wonderfull gifts. This spirit is after the same manner in the body of man. For as the powers of the soul are communicated to the members of the body by the spirit, so also the vertue of the Soul of the World is diffused through all things by quintessence.[120]

Spiritus or quintessence was thus the vital link, the 'medium' whereby occult celestial forces were infused into matter, bestowing on it 'wonderfull gifts' of magic power.

It is in this role – as alchemical *spiritus* – that Our Mercury looms so large in Dee's *Monas*. Indeed Agrippa himself wrote of *spiritus*: 'in it are all generative and seminary virtues, for which cause the Alchymists endeavour to separate this spirit from Gold and Silver.'[121] In his introductory words on the meanings of the monad, Dee writes:

> This our hieroglyphic monad possesses, hidden away in its innermost centre, a terrestrial body. It [the monad] teaches without words by what divine force that [body] should be actuated. When it has been actuated, it is to be united (in a perpetual marriage) to a generative influence which is lunar and solar, even if previously, in heaven and elsewhere, they were widely separated from that body.[122]

Here we see Our Mercury as this kind of magical 'go-between', the vector of celestial influences into matter. For the 'terrestrial body' is earth, alchemical Raw Stuff, and the 'divine force' (*divina potentia*) which must 'actuate' it is the Mercurial spirit. The effect of this alchemical spiritualization will be to imbue matter with a 'generative influence' (*influentia gonetica*) from the planets. Our Mercury 'draws down the life of heaven': celestial and terrestrial, previously 'widely separated', are now 'united in a perpetual marriage'. Elsewhere Dee calls Mercury 'the celestial messenger.'[123] This reminds us of the classic role of the Roman god Mercurius – the messenger of the gods, '*deorum hominumque interpres*'.[124] This is part and parcel of Mercury's role as *spiritus*: it bears down a supernatural 'message', an influx of heavenly powers into earthly matter.

Dee calls this marriage a '*Gamaaea*', which word he explains 'as the earth of marriage, or as the terrestrial sign of a union performed in the realm of influences'.[125] This suggests not only the monad as a 'terrestrial sign' of a supernatural event, but also the alchemical process itself as such a 'sign'. The *opus* is characterised as a kind of rite, the chemical events within the vessel being a visible enactment of an essentially invisible, spiritual occurrence. It is in this sense that Dee calls alchemy '*astronomia inferior*', or terrestrial astronomy, i.e., an earthbound and palpable version of celestial events and patterns. For instance:

> While we were once contemplating the motions, in theory and in the heavens, of the celestial messenger, we were taught that the figure of an egg adds to the scheme. For it is well known to astronomers that he [Mercury] on his course in the ether performs an oval orbit. And this said should be enough for the wise. Behold our interpretations of that celestial counsel (set forth hieroglyphically).... May the most wretched alchemists hence take admonishment and learn to recognize their various errors.[126]

Here Dee draws an analogy between the 'oval orbit' of planetary Mercury and the behaviour of alchemical Mercury. Within the oviform vessel – the 'Philosophers' Egg' – the Mercurial spirit traces its own orbit. Liberated from the embrace of matter, it ascends as volatile vapour to the top, or 'heaven', of the vessel. There it condenses, and descends in the form of drops – the 'dew of heaven' in Dee's biblical epigram – to regenerate the mortified substance below. This is the circular journey of *spiritus* within the microcosmology of the Hermetic vessel: an ascent and return, a fetching of vital celestial influences from 'above'. As the *Emerald Table* puts it: 'It ascendeth from the Earth into Heaven, and againe it descendeth into the earth, and receiveth the power of the superiours and inferiours. So shalt thou have the glorie of the whole world'.[127] The 'oval orbit' of alchemical Mercury – the successive freeing and fixing of volatile spirit – is thus a symbolic enactment of the magician's desire to 'draw down the life of heaven'. Through the agency of spirit, the Stone is imbued with supernatural gifts. The 'most wretched alchemists', labouring after more material rewards, are invited to mend their ways and perceive the true status of alchemy as 'chemical theatre' – a manipulation of cosmic powers in chemical guise, an emblematic performance within the tiny glass theatre of the Hermetic vessel.

*   *   *

Dee may possibly have had that passage from the *Emerald Table* in mind when speaking of the 'oval orbit' of alchemical Mercury. He certainly quotes from it elsewhere in the *Monas*,[128] though he would have known it only in Latin, the earliest English translation being in the *Mirror of Alchimy*, published over thirty years after the *Monas*. The *Emerald Table* purported to be 'the wordes of the secrets of Hermes, which were written in a Smaragdine Table [i.e. a tablet of emerald], and found betweene his hands in an obscure vaute, wherin his body lay buried.'[129] It is one of the classics of alchemical literature, a paramount example of the Hermetic strain in alchemy. This Hermetic element plays a vital role in the alchemy which flourished in the later sixteenth century.

The term 'Hermetic' refers to the legendary Egyptian sage and magician, Hermes Trismegistus ('thrice great Hermes'). In its earliest form, Hermeticism was essentially a religion or mystic philosophy, a curious blend of Greek neo-Platonism, Christian gnosticism and Egyptian magic.[130] Its flourishing in the Hellenistic Middle East in the first centuries AD is intimately associated with the rise of alchemy in the same place and period: the earliest extant alchemical writings – those, for instance, of Zosimos of Panoplis, *circa* AD 300 – are suffused with Hermetic doctrine.[131] This accounts for the frequency of the term 'Hermetic' in considerations of alchemy. Hermes Trismegistus was called the founding father of alchemy, and a variety of texts claimed him as their author, the *Emerald Table* and the *Golden Treatise*[132] being perhaps the most celebrated. Alchemists were the 'sons of Hermes', or the 'Hermetic tribe';[133] their vessel was the *vas Hermetis* or 'vase of Hermes'; from their injunctions that this vessel should be airtight ('luted with the clay of wisdom') comes the still current term 'hermetically sealed'. Finally, since the Greek god Hermes corresponds to the Roman Mercurius, Latin texts referred to Hermes Trismegistus as Mercurius Termaximus (or Triplex). As the teachings of Hermes were the life-blood of alchemical wisdom, so the spirit of Mercury was the life-blood of chemical matter: these were twin forms, philosophical and chemical, of a single sought-for revelation.

The alchemical and magical writings of Hermes Trismegistus were well-known in medieval Europe, but the rediscovery of Hermeticism as a religion was a phenomenon of the Renaissance. In 1460, an ancient Greek manuscript containing fourteen mystical Hermetic tracts was unearthed in Macedonia by an agent of Cosimo de' Medici. Brought back to Florence, it was translated into Latin by Marsilio Ficino and published in 1471 under the title *Pimander* (Poimandres, the 'shepherd man', being the name of the Hermetic god). This publication, which went through sixteen editions before 1600, sparked off a huge Hermetic revival which deeply influenced the development of Renaissance magic. Because Hermes Trismegistus was accepted as an historical figure – one who, in Sir Walter Raleigh's words, 'lived at once with, or soone after, Moses'[134] – his religion was deemed an inspired precursor of both Platonism and Christianity. 'There is one ancient theology,' enthused Ficino, 'taking its origin in Mercurius and culminating in the divine Plato.'[135] The rediscovery of the religious *Hermetica* thus provided a redeeming theological context for the potent magic long associated with Hermes and the Egyptians. Alongside the devotional texts of the *Pimander*, Ficino published a new translation of the *Asclepius*, an account of

Egyptian magical practices and 'celestial rites' whereby the 'continual effluvia' from heavenly bodies could be channelled into matter. All this, dangerously pagan by medieval standards, could now be vindicated as part of the 'one ancient theology'. The characteristic note of Renaissance magic is an insistence on its godliness, its rejection of diabolic pacts and 'black' arts. The magician was, in the words of Cornelius Agrippa, '*divinorum cultor & interpres*, a studious observer and expounder of divine things'.[136]

Hermeticism was tailor-made for the Renaissance in its whole emphasis on the innate divinity of man himself. This rings out in the *Asclepius*:

> O Asclepius, man is a great miracle, a Creature born to be reverenced and honored, being after the Nature & Image of God, as though he were a God. . . . O therefore the more temperate the Nature of man is, and comes nearest to God and the divinity, the more he despiseth that part of his whereby he becomes earthly.[137]

In similar vein, the first tract of *Pimander* concludes that 'man above all that live upon earth is double; mortal because of his body, immortal because of "the substantial man" '.[138] The 'substantial man' (or better, 'essential man' – the Greek word is οὐσιώδη) is a divine spirit locked in the embrace of physical nature. Renaissance magic is not only a 'drawing down' of influences but a 'mounting up' of the magician, the liberation and ascent of the 'essential man' within him. The magician, writes Agrippa, 'has co-habited with the elements, mounted higher than the heavens, elevating himself above the angels to the archetype itself, with whom he then becomes co-operator and can do all things.'[139] This is the thrust of Dee's mathetic and astral magic too: 'by good meanes mount above the cloudes and sterres.'

All this had a deep effect on the status of alchemy, the art so closely linked with the names of Hermes. Increasingly through the sixteenth century, alchemy is embraced into the ever-widening repertoire of Renaissance magical systems (the ancient tradition of the Cabbala was similarly annexed, largely through the agency of Pico della Mirandola). Trithemius spoke of the secret power essential not only for the magus to give power to images but for the chemist to 'control spirits'.[140] His pupil Agrippa spoke similarly of the 'one thing' without which 'neither Alchymie nor Naturall Magick can attain to their compleat end.'[141] *Neque alchymia neque naturalis magia*: this is the characteristic partnership, with Cabbala to make up the trinity.[142] And this is clearly the alchemy of Dee's *Monas*, with its alchemical talisman being 'mathematically, magically, cabbalistically and anagogically explained'.

Dee's *Monas* crystallizes the new status of alchemy as part of the complex, philosophical, Hermetic-Platonic magic which loomed so grandly over sixteenth-century Europe. And with that new status came a vital new emphasis. For just as the Hermetic magus asserted the self as the source of powers and sought to free the 'essential man' within him, so the new Hermetic alchemist asserted the self as the true arena, and object, of transformation. This Dee appears to affirm in the thirteenth theorem of the *Monas*. During the 'work of rehabilitating by fire' (i.e. the alchemical *opus*), he writes,

> there becomes at length apparent that other Mercury – who is indeed the uterine brother of the first – when the lunar and solar magic of the elements is completed. . . . He is (by the will of God) that most famous Mercury of the Philosophers, the microcosm, and Adam.[143]

Here Our Mercury, the quintessence drawn from Luna and Sol, is not only called a 'microcosm' of the universal *anima mundi*. He is also 'Adam', surely a reference to Adam Kadmon (Hebrew, 'primaeval man'), the cabbalistic name for that divine 'essential man' trapped within mortal nature. The *opus* draws forth a spirit from ourselves: we are the Raw Stuff in which Our Mercury, our hidden and magical spirit, can be arduously revealed. The same message is elliptically expressed when, after an obscure passage of alchemical symbolism featuring eggs, eagles and scarab beetles, Dee remarks: 'Here (O King) I am not attempting to act the part of Aesop, but that of Oedipus.'[144] His 'magic parable', in other words, is no mere fable. It is a riddle, like that of the Sphinx, which has but one answer: man.

Though Dee was an enthusiastic experimenter and distiller, there is scarcely a word of chemistry in the *Monas*. It is no esoteric handbook, like the *Mirror*; it has none of the smoky redolence that clings to Ripley's *Compound*. Chemistry has become for Dee totally emblematic. The *opus* is a mirror where chemical change reflects unseen alterations within the self, a theatre where the ascent and descent of volatile *spiritus* enacts the mystic transports of the magus who 'mounts above the clouds and stars' to fetch back archetypal powers. So Dee says of the successful *Gamaaea* (the *opus* as 'terrestrial sign' of supernatural event):

> When that advance has been made, he who fed [the monad] will first himself go away into a metamorphosis and will afterwards very rarely be held by mortal eye. This, O very good King, is the true invisibility of the magi which has so often (and without sin) been spoken of, and which (as all future magi will own) has been granted to the theories of our monad.[145]

The transmutation of matter and the metamorphosis of the magus: a chemical enactment of mystic themes. This is the keynote of Dee's alchemy.

*   *   *

The influence of Dee's *Monas* in European alchemical circles was considerable – less so in England, where Dee complained of his 'disdainfull and sklanderous cuntrymen' and their 'strange and undue speeches devised of that hieroglyphicall writing'.[146] Among the Europeans, a paramount example of Dee's type of alchemy is found in the work of Heinrich Khunrath, alchemist and physician of Hamburg. Dee in fact met Khunrath, at Bremen in June 1589, on his homeward journey from Bohemia.[147] The influence of the *Monas* is discernible in both of Khunrath's major works – the *Von Hylealischen Chaos* (1597), in which the Stone is styled the 'plusquamperfect Catholic monad',[148] and the *Amphitheatrum Sapientiae Aeternae* (1595), which quotes from the *Monas* on the subject of Cabbala as well as citing Dee's *Propadeumata Aphoristica*.[149]

Khunrath is claimed, as Waite puts it, 'as a hierophant on the psychic side of the *magnum opus*'.[150] The full title of his *Amphitheatrum* gives some idea of his peculiar brand of spiritual alchemy: 'The Amphitheatre of the only true eternal wisdom, the Christian-Cabbalistic, Divine-Magical, Physical-Chemical, Three-in-One Catholicon.' The title-page shows an engraving of Khunrath surrounded by the words '*Non, certe, haec sine numine Elohim*' ('None of these things without the will of God'); a mountain labelled 'Mercurius' surmounted by a shining light full of Hebrew characters; and two columns with the symbols for Magnesia and Sulphur at the top and the words '*Id quod inferius sicut quod superius*' (the Hermetic 'As above, so below') at the bottom. All this sums up the ingredients of Khunrath's alchemy – mystical, cabbalistic, Hermetic: an alchemy integrated, like that of the *Monas*, into a comprehensive system of penetration and revelation. The title again suggests that idea of alchemy as a kind of 'theatre', an emblematic chemical enactment of magical themes: the *opus* itself as an 'amphitheatre of eternal wisdom'.

The texts amplify the mystical promise of this title-page. The alchemist can transmute 'our Chaos of Nature' by virtue of a 'secret divine vision and revelation'.[151] The transforming Mercurial spirit is called '*vapor virtutis dei*' ('the vapour of God's virtue') and '*ruach Elohim*' ('the spirit of God', in cabbalistic language), and the Stone is matter 'filled, animated and impregnated . . . with the fiery spark of *ruach Elohim*, the spirit, breath, wind or blowing of the triune God'.[152] The ardours of the *opus* – 'study, meditate, sweat, work, cook' – culminate in religious ecstasy:

> So will a healthful flood be opened to you, which comes from the heart of the Son of the Great World (*filii macrocosmi*), a water which he pours forth from his body and heart to be a true and natural *aqua vitae*.[153]

This translation of *aqua vitae* – originally pure alcohol distilled from wine, hence the 'quintessence' of wine[154] – into the blood of Christ is typical of Khunrath's marvellously fevered text. The interplay of chemical and spiritual is illustrated in the *Amphitheatrum* in an engraving of the alchemist in his 'laboratory-oratory'. This is a vast hall, stretching away into the distance. On one side, in the circular *oratorium*, the alchemist prays before an altar on which are cabbalistic letters and geometric symbols and various godly admonitions: '*Ne loquaris de deo absque lumine*' ('only the illuminated may speak of God') and '*Hoc hoc agentibus nobis aderit ipsi deus*' (roughly: 'God will stand by us in our works'). On the other side is the *laboratorium*, complete with furnace, retorts, bellows, etc., and an array of vessels labelled '*ros caeli*', '*azoth*', '*hyle*', '☿', etc. Separating these two zones is a long table, filled with musical instruments, writing materials, a pair of scales and a rubric to the effect that 'sacred music disperses sadness and malignant spirits'. The engraving neatly represents the double nature of alchemy: chemical labour in the *laboratorium*, ardent prayer in the *oratorium* ('oratory' perhaps including alchemical literature). These twin aspects are perhaps united by the goal of celestial harmony represented by the musical instruments and scales.

Another exponent of the explicitly mystical alchemy so characteristic of the late sixteenth century is the Belgian, Gerhard Dorn. In a series of tracts published in the anthology *Theatrum Chemicum* (1602), he draws unequivocal analogies between alchemical and psychic transformation. The chemical art of 'freeing the spirit from its fetters' teaches how 'the mind can be freed from the body'. For

> if a man knows how to transmute things in the greater world ... how much more shall he know how to do in the microcosm, that is, in himself, the same that he is able to do outside himself, if he but know that the greatest treasure of man dwells within him and not outside him.[155]

This 'treasure' – the inner Stone – is that hidden divinity within man which the Hermetic magus sought to liberate. It is, says Dorn, 'in us and not of us, but of Him to whom it belongeth, who has deigned to make us His dwelling place .... In this especially we are made like unto Him, that He hath given us a spark of His light.'[156] Dorn also calls this divine 'spark' within the self an 'invisible sun', i.e. an inner 'philosophical gold'. In his specifically Hermetic *Physica Trismegisti*, he writes of Sol:

> As the fount of life of the human body, it is the centre of man's heart, or rather that secret thing which lies hid within it, wherein the natural heat is active. Therefore Sol is rightly named the first after God, and the father and begetter of all, because in him the seminal and formal virtue of all things whatsoever lies hid.[157]

Entwined in the symbol of the sun, the alchemical pursuit of *aurum nostrum* and the mystic quest for enlightenment become one and the same *opus*: 'transmute yourselves from dead stones into living philosophical stones'.[158]

This stress on the psychic and spiritual implications of alchemy is one clue to its huge popularity in the late sixteenth and early seventeenth centuries. These implications were always present in alchemy, but coded and submerged in esoteric symbolism, obscure chemical instruction and smoky medieval language. The 'new alchemy' is no less esoteric, but is explicit enough in its theme of human transmutation, its abstraction of chemical procedure to a ritual or emblematic level. Dee's *Monas* is formative, and the texts of Khunrath and Dorn typical, of this new reading of alchemy as a philosophy of transformations, whose chemical terms expound a spiritual system.

However one approaches alchemy, one is met with paradox and subterfuge. I have tried in this chapter to offer some composite image of alchemy by briefly surveying three English alchemical texts of Shakespeare's day. Each of them shows alchemy in a different guise – transmutations, elixirs, solutions and congelations in the *Mirror*; the arduous chemical journey of the *magnum opus* in the *Compound*; the strong magical and mystical tone of sixteenth-century 'Hermetic alchemy' in the *Monas*. One can say that these texts have a relationship: that the first two belong to an ancient tradition which the third did much to vindicate. The contemporary insistence on alchemy's spiritual intentions threw new light on the old adepts and their dark adages. The 1591 edition of the *Compound* illustrates this quality of revival, for it came complete with 'prefatory verses' by none other than Dr Dee, who praised its 'rare effects' and affirmed that 'the learned' will 'delight therein, And their delight will draw them on to skill'.[159] No doubt Dee found, beneath its opaque surface, the secrets of spiritual alchemy – in Ripley's beautiful expression of the Stone's ubiquity, for instance:

> Foules and fishes to us doth it bring,
> Every man it hath, and it is in every place,
> In thee, in me, and in each thing, time and space.[160]

A rich medieval English tradition revived; a strong contemporary identity entwined with the aspirations of Renaissance magic – these are two interacting factors in the alchemy of Shakespeare's day. To complete the picture, we must explore a third crucial factor: the role of alchemy in certain medical developments in the sixteenth century. It is time to bring on Paracelsus to the stage of the chemical theatre.

· 3 ·

The Rise of 'Chymicall Physick'

Paracelsus

'Wie ein Ruhmen Bestellter
ging er hervor wie das Erz aus des Steins
Schweigen.'
Rainer Maria Rilke – *Sonnete an Orpheus*

5 Paracelsus at age 45. Engraving by Augustin Hirschvogel (1538)

Paracelsus was a shock from which European medicine has never quite recovered. His revolutionary medical teachings – suffused with magic, alchemy and astrology – echoed across sixteenth-century Europe, polarizing opinion into traditional 'Galenist' and radical 'Paracelsist' camps. The controversy hit England in the closing decades of the century. Paracelsus has been called the 'Luther of medicine', but his more exact contemporary and soul-mate was Rabelais.[1] The *'espoventables Faictz et Prouesses'* of Pantagruel are most nearly incarnate in the catastrophic and prodigious life of Paracelsus.

Philippus Aureolus Theophrastus Bombastus von Hohenheim was born in Einsiedeln, near Zürich, in late 1493, the son of a doctor and scion of the Swabian family of Bombasts from Hohenheim, near Stuttgart. The pseudonym Paracelsus, which first appears on an astrological *practica* published in 1529, was a characteristic swagger, meaning 'surpassing Celsus' – Celsus being the Roman physician of the first century AD, whose medical works were rediscovered in the Renaissance and, like the *Hermetica*, published at Florence in the 1470s.[2] The prefix 'para' crops up throughout Paracelsus' work: two of his major works are the *Paragranum* ('Beyond the Grain') and the *Opus Paramirum* ('Work Beyond Wonder'). 'Beyond' was a kind of war-cry for him.

He was educated, he states, in *'adepta philosophia'*.[3] Among his teachers was Johannes Trithemius, the scholar and magician whose *Steganographia* (published in 1606, but influential long before) was a key work in Renaissance Cabbalism. Paracelsus' interest in the 'Gabal', as he called it, probably reflects Trithemius' influence.[4] The other side of his education was his apprenticeship in the mines of the Tyrol. From early childhood, he says in the *Grosse Wundarznei* ('Great Surgery'), the transmutation of metals fascinated him, and he saw the mines as Nature's own alchemical laboratories: 'Where the minerals lie, there are the artists. If one is to search for artists in the separation and preparation of Nature, he must look for them where the minerals are found.'[5] These twin poles of his education – on the one hand 'pansophic' and occult, on the other practical and metallurgical – laid the foundation for his own philosophy, with its fusion of disciplines, its restless and ransacking temper, its strange blend of arcane and empirical.

All his life Paracelsus was a wanderer. Of his early wanderings little is known, though perhaps the thorny, Grimmisch landscape of Dürer's engraving, *Knight, Death and Devil* (1513), tells us something of their texture and atmosphere. Perhaps as early as 1509, aged about 15, Paracelsus was on the road, a vagrant journeyman-scholar attending one university after another, possibly (on his own unsubstantiated evidence) graduating as a doctor at Ferrara in 1516. These were the years of his medical apprenticeship, drawn as much from rustic tradition and witches' lore as from recognized academic sources like Manardus and

Leonicinus at Ferrara. He claims to have crossed and recrossed Europe, performing cures from Stockholm to Naples.[6] In 1522 he was in service as a surgeon in the Venetian army. In 1524 he was at Salzburg, arrested and narrowly escaping death for taking part in a peasant rebellion. The peculiar ambiance of his life already surrounds these fragments.

In 1527 came a brief and notorious interlude. His healing powers had begun to earn him influential friends – among them Erasmus and the publisher Frobenius[7] – and in March that year Paracelsus was appointed City Physician at Basle. His impact was immediate and explosive: he issued an iconoclastic manifesto, the *Imitatio*, denouncing the traditional Greek medical authorities, Galen and Hippocrates; he burned the medical *Canon* of Avicenna on a bonfire in the city square; he styled himself the 'Monarch of Arcana', a physician 'instructed by Nature', not a 'wormy and lousy sophist' like the orthodox doctors.[8] His pupil and amanuensis in Basle, Johannes Oporinus, later recalled him: always drunk but always lucid, dictating his lectures late into the night, working all day at the furnace, throwing himself down to sleep fully dressed and clutching his sword.[9] This sword, supposedly given him by a hangman, is an indispensable part of the Paracelsus legend. In the pommel he kept a supply of his '*arcanum*', or secret remedy. In Jonson's *Volpone*, Nano the dwarf sings of 'Paracelsus with his long sword',[10] and Samuel Butler writes in *Hudibras* (1663):

> Bumbastus kept a devil's bird
> Shut in the pummel of his sword
> That taught him all the cunning pranks
> Of past and future mountebanks.[11]

Not surprisingly, the medical establishment closed ranks against the Monarch of Arcana. Bitter lampoons against 'Cacophrastus' and the 'mad bull' began to appear, and early in 1528 Paracelsus made a hurried and secret exit from Basle, pursued by a trumped-up indictment for contempt of court.

From then until his death in 1541, Paracelsus drifted, preached, healed and wrote, constantly pitted against the authorities. It is typical that, when he published his *Von der Frantzosischen Kranckheit* (1529) criticizing the use of guaiac wood to treat syphilis, the work was immediately banned at the instigation of the Fugger family, who owned the monopoly on importing guaiac from America. In 1530, at Beratzhausen, he composed the great *Paragranum*, founding his medicine on four 'pillars' (*columnae*): Philosophy, Astronomy, Alchemy and Virtue. This last word we shall often meet in its Paracelsist sense, which has more in common with Latin *virtus* (manliness, strength, excellence) than with our morally flavoured definition. 'Virtue' is the hidden power, spiritual and concentrate, contained in all bodies – plants, metals and men – and having specific curative effects when extracted. It is closely connected to the alchemical quintessence, to which Paracelsus often likens it.[12] The following years were a storm of literary activity. As one acquaintance[13] put it, '*subinde, subinde scribit*': he was ceaselessly, ceaselessly writing. The *Opus Paramirum* (1531), the *Grosse Wundarznei* (1536), the *Philosophia Sagax* (1537), the nine books *De Natura Rerum* (1537), and the *Labyrinthus Medicorum Errantium* (1538) are among the great works of this decade.[14] Towards the end he regained something of his former reputation, and attended patients of high rank. At Vienna he was admitted to an audience with King Ferdinand, and he was in Salzburg at the request of Bishop Ernst of Wittelsbach when he died, on 24 September 1541, bequeathing his few belongings to the poor. His epitaph described him as the 'distinguished Doctor of Medicine, who with wonderful art cured dire wounds, leprosy, gout, dropsy and other contagious diseases'.[15] His own motto was his true epitaph: '*Alterius non sit qui suus esse potest*' – 'Be not another's if you can be your own.'

The hobo's life Paracelsus always chose is the great clue to his revolutionary medical teaching. It fostered in him an intense empiricism, a devotion to what is learnt from '*Erfahrung*' ('experience', with an echo of *fahren*, to journey) rather than from '*logica*' or from adherence to standard orthodoxies.

> He who wishes to explore Nature must tread her books with his feet. Writing is learnt from letters, Nature however from land to land. One land, one page. Thus is the *Codex Naturae*, thus must its leaves be turned.[16]

The physician must address himself to Nature, and learn to 'overhear' (*ablauschen*) the secret knowledge innate in natural things:

> Nature is the physician, not you. From her you take your orders, not from yourself. She composes, not you. See that you learn where her pharmacies are, where her virtues are written and in what boxes they are kept.[17]

For Paracelsus, Nature was active, dynamic, proffering. Its medicine 'grows unbidden from the earth even if we sow nothing. . . . The herbs and roots speak with you, and in them will be the power you need.'[18] Cognition was an activity of the object itself – a '*Zuwerffen*': the object 'throws' its meaning out, 'the tree gives the name tree without the alphabet'.[19] To know Nature – its meanings and medicines – was to be illuminated by the *lumen naturae*, the light of Nature, a key Paracelsist phrase.

* * *

Paracelsus' medicine was deeply influenced by alchemy, both in theory and practice. In the *Paragranum* he wrote:

> The third fundamental part, or pillar, of true medicine is alchemy. Unless the physician be perfectly acquainted with, and experienced in, this art, everything he devotes to the rest of his art will be vain and useless. Nature is so keen and subtle in her operations that she cannot be dealt with except by a sublime and accurate mode of treatment. She brings nothing to the light that is at once perfect in itself, but leaves it to be perfected by man. This method of perfection is called alchemy.[20]

Alchemy was particularly applied to the distillation and extraction of medicines, both from herbs and, more controversially, from metals and minerals. These preparations were the Paracelsist *arcana*, concentrated essences drawn from substances and containing in them the invisible spiritual 'virtue' of the substance. What medicine needs, he wrote, is 'to extract, not to compose: it lies in knowledge of what is inside, not in composing and patching up pieces'.[21] The preparation of *arcana*, the isolation of curative 'virtues', was thus an alchemical separation of spirit from body:

> Many, by means of alchemy, hunt after a fifth essence, which means nothing else than that the four bodies shall be separated from the *arcana*.... If the sum total of the matter lies in *arcana*, it follows that the foundation of all is alchemy, by which the *arcana* are prepared.[22]

An *arcanum* says Paracelsus, is 'a chaos [i.e. a gas], clear, pellucid, and in the power of a star'.[23] This star, the Paracelsist *astrum*, introduces the astrological component of his medicine, closely connected to the preparation of *arcana*. To Paracelsus, the stars presided over every action, substance and anatomy. There was, corresponding to the macrocosmic heavens, a human '*corpus sydereum*', or 'firmamental body'. 'In every human being, there is a special heaven, whole and unbroken.'[24] Thus every disease, and every remedy, had its particular astral identity: 'the higher stars weaken and cause death, but they also heal; if any of these effects is to be produced, it cannot be done without the stars.'[25] The alchemical preparation of *arcana* is precisely related to the presiding influence of the *astra*. Just as the alchemist liberated the Mercurial spirit to make it a vehicle for the influx of celestial powers into matter, so the physician prepared the *arcanum*, whose volatile spiritual properties made it a fit receptacle for astral influences:

> Since medicine is worthless save insofar as it is from heaven, it is necessary that it shall be derived from heaven. And this bringing

down from heaven means ... the abolition and elimination of every earthly element which exists in it. Heaven does not rule except these earthly elements be separated from it. ... So medicine must be reduced to air, that it may readily be ruled by the stars. Can a stone be lifted up by the stars? No, unless it be volatilized.[26]

Since disease and remedy were often ruled by the same star, there was an intimate kinship between them. In this, despite its arcane astrological expression, lies one of the essentially modernizing aspects of Paracelsist physic: its recognition of specific treatment. Orthodox 'Galenist' therapy was non-specific. All disease was interpreted as an upset in the balance of the four physiological 'humours' — blood, choler, phlegm and black bile — and all remedy sought to restore that balance, by counteraction or by evacuations such as blood-letting, sweating and vomiting. To Paracelsus, however, each disease was a specific entity — 'a living thing, a seed'[27] — within the body, presided over by a specific star, and healed by a specific *arcanum*. Remedy and disease are 'attuned to each other like lock and key'.[28] He writes in the *Opus Paramirum*:

When the doctor says, 'Marcasite is good for this', he must know beforehand that the marcasite is in the world and that it is in the human microcosm. This is how the philosopher speaks. If he wants to speak as a physician, however, he must say: 'This marcasite is man's disease, hence it cures him.'[29]

This is the strongly homoeopathic element in Paracelsist physic. Like cures like: *'wie die Kranckheit ist, also ist auch die Artzney'* — 'as the disease, so also the physic'.[30] Here too alchemy plays a vital role, for the disease was identified chemically as well as astrologically — the marcasite mentioned above is iron pyrites. To combat a harmful bodily chemical, an *arcanum* from that chemical was required, and the advocacy of medicinal preparations from metals and minerals — notably mercury, antimony, sulphur, vitriol and tartar, as well as the more traditional cordials of gold and pearl — was one of the most controversial aspects of Paracelsist physic. Many of these were drastic, caustic remedies, violently expelling the disease: Paracelsus asserted a close relationship between remedy and poison in such substances as arsenic.[31] The skilful chemical physician, moreover, could draw several *arcana* out of a single substance, as he refined it towards 'maturity'. Vitriol, for instance: in its first 'budding time', it acts as a potent laxative; in its 'second time', as colcothar, it is styptic and assists scab formation ('colcothar mends the hole because colcothar is the salt that makes the hole'); in its oily, 'leafy' stage, it is a remedy for epilepsy; finally, its 'fruit' is a refreshing tonic.[32] The alchemist presides over the 'growth' of metals and minerals, extracting 'virtues' at every stage: 'As the hen by incubation converts the world outlined in the shell into a chick, so alchemy matures the *arcana* which are contained in the philosophical physician.'[33]

In this sense, Paracelsus saw alchemy at work in all things, in continuous natural processes of change and growth, separation and transformation. He spoke of Vulcan – the Roman god of fire, smiths and smelters – as a kind of cosmic alchemist, working on the formless *prima materia* (or '*Iliaster*', as he called it) and separating it into diverse qualities and species. There was also the '*Archeus*', which was the '*inwendig Vulcan*', the inner Vulcan, which 'directs everything into its essential nature'.[34] There was an '*archeus terrae*' at work in the mines, regulating the 'mineral fire in the mountains',[35] and an '*archeus stomachi*' within man, digesting and converting food: 'I place a smelting-works in the microcosm [i.e. man], and a smelter therein called *Archeus*.'[36] All this was Nature's alchemy, in whose footsteps the alchemist-physician must follow. In the *Labyrinthus Medicorum Errantium*, Paracelsus wrote:

> Alchemy is an art and Vulcan is the operator therein. Whoever is a Vulcan, he has power in this Art. Whoever is not a Vulcan has no power herein. ... All things are created with this view, namely, that they should be placed in our hands, but not altogether perfect. Wood grows to its proper end, but not to coal; clay is created, but a vessel is not formed from it. ... God created iron, but not ... as a horseshoe, a sickle, or a sword. These modifications are entrusted to Vulcan, and so this art is good. ... This, indeed, is alchemy, which directs to its final end everything which has attained some intermediate end. ... This only is alchemy, which by preparation through fire, separates what is impure and draws out what is pure. Though all fires do not actually burn, still they are fires and remain fires. So also there are alchemists of wood, such as carpenters ... or sculptors, who take away from the block of wood whatever does not form part of the contemplated statue. So too there are alchemists of medicine, who take away from medicine what is not medicine.[37]

This wide-ranging conception of alchemy is further dilated in the *Paragranum*, where to the carpenter and sculptor are added the baker, the wine-maker and weaver. 'Whatever is poured forth from the bosom of Nature, he who adapts it to that purpose for which it is destined is an alchemist.'[38] 'Alchemy' is almost Paracelsist for 'technology'.

*   *   *

This is a brief digest of the alchemical elements in Paracelsist medical theory. We see how central alchemy is to the whole enterprise. It provided the chemical techniques for distilling and extracting the virtue-laden *arcana* from herbs, metals and minerals: quintessential medicine suffused with astral influences. It furnished a whole range of chemical substances which Paracelsus pressed into the pharmacopoeia. It fuelled his perception of an organic, operative Nature, with its Vulcans and *Archei*, the Nature whose medicine 'grows unbidden' from the earth. The physician, said Paracelsus, is 'the means by which Nature is put to work',[39] and again it is alchemy which defines that relationship: 'as the *Archeus* in the earth shapes, cooks and makes, so shall the physician be another *Archeus*.' Overall, alchemy was the prime Paracelsist technology, the means of adapting and perfecting Nature's *data* – things given – into active therapeutic form. It was his *access* to Nature. Alchemy begins to sound like a modern 'science', and although Paracelsus was steeped in the magical and esoteric traditions of the Renaissance, his paramount effect was to realign alchemy towards a practical, tangible purpose:

> Nicht als die sagen ALCHIMIA mache Gold mache Silber: Hie ist das fürnemmen mach ARCANA, unnd richte dieselbigen gegen den Kranckheiten.
>
> Not, as people say, alchemy to make gold or silver: here the purpose is to make *arcana*, and direct them against diseases.[40]

Metal-brewing Paracelsians

Paracelsus wrote of the physicians of the future: 'They will be *Geomantici*, they will be *Adepti*, they will be *Archei*, they will be *Spagyri*, they will possess the *Quintum Esse*.'[41] First, however, they had to be interpreters. Paracelsus' prodigious literary output was a chaos – scattered fragmentary manuscripts, part Latin, part German, peppered with hybrid neologisms – and the first generation of his disciples devoted themselves to editing, glossing, publishing and generally untangling the master's writings. The only important work of Paracelsus that was actually published during his lifetime was the *Grosse Wundarznei* of 1536. A few specialist medical tracts had also appeared, but none of the great philosophical works. During the decades after his death, a steady stream of texts appeared, in Latin, mainly from the great publishing centres of Basle and Frankfurt. Among these were the *Labyrinthus Medicorum Errantium* (1553), the *De Vita Longa* (1562) edited by Adam von Bodenstein, the *Archidoxis* (1569), the *Philosophia Sagax* (1571) edited by Michael Schuetz (Toxites), and the *De Natura Rerum* (1573). It was not until 1589, nearly fifty years after Paracelsus' death, that the first definitive Collected Works appeared: the mammoth ten-volume

6 Chymist at work. From Conrad Gesner, *The Practise of the New and Old Phisicke* (1599)

quarto *Bücher und Schrifften*, edited by John Huser and printed by Conrad Waldkirch of Basle.

By this time, the Paracelsist movement was at its height, and a number of original works had appeared under its banner. One of the earliest was the *Idea Medicinae Philosophiae* (1571) by the Dane Peter Soerensson (Severinus), who echoed Paracelsus' exhortation to learn from the great *Codex Naturae*:

> Sell your lands, your house, your clothes and your jewelry; burn up your books. On the other hand, buy yourselves stout shoes, travel to the mountains, search the valleys, the deserts, the shores of the sea, and the deepest depressions of the earth; note with care the distinctions between animals, the differences of plants, the various kinds of minerals, the properties and mode of origin of everything that exists. Be not ashamed to study diligently the astronomy and terrestrial philosophy of the peasantry. Lastly, purchase coal, build furnaces, watch and operate with the fire without wearying. In this way and no other, you will arrive at a knowledge of things and their properties.[42]

In this work, wrote an enthusiastic contemporary, Severinus 'hath opened the invincible foundation of Paracelsus physicke'.[43]

Another early contributor to the Paracelsist cause was Leonhardt Thurneisser, a roving physician, Cabbalist and alchemist, whose *Quinta Essentia* (1574) praised the curative virtues of the 'highest subtlety, strength and effect . . . of Medicine and Alchemy'.[44] In France the major figure was Joseph Duchesne (Quercetanus), a Calvinist physician who studied at the Paracelsist stronghold of Basle. A number of his chemical-medical works were available to the English reader of Shakespeare's day, through the translations of John Hester (see below, pp. 66f) and Thomas Tymme. Gerhard Dorn, already mentioned as a prominent 'spiritual alchemist', was another committed Paracelsian. He flourished as a physician in Frankfurt, wrote commentaries on Paracelsus' *Archidoxis* and *De Vita Longa*, and compiled a *Dictionarium Paracelsi* (1583; translated, in 1650, as the *Chymical Dictionary*).

These first literary broadcastings of Paracelsist theory were accompanied by a growing acceptance of actual medical practice involving chemically prepared metallic and mineral remedies. These remained hotly controversial – condemned by official medical faculties and denounced in counterblasts like Thomas Erastus' *Disputationes* (1572) – but nevertheless many of the authors mentioned above were practising physicians who held, at one time or another, influential posts. Von Bodenstein was physician to Otto, Count Palatine; Thurneisser to the Elector of Brandenburg; Severinus to King Frederick of Denmark; Quercetanus to Henri IV of France. Later powerful figures were Oswald Croll, physician to

Christian of Anhalt; Michael Maier, physician and private secretary to Rudolf II; and Turquet de Mayerne, who, outlawed by the medical faculty of Paris, set up practice in London in about 1607, became court physician to James I and Charles I, and was the chief mover of the official *London Pharmacopoeia* of 1618.⁴⁵

*   *   *

The Paracelsist enthusiasm that swept through Europe from the 1560s onwards was gingerly received in England. The earliest mention of Paracelsus by an Englishman was in 1562, in a pamphlet on spa waters by William Turner.⁴⁶ The leading authority on the transmission of Paracelsism into England, Professor Debus, has characterized its reception as a 'compromise': a growing acceptance of chemical therapy contrasted with a deep suspicion of the violent iconoclasm and arcane mysticism of 'out-and-out' Paracelsism. Typical of this 'middle path' is the Elizabethan surgeon, George Baker, who wrote in 1576 that 'the vertues of medicines by Chimicall distillation are made more vailable, better, and of more efficacie than those medicines which are in use and accustomed',⁴⁷ but insisted that a proper knowledge of Galen and Hippocrates – the orthodox authorities that were Paracelsus' *bêtes noires* – was essential.

The first fully-fledged Paracelsist work in English appeared in 1585, under the authorship of 'R. B. Esquire', probably one Robert Bostocke.⁴⁸ Its full title reads:

> The difference between the auncient Phisicke, first taught by the godly forefathers, consisting in unitie, peace and concord: and the latter Phisicke proceeding from Idolaters, Ethnickes and Heathen: as Gallen, and such other, consisting in dualitie, discorde and contrarietie.

Paracelsism is thus equated with ancient, pristine wisdom. The emphasis is drawn away from Paracelsus as innovator, to suggest that his chemical medicine has, like alchemy itself, a pedigree of profound antiquity: in the text, Bostocke cites Hermes Trismegistus, Pythagoras and Democritus as among the 'godly forefathers' who practised this medicine. It consists in 'unitie, peace and concord' in the sense of curing 'like with like' – matching the *arcanum* to a specific disease – rather than the Galenic 'contrarietie'. The new physic is explained as a blend of godliness and empiricism: 'the Chymicall Phisition in his Phisicke first and principally respecteth the worde of God, and acknowledgeth it to be his gifte; next he is ruled by experience.'⁴⁹ Bostocke smartly weaves in an anti-Catholic bias, comparing the spiritual virtues of the pure *arcana* with the true doctrine of Christ, and the 'corporal and grosse medicines' of the Galenists with the Romish religion, which is 'mixed with impurities' and 'standeth in outward ceremonies and traditions, corporal exercises which be less to the workes of the spirite'.[50]

Another early champion of the Paracelsist cause in England was the physician, Thomas Moufet, or Moffett, who had studied at Basle and met the Dane Severinus at Elsinore in 1582. His *De Jure et Praestantia Chemicorum Medicamentorum* (1584) predates Bostocke's *Difference* by a year, but it was written in Latin, published at Frankfurt and appears to have had no English edition. His personal influence was, however, considerable. A fellow of the Royal College of Physicians from 1585, he pioneered its guarded acceptance of the new 'chymicall physick': the proposed Pharmacopoeia of that year was not completed, but laid the foundations of Mayerne's of 1618.[51] Moffett's patients included Francis Drake and Francis Walsingham, and he was chief physician of the Herbert family – Henry, second Earl of Pembroke, and his wife, Mary, sister of Philip Sidney. In their service he resided at Wilton House, and it was for the later third earl – William Herbert, identified by some as the 'Mr W. H.' of the Shakespeare *Sonnets* dedication[52] – that Moffet wrote his biography of Sidney, *Nobilis*, in which he praised Sidney's pursuit of 'chemistry, that starry science, rival to Nature'.[53]

The man who perhaps did most to disseminate the new 'chymicall physick' was the distiller, John Hester, who prepared and sold his chemical remedies 'at the signe of the Furnaises' (or 'Stillitorie') at Paul's Wharf on the north bank of the Thames.[54] One of the broadsheets he posted round the city to advertise his wares is preserved at the British Museum. It announces:

> These Oiles, Waters, Extractions or Essences, Saltes, and other Compositions; are at Paules Wharfe ready made to be solde, by IOHN HESTER, practisioner in the arte of Distillation; who will also be ready for a reasonable stipend to instruct any that are desirous to learne the secrets of the same in few dayes, &c.[55]

Over two hundred preparations are listed, ranging from traditional herbal extracts to the controversial chemical remedies, including *essentia perlarum* (essence of pearl), *creta vitrioli* (a kind of fuller's earth), *balsamum sulphuris* (balm of sulphur), *saccharum plumbi* (sugar of lead, or lead acetate), and the two staples of Paracelsist therapy – *mercurius sublimatus* (sublimed mercury) and *vitrum antimonii* (the 'antimoniall cup').[56] Some of his potions are quite bizarre – *sal cranii humani* (salt of human skull) and *aqua spermatis ranarum* (water of tadpoles), for instance. He also offered more practical preparations, like 'divers and sundrie vernishes' and 'strange and terrible fireworks'.

Hester is typical of this new type of alchemist – the practical and medicinal 'chymist' – that flourished in late sixteenth-century England under the aegis of Paracelsist physic. He was also a prolific translator.

His earliest extant works are two translations (1579 and 1582) of the Bolognese physician Leonardo Fioravanti.[57] From 1590 onwards, he produced a steady stream of Paracelsist translations, including Quercetanus' *Sclopotarie* (1590) and *Originall and Causes of Mettalles* (1591), the pseudo-Paracelsist *Hundred and Foureteene Experiments and Cures* (1596), and works by Philip Hermann, Barnardus Penotus and Isaac Hollandus. These are generally technical texts, full of chemical procedure and recipes, together with directions for their medical application – 'one scruple of *laudanum precipitatum* in the water of plantaine' for the patient that 'spit blood',[58] oil of white amber for that 'rising of the mother' (i.e. hysteria) which Shakespeare's Lear suffers from,[59] and so on. In the preface to the *Key of Philosophie*,[60] however, Hester delivers an illuminating insight into the world of the 'chymist'. He relates how he was 'casting about' for a career to follow, and found none to his liking,

> till at length in the middest of this muse, I met in my minde with two such minions, as in my conceit were the only Paragons of the rest: the one gallant and gorgeous, garnished with gold and silver, bedect with iewels, sole Ladie and Governesse of all the rich mines and mineralles that are in the bowels of the earth: the other sweet and odiferous, adorned with flowers and hearbes, beautified with delicate spices, sole Lady and Regent of all the pleasant things that grow uppon the face of the earth. These I vowed to serve and honour, even to the losse of life and limme. . . . Divers and sundrie their affaires have they imployed mee in, in the which I have faithfully, painefully, and chargeably applied myselfe, and attained by their instructions (to mine own destruction almost) manie their hidden secretes.[61]

The metaphor is one of seduction and servitude. The Ladies of Minerals and Herbs – Hester's 'minions' in the sense of French *mignonnes*, delicate beauties or darlings – are twin guises of Lady Alchymia herself. Alchemy is a kind of enticement, a beckoning promise of 'hidden secretes' behind the veils of matter. Secrets, Hester adds, learnt arduously: knowledge 'digged out of harde stones, blowne out from hote fire, raked out from foul ashes, with great cost and greater travaile'. A sense of strenuousness, even of risk and danger, is something stressed in all alchemical writings. They were lonely pioneers down 'this perilous passage to Science'.[62]

By the 1590s, through the enthusiasm of men like Bostocke, Moffett and Hester, 'chymicall physick' had come to roost in England. By the turn of the century, a number of Paracelsist physicians were successfully practising in London. There was Simon Forman, distilling 'strong waters' to cure the plague, diagnosing astrologically ('as Saturn doth cause the black plague, so Mars doth cause the red')[63] and fighting a running battle with the Royal College of Physicians. There was Francis Anthonie, a goldsmith's son who set up practice in London in 1600, marketing his *aurum potabile* (cordial of gold) at 5 shillings per ounce.[64] He was fined £5 and imprisoned for eight months for practising without a license. In 1616 he published his delightful *Apologie of Aurum Potabile*, full of invective against the 'pasqualling libertie of traducing slanderers' and containing many 'irrefragable testimonies' from satisfied customers, among them Thomas Kisley, secretary to the Earl of Southampton, and James Mosan, 'primate physician' to the Landgrave of Hesse.[65] Other practising Paracelsians included the Hermetic philosopher and Rosicrucian apologist, Robert Fludd[66]; the French exile, Turquet de Mayerne; John Dee's son, Arthur; and, of course, Sir Walter Raleigh, distilling his 'Great Cordial' and 'Guiana Balsam' in the Tower.

*   *   *

The new 'chymicall physick' was more than a challenge to orthodox medicine: it entailed a drastic reappraisal of the purpose and value of alchemy. Alchemy, Bostocke asserts in the opening pages of the *Difference*, is the very name by which Paracelsist physic is to be known:

> The true and aunciant phisicke which consisteth in the searching out of the Fountaines of Nature . . . is called by aunciert name Ars sacra, or magna, & sacra scientia, or Chymia, or Chemeia, or Alchimia, & mystica & by some of late Spagirica ars.[67]

Alchemy is, as Paracelsus himself insisted, the essential method of discovering, defining and controlling the therapeutic secrets of Nature. 'We have', claims Barnardus Penotus in Hester's translation, 'brought into physicke essences, oiles, balmes and saltes, all which the Alchymists schooles have found out. And how great light is come into physicke, only by true distillation it is knowen to all men.'[68] The key-note of Paracelsist alchemy is practical purpose. Practice, says Penotus, is 'the plaine and manifest workemistresse, established even by trueth it selfe'.[69] Practice meant partly the proof of results: to the carping traditionalist that 'backbitest Paracelsus', Penotus asks how can 'thy trifling wordes deface the worke it selfe, or with thy theoricke refell the practise?'[70] More than that, it meant the physician's operative participation in chemically preparing his medicine. Orthodox doctors are challenged as aloof and theoretical:

> Have you tasted the most sharpe salt? Or the most sweet oile? Or the balme of that most delicate liquor? All those being hidden in everie thing that is created, you have not once perceived. The metalline spirits, in whom physicke doeth consist, by no meanes

can be founde out, neither what force they have or fellowship with man's nature, but only by fire. . . . But thou didst never handle coales . . . therefore being ignorant in metalline physicke, thou canst not so much as once ghesse what it is, and therefore doest judge of things unknowen, as the blinde man doeth of colours. . . . Hereby it appeareth that ye nourish a secret ignorance of naturall things in your selves. . . . How vaine is this reasonable Phisitian, which prepareth his medicines with reason and not with the hand.[71]

The new physician's central role was as alchemist: the Paracelsist Vulcan, distilling, extracting, separating, purifying. Give yourself, Penotus concludes, 'unto the studie of that sensible and metalline philosophie called Chymia', and learn by 'long travell' to draw forth spirits and balms from sundry matters.[72]

These words, published in London in 1596, are typical of the Paracelsist vindication of alchemy. Apprentice to Vulcan and Nature, toiling experimenter, revealer of the secret healing 'virtues' in matter: this was the new image of the alchemist as medicinal 'chymist'. This partly entailed a demystification of alchemy, a deliberate evasion of its esoteric and magical connotations. Hester, for instance, is guarded about metallic transmutation. He has 'seene many wonderful things' performed by alchemy, but thinks it 'unreasonable that a man in so shorte a time should doe that thing the which Nature doeth in manie yeeres'.[73] In this sense, the Paracelsist current in sixteenth-century alchemy is precisely complementary to the extreme mystical developments described in the previous chapter, the spiritual Hermetic alchemy of Dee, Khunrath and Dorn. The 'chymist' and the 'philosopher' are certainly two increasingly divergent identities: the embryonic scientist and the declining magician. And yet these years around 1600 were the heyday of alchemy precisely because *both* these identities were alive and active, and both insisted that the true province of alchemy was man himself, whether the healing of his body or the transforming of his self.

Some literary reactions

The rise of 'chymicall physick' in England in the last years of the sixteenth century was quickly picked up by literary antennae. In the *Arte of English Poesy* (1589), for instance, George Puttenham describes the 'poeticall lamentations' of antiquity, in which the poet sought to 'remove or appease' sorrow

> not with any medicament of contrary temper, as the Galenistes use to cure (*contraria contrariis*) but as the Paracelsians, who cure (*similia similibus*) making one dolour to expell another, and in this

case, one short sorrowing the remedie of a long and grievous sorrow.[74]

Elegiac poetry is, Puttenham suggests, a kind of Paracelsist homoeopathy. It cures like with like, sorrow with lamentation: it makes poetic grief expel actual grief. 'As the disease, so also the physic,' said Paracelsus. In a similar vein, John Donne wrote in his verse letter to Henry Wotton (*c.* 1597):

> Onely in this one thing, be no Galenist: to make
> Courts hot ambition wholesome, do not take
> A dramme of Countries dulnesse; do not adde
> Correctives, but as chymiques, purge the bad.[75]

Again Paracelsist or 'chymique' medicine is associated with drastic purgative action. The disease, 'hot ambition', will not be cured *contraria contrariis* by administering a counteracting 'cool' quality, a dram of rustic sloth, to restore the balance or 'temperament' of the humours. It must be chased out by the potent penetrating spirit of its *arcanum*. '*Arcanum* is health, disease contrary to health: these two expel each other.'[76] Shakespeare plays on the same idea of *similia similibus* in *Macbeth* (1606): 'Let's make us med'cines of our great revenge, to cure this deadly greefe.'[77] The idea of *arcanum* as chemical spirit provides Donne with a more metaphysical metaphor in 'The Crosse', probably among his earliest Divine Poems:

> Materiall Crosses then, good physicke bee,
> But yet spirituall have chief dignity.
> These for extracted chimique medicine serve,
> And cure much better, and as well preserve;
> Then are you your own physicke, or need none,
> When Still'd, or purg'd by tribulation.[78]

We shall find this alchemical concept of purification – the self 'still'd [i.e. distilled] or purg'd by tribulation' – at the heart of one of Shakespeare's greatest plays.

If the new chemical medicine offered potential for metaphor, it also provided a target for satire. In his *Terrors of the Night* (1594), Tom Nashe gives a jaundiced account of the rise of the Paracelsist physician to high places.[79] First he must get 'a Library of three or foure old rustie manuscript books', and 'rake some dunghil for a few durtie boxes and plaisters', and 'of tosted cheese and candles endes temper up a fewe oyntments and sirrups'. Next he sets up in practice, 'not in the hart of the Cittie . . . but in the skirtes and out-shifts steale out a signe over a Coblers stall, lyke *Aqua vitae* sellers and stocking menders'. Now 'this mountebank' must insinuate his name in the right circles: he makes an arrangement with some 'needie Gallaunt' who

> straight trudgeth to some Noble-mans to dinner, & there enlargeth the rumor of this new Phisition, comments upon everie glasse and violl that he hath, rayleth on our Galenists, and calls them dull gardeners and hay-makers in a mans belly.

The nobleman is 'inflamed' and the vaunted physician is summoned:

> To court he goes; where being come to enterview, hee speaks nothing but broken English like a French doctor, pretending to have forgotten his naturall tung by travell.... Suffiseth he set a good face on it, & will sweare he can extract a better Balsamum out of a chip than the Balm of Iudaea: yea, all receipts and authors you can name he syllogizeth of, & makes a pish at, in comparison of them he hath seen and read: whose names if you aske, hee claps you in the mouth with halfe a dozen spruce titles, never til he invented them heard of by any Christian. But this is most certaine, if he be of any sect, he is a mettle-bruing Paracelsian, having not past one or two Probatums for al diseases.

The continental image, the chemical prowess, the disparaging 'pish' against traditional medicine, the *balsama* and metallic brews: all the trappings of 'chymicall physick'. His 'deepe prescience' is soon found wanting, however, and the 'hungry druggier' is back on the streets:

> This is the Tittle est amen of it : that when he waxeth stale, and all his pispots are crackt and wil no longer hold water, he sets up a coniuring schoole, and undertakes to play the baud to Ladie Fortune.

He becomes, in other words, someone much like Dr Subtle, Jonson's bawdy alchemist and cunning-man of some fifteen years later, who is also boasted as

> a rare physitian, ...
> An excellent Paracelsian! And has done
> Strange cures with minerall physicke. He deales all
> With spirits, he. He will not heare a word
> Of Galen or his tedious recipes.[80]

Jonson could hardly resist the comic potential of Paracelsism, and when Volpone dons the disguise of Scoto the mountebank and sets up his platform in a Venetian *piazza*, the scene is set for burlesque. Italy was famed for its mountebanks – the word is of Italian origin (*montare in banco*, to mount a platform), as is 'charlatan' (*cialare*, to prattle) – and Jonson may well have found a prototype for Scoto's grandiose 'drug lecture' in John Hester's translations of the Italian physician, Leonardo Fioravanti. Where Scoto trumpets the virtues of his miraculous *'Oglio'* – 'this precious liquor . . . this blessed *unguento*'[81] – Hester's Fioravanti similarly extols his *'Oleum Philosophal'*, his *'magno licore'* and *'unguento magno'*.[82] Amidst these Italianate flourishes, Jonson places more specifically Paracelsist echoes. Scoto begins his tirade against false physicians: 'These turdy-facy-nasty-paty-lousy-farticall rogues, with one poore groats-w rth of unprepar'd antimony, finely wrapt up in severall scartoccios, are able very well to kill their twentie a weeke'.[83] This catches the tone of Paracelsus' own denunciations of the 'wormy and lousy sophists' and their 'dirty drugs'.[84] Again Hester might provide a source. The dangers of 'metalline tinctures' that 'have not been rightly prepared', and of 'collier phisitions' that 'kil al men . . . with their venemous medicines',[85] are warned against in the *Hundred and Foureteene Experiments and Cures*. Scoto speaks of 'unprepared antimony': Hester's text also asserts that *'vitrum antimonii'* is lethal unless 'prepared and purged' from its 'venome and poison'.[86] Scoto's spiel is typically Paracelsist in its boasts of chemical prowess: 'Honourable gentlemen, I will undertake (by vertue of chymicall art) out of the honourable hat that covers your heade, to extract the foure elements . . . and returne you your felt without burne or staine.' His miraculous *'oglio'* – which he compares to 'Raymund Lullies great elixir' – is, he claims, compounded out of 'sixe hundred severall simples' and its composition plumbs deep alchemical secrets:

> Many have assay'd, like apes, in imitation of that which is really and essentially in mee, to make of this oyle; bestow'd great cost in furnaces, stilles, alembeks, continuall fires, and preparation of the ingredients. . . . But when these practitioners come to the last decoction, blow, blow, puff, puff, and all flies *in fumo*.[87]

*   *   *

Friar Laurence's phial of 'distilling liquor'[88] plays a vital role in the story of Shakespeare's *Romeo and Juliet* (*c.* 1595). When we first meet the Friar, collecting 'baleful weeds and precious juiced flowers', his words expressed a richly Paracelsist perception of Nature. He praises the earth as 'Nature's mother', bringing forth 'children of divers kind':

> We sucking on her naturall bosome find
> Many for many virtues excellent:
> None but for some, and yet all different.
> O mickle is the powerfull grace that lies
> In plants, hearbs, stones, and their true qualities:

> For nought so vile that on the earth doth live
> But to the earth some speciall good doth give.[89]

This is Paracelsus' Nature, full of 'virtues' and medicines which 'grown unbidden from the earth'. The *arcana* of 'chymicall physick' are beautifully expressed by Shakespeare as the 'powerfull grace that dwells In plants, hearbs, stones'. The Friar observes the relationship between poison and remedy, as twin aspects of the potent virtue in substances:

> Within the infant rind of this weake flower
> Poyson hath residence, medicine power.

So Paracelsus observes of arsenic: 'As long as it lives, poison and remedy are close together: when its poisonous substance is subdued, it loses its power of physic.'[90] And Raleigh wrote of how the 'skilful and learned Chymist' could 'draw helpful medicines out of poison, as poison out of the most helpful herbs and plants (all things having in themselves both life and death)'.[91] From this duality Friar Laurence draws a human analogy:

> Two such opposed Kings encampe them still
> In man as well as hearbes, grace and rude will.

The more philosophical Paracelsians made similar connections, as Thomas Moffett, who concludes from the alchemical principles of Paracelsus that 'man also is of a doubly principled nature, the one being volatile and the other fixed'.[92] Grace and rude will, volatile and fixed: the duality of body and spirit is a recurrent theme in *Romeo*, and Friar Laurence – distiller of 'virtues' and practiser of 'holy phisicke'[93] – injects an alchemical element into this theme.

Distillation – the key 'chymicall' technique, or at least the most readily identifiable of many extracting and purifying techniques – provided Shakespeare with a fund of metaphors. In the *Sonnets* (*c.* 1593–5), variations are played on a simple 'household' form of distillation: the extraction of rosewater from roses. It is a counteracting of material transience:

> Then were not summers distillation left,
> A liquid prisoner pent in walls of glasse,
> Beauties effect with beauty were bereft.[94]

To this is compared the getting of children as a means of extracting and preserving one's 'essence':

> Then let not winters wragged hand deface
> In thee thy summer ere, thou be distil'd.

> Make sweet some viall, treasure thou some place
> With beauties treasure ere it be selfe-kil'd.[95]

Poetry itself is styled a distillation. The beauty of the 'lovely youth' is transient, but 'when that shall vade, my verse distils your truth',[96] preserves it in the vessel of the sonnet. This is reminiscent of Marlowe's lines in *Tamburlaine* (c. 1587) about poets and 'the heavenly quintessence they still from their immortal flowers of poesy'.[97] By the time of *Henry V* (c. 1599) the implications of distillation are more alchemical:

> There is some soule of goodnesse in things evill,
> Would men observingly distill it out.[98]

This suggests the drawing out of precious spirit from vile matter, or of remedial *arcana* from poisonous substance: 'out of the gross and impure one comes forth the exceedingly subtle one,' asserts Khunrath.[99] The phrase 'observingly distil' is a neat epitome of the alchemist's experiments, the sense in which his 'speculative' chemistry was literally a matter of watching (Latin *speculari*, to watch or observe) and of drawing metaphysical inference from physical observation.

In *All's Well That Ends Well* (c. 1602), Parolles draws a lascivious analogy as he tells Helena she must wait before consummating her marriage to Bertram. She must accept

> a compell'd restraint;
> Whose want and whose delay is strew'd with sweets,
> Which they distill now in the curbed time,
> To make the comming houre oreflow with ioy
> And pleasure drowne the brim.[100]

Here abstinence itself – 'want' and 'delay' – plays the distiller, working the scented flowers ('sweets') of sexual desire into more and more potent form. With the delicious pun on the 'coming hour', Parolles imagines this quintessential brew of sexual juices overflowing the brim of the vessel. In *Venus and Adonis* Shakespeare wrote of the 'stillitorie of thy face': Parolles locates Helena's 'stillitory' elsewhere. *All's Well* is the only one of Shakespeare's plays that mentions Paracelsus by name – 'the artists . . . both of Galen and Paracelsus'[101] – adding his presence to the motif of 'healing the King' which the play so sardonically treats. The promises of 'chymicall physick' are no doubt lurking in Lafew's claim that

> I have seen a medicine
> That's able to breathe life into a stone,
> Quicken a rocke, and make you dance Canari
> With sprightly fire and motion.[102]

Arcanum merges into elixir, the 'universal medicine' that breathes spirit into matter: this too gets its mention in the play, styled with Shakespearean panache 'the tinct and multiplying medicine'.[103]

The most technical of Shakespeare's distillation metaphors occurs in *Macbeth* (c. 1606). Lady Macbeth will get Duncan's guards so drunk that

> Memorie, the warder of the braine,
> Shall be a fume, and the receipt of reason
> A lymbeck only.[104]

The 'receipt' is the receiver at the bottom of the still, where the pure distillate collects as it condenses. The 'lymbeck', or alembic, contains the original impure liquid to be distilled. The brain is compared to a still, in which the crude material of 'sensory data' is purified and separated, and its essence collected in the 'receipt of reason'. Alcohol will so befuddle this process that, where the pure and rational distillate should be, there will only be the crude and undigested liquor of experience. The guards will remember nothing of the murder of Duncan: it will be murky and meaningless in their minds.

These are only a few of the metaphors Shakespeare drew from distillation.[105] He knew his way around 'chymicall' techniques and, in plays like *Romeo* and *All's Well*, this connects with an awareness of the new Paracelsist medicine so closely associated with chemical practice. We are only scratching the surface here. In Shakespeare's later work, Paracelsist principles contribute richly to a central motif of healing and regeneration. In *Pericles* (c. 1608), for instance, the physician Cerimon delivers what is virtually a Paracelsist manifesto as he prepares to restore Queen Thaisa. He speaks of the 'virtue and cunning' that makes 'man a god', and of his physic as a 'secret art' deploying the 'blest infusions That dwells in vegetives, in metals, stones'.[106] We shall return to this later interest in another chapter.

*   *   *

The strange themes and theories of Paracelsism are all that remain to us, but the chymists, distillers and metal-brewers of the 1590s were a living presence. John Hester's distillery at Paul's Wharf was a moving part of the busy machinery of Shakespeare's London, adding its fumes to the none-too-fragrant air of Thameside. At least one literary figure, Gabriel Harvey, was a client and friend of Hester: Hester's broadsheet in the British Museum has handwritten annotations by Harvey, and in *Pierces Supererogation* (1593), Harvey pauses to deliver a brief elegy:

> Poules Wharfe honour the memorye of oulde Iohn Hester, that would not sticke with his frende for twentye such experimentes, & would often tell me of *A Magistrall Unguent* for all sores.[107]

Sir Walter Raleigh may have been another customer. His interest in distillation is well known, and Hester's dedication of the *Hundred and Foureteene Experiments* to him might suggest an acquaintance[108].

On his death in 1593, Hester's distillery was purchased by one James Fourestier, or Forester: an interesting figure. Earlier that year Forester had narrowly escaped with his life, for on 21 March he had been indicted along with the Puritan extremists Henry Barrow and John Greenwood for publishing works to 'cry down the Church of England and the Queen's prerogative'. While confined in the Fleet prison, Barrow and Greenwood had composed a series of militant pamphlets which they smuggled out in 'slips and fragments'. For his part in transcribing one of these, Forester found himself in the dock at the Old Bailey. He repented his 'sharpe maner of writing' and was spared: Barrow and Greenwood were hanged at Tyburn that April.[109] Again a whiff of subversion hangs over the alchemist: we remember the malcontent Poole, the coiner full of 'traitorous speeches' whom Marlowe found so entertaining in Newgate prison.

In 1594, Forester published *The Pearle of Practise*, a collection of Hester's chemical recipes 'garnished and brought into some methode.'[110] This book has two rather tantalizing connections with Shakespeare. To begin with, it was printed by Richard Field, 'dwelling in the Black-friers', a friend of Shakespeare's perhaps from boyhood, a fellow Stratford man seeking his fortune in the city.[111] It was Field who printed both Shakespeare's early narrative poems, *Venus and Adonis* and *The Rape of Lucrece*, registered at Stationers' Hall in April 1593 and May 1594 respectively. Dr Rowse has observed of these works that 'the proofs were very well corrected, and no doubt the poet was in and out of Blackfriars'.[112] So Shakespeare was 'in and out' of Field's printing house at the very time that Forester's *Pearle of Practise* was on the presses. Forester himself may often have been there, for he lived nearby: his preface to the *Pearle* is signed 'from my studie in the Blackefriers, the 19 of Ianuary [1594]'.[113] This proves absolutely nothing, but at least populates a corner of the London 'around' those early Shakespeare poems. One might glimpse the street-sign of Hester's distillery – called equally 'the Stillitorie' and 'the Furnaises' – in such metaphors from *Venus* as 'the stillitorie of thy face' and 'forth againe as from a fornace vapors doth he send'.[114] Or one might hear some earnest lauding of chymicall balms and salves – Hester 'would often tell me of a Magistrall Unguent,' recalled Harvey in 1593 – in Shakespeare's 'balme, Earths soveraigne salve' and 'drop sweet Balme in Priam's painted wound'.[115]

But a shared printer was not the most interesting connection between Forester and Shakespeare in 1594. Forester's *Pearle* was dedicated to 'the famous Mecaenas of all good learning, his honorable good patrone, Sir George Carey'.[116] George Carey, who also had a house at Blackfriars, was another person in Shakespeare's orbit, for it is from June of this year, 1594, that we have the first record[117] of the newly formed Lord Chamberlain's Men, the theatre company with which Shakespeare was to be associated, in one form or another, throughout his career. Their patron, the Lord Chamberlain, was George Carey's father – Henry, first Lord Hunsdon – upon whose death in 1596 George himself became their patron. Carey's patronage of Forester is by no means the only instance of his interest in Paracelsism and related pursuits. As early as 1587 he was playing host to Simon Forman, occultist and physician, at Carisbrooke Castle. Though Forman's reputation was not yet established, his interests certainly were. His diary for that year notes that he 'began to distill many waters' (11 May), that he met his 'scryer' John Goodridge (28 August), and that Goodridge 'first saw' spirits (4 November): a year of alchemical and angelic pursuits, rounded off by a visit to George Carey in December.[118] Another guest of Carey's – at Christmas 1592 'and a great while after' – was Tom Nashe: his *Terrors of the Night* (1594) was dedicated to Carey's daughter, the 'high wonder of sharpe wit and sweete beautie, Mistres Elizabeth Carey'.[119] It is conceivable that Nashe's witty account of the metal-brewing Paracelsian's insinuation into courtly circles referred to Carey's favouring of such as Forman and Forester. Such insolence towards a patron would be typical of Nashe: perhaps he was making amends for it when he wrote, in the same work, that

> of the divinest quintessence of mettals and of wines are many of these spirits extracted. It is almost impossible for anie to bee encumbered with ill spirites who is continually conversant in the excelent restorative distillations of wit and of Alcumie.[120]

Another reference to Carey from this year, 1594, is found in George Chapman's *Shadow of Night*, a work full of vaguely occult Platonic and Hermetic themes. In the preface, Chapman praises 'most ingenious Darbie, deepe searching Northumberland and skill-imbracing heire of Hunsdon'.[121] This places Carey in the company of Ferdinando Stanley, Earl of Derby, and Henry Percy, Earl of Northumberland, both famed as devotees of the occult, and Northumberland – the 'Wizard Earl' – as a distiller and alchemical enthusiast. What was the 'skill' which Carey embraced? Surely the one which is common to Forman and Forester, the *Terrors of the Night* and the Wizard Earl: the 'restorative distillations' of medicinal chymistry. By 1600, Carey was suffering from syphilis: Forman was again consulted (May 1601),[122] but despite the help of his 'strong waters', Carey died in April 1603. A court-lampoon summed him up thus:

> Chamberlain, Chamberlain,
> He's of her Grace's kin,

> Fool hath he ever been
> With his Joan Silverpin.
> She makes his coxcomb thin
> And quake in every limb:
> Quicksilver is in his head,
> But his wit's as dull as lead.[123]

The final couplet refers to mercury, the Paracelsist prescription (perhaps Simon Forman's) for syphilis.[124] The joke is clearly alchemical: the volatile spirit of Mercury has gone to his head but not transmuted his leaden wit. It was a common belief that laboratory fumes turned alchemists mad: Mammon in *The Alchemist* promises to 'repair this braine, Hurt wi' the fume o' the mettalls'[125] and in his *White Divel* (1612) John Webster refers to 'a gilder that hath his braines perished with quicksilver'.[126] Perhaps 'quicksilver is in his head' also suggests Carey besotted and befuddled by his pursuit of Lady Alchymia.

James Forester opens up interesting avenues. He shared printer and patron with Shakespeare, and the patron appears to have pursued alchemy — at least its practical Paracelsist forms — with some enthusiasm. This proves nothing about Shakespeare but suggests how the chymicall and literary circles of late sixteenth-century London might intersect. The distillation metaphors in the *Sonnets* and *Venus*, the Paracelsist traits of Friar Laurence's 'holy physic', all roughly dateable to 1593–5, are perhaps traces of that intersection. There are traces too in a play closely associated with George Carey, *The Merry Wives of Windsor*.

In 1597, George Carey received the Order of the Garter. The speech of the 'Fairy Queen' in the last act of the *Merry Wives*, with its unmistakable references to the Garter ceremony,[127] is now generally accepted as evidence that the play was commissioned by Carey to celebrate his investiture. It was probably first performed during the Garter Feast at Whitehall on 23 April 1597.[128] And George Carey would surely have enjoyed the character of Dr Caius. The *dramatis personae* describes Dr Caius as a French physician, and his Frenchness is the staple of his comic presence: instead of 'by my troth' he says 'by my trot' ('trot', prostitute); he joins two others saying 'I shall make-a the turd'; and so on. Despite his ridiculousness, 'Master Doctor Caius, the renowned French physician' is much sought after. He has 'friends potent at court' and promises the host of the Garter Inn to procure for him 'de good guest: de earl, de knight, de lords, de gentlemen, my patients'.[129] There is, however, some doubt about his medical skills: according to Sir Hugh Evans, he knows as much as a 'mess of porridge' about 'Hibocrates and Galen'.[130] Heavy French accent, fashionable court clientele, ignorance of orthodox medical authorities: Dr Caius sounds very like Tom Nashe's 'mettle-bruing Paracelsian' who 'speaks nothing but broken English like a French doctor' as he worms his way into noble favour. In Dr Caius, Shakespeare gently guys – as Nashe had more scurrilously – his patron's Paracelsist leanings.

Caius is a peripheral figure, but there is also a Paracelsist element in the central comic action of the play, Sir John Falstaff's preposterous wooing of the eponymous merry wives, Mistresses Ford and Page. They determine to humiliate him – 'to entertain him with hope till the wicked fire of lust have melted him in his own grease'.[131] The gross Falstaff – he is called 'unwholesome humidity', 'gross watery pumpion', 'greasy knight', 'unclean knight', 'hodge pudding', etc. – is to be melted down by fire to purge him of his grossness. To this end the wives trick Falstaff into making an undignified escape from a jealous husband, being bundled into a 'buck-basket' (laundry basket), and covered up with 'foul linen'. Falstaff recalls the ordeal:

> Yes: a Buck-basket! Ram'd mee in with foule shirts and smockes, socks, foule stockings, greasie napkins, that, Master Broome, there was the rankest compound of villanous smell that ever offended nostrill. . . . To be stopt in, like a strong distillation, with stinking cloathes that fretted in their owne grease: thinke of that, a man of my kidney; thinke of that, that am as subiect to heate as butter, a man of continuall dissolution and thaw: it was a miracle to scape suffocation. And in the height of this Bath, when I was more then halfe-stew'd in grease (like a Dutch dish) to be throwne into the Thames, and coold glowing-hot in that serge like a horseshoo: thinke of that![132]

The whole thing is a piece of comic chymistry. Falstaff is gross, unwholesome, corrupt matter. He is placed in the vessel, the 'buck basket', and immersed in a rank 'compound' of foul linen. Thus 'stopped in' (hermetically sealed), the mixture ferments and frets: as Hester puts it, 'they putrifie in a moist heate'.[133] The Falstaffian matter, being 'subject to heat', melts: he suffers 'dissolution and thaw'. In 'this bath' (the caustic chemical *balneum*) he undergoes, as he says, a 'strong distillation'. He is purged and purified, the hapless victim of such chymicall instructions as: 'set it to distill upon the fire with great heate' or 'let it seeth an houre long, and the earth and other filthinesse will separate it selfe, which you shall scumme off'.[134] It is a witty *reductio ad absurdum* of 'chymicall physick': not to be cured *by* distilled liquor but to be turned *into* distilled liquor. Falstaff's remedial punishment – the wives refer to it as his 'medicine'[135] – is to become a Paracelsist remedy, a 'strong water' of the kind Dr Caius distilled for 'de gentlemen, my patients' or Simon Forman for the Garter knight, Sir George Carey. Perhaps the absurd was ready and waiting for Shakespeare. In Forester's *Pearle of Practise*, which began this brief excursion, we find the claim that 'I have made the Quintaessence of mans blood, rectified and circulated, with which I have done most wonderfull cures'. And, even more appropriate for Falstaff melted in his own grease: 'The fat of a man is (as every man knoweth) hote and penetrative, and mollifying if you annoint the parts therewith.'[136] There perhaps is the source for Shakespeare's bizarre *arcanum*, distilled out of a fat man in a laundry tub.

· 4 ·

A New Chemical Light: Sources and Reflections

An alchemical renaissance

We have now isolated the three staple ingredients of alchemical philosophy in Shakespeare's day, the three 'simples' which make up that potent 'compound'. There is the ancient esoteric tradition of alchemy, typified by the medieval adept with his darkly elaborate chemical procedures for probing, changing and perfecting matter. And there are the two currents of sixteenth-century alchemy, promoting a dual image of the alchemist – the Hermetic mystic-magician, intent on revealing the transforming Mercury, philosophical gold or perfect Stone within man's nature; and the Paracelsist chymist-physician, labouring to extract the healing virtues and *arcana* hidden in 'the bosom of Nature'. Nevertheless, though it is clear that Ripley's *Compound*, Dee's *Monas* and Hester's *Hundred and Foureteene Experiments* are entirely distinct in tone and appeal, it is also true to say that they are different versions of the same system, different routes towards a common ideal of enrichment and wholeness, each insistent on the occult efficacy of spirit, whether defined chemically, mystically or medicinally. So having separated these ingredients we must now, in true alchemical fashion, conjoin them again. Such a conjunction or synthesis is, I suggest, the key-note of certain alchemical texts published in the first years of the seventeenth century. These were mostly European works, notably German and Bohemian, but their effect was quickly registered in England. They are the first texts of what might be called alchemy's 'renaissance': a new and highly refined synthesis of systems, an upsurge that was partly a revival, a potent optimism that promised a pan-European enlightenment around the corner – these are loosely suggested by the term 'renaissance'. Poised at the dawn of a century, these alchemists promised (in the words of one of their titles) a 'new chemical light'.

Perhaps the most typical voice of this alchemical renaissance is Basil Valentine's. His true identity can only be conjectured. The name is an alchemical epigram: *basilius valentinus* means 'the mighty king' (hence gold, the Stone, etc.); or, according to one ingenious exegete, is an anagram of *albus intus latens* (the white one hidden within).[1] Alchemical

7 Landscape with *lumen chemicum*. From Heinrich Khunrath, *Amphitheatrum Sapientiae Aeternae* (1609)

lore presents Basil Valentine as a medieval Benedictine monk of St Peter's, Erfurt, and his works as manuscripts miraculously rediscovered under a 'table of marble'.[2] Internal evidence and unmistakable echoes of Paracelsus make this dating impossible,[3] and it is highly likely that the man who 'edited' and published these allegedly antique texts was in fact their true author. This was Johann Thölde, chemist and salt manufacturer from Frankenhausen, and reputed secretary of the Rosicrucian fraternity.[4] The first of Basil's works to be published was the *Zwölff Schlüssel* ('Twelve Keys') in 1599: this became his best-known text and by 1618 had gone through four editions.[5] Thölde followed up this success with three more texts, published at Leipzig in 1603–4, the most celebrated being the *Triumphwagen Antimonii* ('The Triumphant Chariot of Antimony').[6]

The medieval 'packaging' suggests the works' appeal to alchemical tradition: Basil is presented as an ancient adept, a 'veiled master'. The *Twelve Keys* are those 'wherewith the Doors are opened to the most ancient Stone of our Ancestors, and the most secret Fountain of all Health is discovered'.[7] This reminds us of Ripley's *Compound*: keys to unlock those twelve gates into the castle of wisdom. The echo may be intentional: Ripley's work was much admired in continental circles, and Latin translations of the *Compound* appeared at Prague in 1599[8] and in the *Theatrum Chemicum* of 1602. Like the *Compound*, Basil's *Twelve Keys* presents an unfolding *magnum opus* whose end product is the Stone or Tincture in which are 'all the mineral and metalline forms, yea, and all the qualities and properties of the whole world' and by which 'the cold body of Saturn [i.e. lead] is warmed, and by that heating is changed into the best gold'.[9] But though traditional in format and goal, the *Twelve Keys* is entirely contemporary in its explicitly mystical analogies. Alchemical injunction – 'Take your spiritual water, in which the spirit was from the beginning, and preserve it in a closely shut chamber' – comes complete with religious interpretation: 'if we rightly behold our own souls, then shall we ... effect that which seemeth now impossible to us.'[10] In the fourth Key, the disintegration of matter by means of the secret fire, or *ignis innaturalis*, is described in terms which typify this intertwining of chemical and spiritual meanings:

> All flesh that came from the Earth must be corrupted and return to Earth again, as it was Earth at the first. Then that Earthly Salt begetteth a new generation, by a coelestial revivification, for if it were not first Earth there could be no revivification in our Work. For in the Earth is the balsome of Nature, and is their Salt who sought after the knowledg of all things. At the Day of Judgment the World shall be judged by Fire and ... by Fire be burnt to Ashes, out of which Ashes the Phoenix produces her young. For in

those ashes lye the true and genuine Tartar which must be dissolved, and when that is dissolved, the strongest Lock of the Kings Palace may be opened. After that burning, a new Heaven and a new Earth shall be formed, and the new man shall more gloriously shine forth, than ever he lived in the old world, for he shall be purified. When ashes and sand are well maturated and concocted in the fire, then the artist turneth it into Glass, which afterward will endure in the fire, and in colour like a transparent Stone. . . . This to the ignorant is a great Mistery, but not so in any wise to the experienced Artist.[11]

We see how integrated the different levels are. There is, lurking somewhere there, an actual chemical operation: the earthly salt, tartar and 'glass' (probably a coded reference to green vitriol) are all palpable substances, indeed Basilius is an alchemist often praised as a 'practical manipulator and observer of the first rank', who accurately described 'derivatives of antimony, hydrochloric acid, solutions of caustic alkali, the acetates of lead and copper, gold fulminate, and many other salts'.[12] Obviously this particular passage would not impress a modern chemist, but chemical practice forms the basis for its more esoteric elements – the traditional symbols of regenerated Phoenix and locked royal chamber, the boast of initiation into a 'great mistery', and the analogies with religious experience. Earthy matter is called 'flesh' and the action of fire likened to a conflagration consuming the great world. The promise of spiritual alchemy is overtly made – 'the new man shall more gloriously shine forth . . . for he shall be purified' – and then the passage shades back into the world of chemical substance: sand, ashes and crystalline vitriol. The tortuously abstract alchemy of Dee's *Monas* and the opaque but richly concrete alchemy of Ripley's *Compound* find common ground in the *Twelve Keys*: something is lost from each, but the synthesis is made.

There is, to complete the picture, the strong Paracelsist flavour of Basil's alchemy. In the first Key, he compares the purifying of matter to a healing physic:

As Physicians cleanse and purifie the inward parts of the body, by means of their medicines, expelling all impurities from thence, so. . . our Masters require a pure and undefiled body, which is not adulterated with any spot or strange mixture. For the Addition of another thing is a Leprosie to our metals.[13]

His specifically Paracelsist work is the *Triumphwagen Antimonii*, a work in praise of the 'potency, virtues, powers, operation and efficiency' of antimony as 'chymicall physick'. In its denunciation of 'ignorant medicasters', its discourse on 'out of a remedy a poyson and out of a poyson a remedy'[14], and its lengthy directions on how to become 'a true Antimonial Anatomist',[15] the *Triumphwagen* shares many features with the chymicall works of Hester. But the whole tone is different – the alchemical personification ('It is, I, Antimony, that speak to you'); the devotional words which compare antimony to 'an infinite circle and painted with all sorts of colours' and conclude that 'one man's life is too too short perfectly to understand all its mysteries':[16] these are quite foreign to the industrious tone of Hester's work. So too is Basil's contention that the Philosophers' Stone itself can be composed of antimony. Once again the key-note is synthesis: the practicalities of the chymist allied (in the true spirit of Paracelsus himself) to more esoteric formulations. The laboratory work of Hester and the exclusively metaphysical Paracelsism of Gerhard Dorn[17] each find their complement in the *Triumphwagen*. Exactly contemporary with Basil's work are two texts by the French Paracelsian, Quercetanus: *De Priscorum Philosophorum Verae Medicinae Materia* (1603) and *Ad Veritatem Hermeticae Medicinae* (1604). These display the same blend of practical and esoteric Paracelsism which characterizes the *Triumphwagen*, as the title of their English translation – *The Practise of Chymicall and Hermeticall Physicke* by Thomas Tymme (1605) – makes clear. The practical bias of 'chymicall physick' is balanced with lengthy discourses on the 'Soule of the World', on 'astral seedes and spiritual beginnings' and various other metaphysical themes comprehended by the term 'Hermetic physic'.

Another important text from those years is the *Novum Lumen Chemicum* ('New Chemical Light') by the Pole, Michal Sendivoj (Sendivogius; occasionally, Sensophax). This was published at Prague in 1604 and, like the *Twelve Keys*, went through several editions in quick succession: the 'Wizard Earl' of Northumberland's library included the 1614 edition.[18] The *Novum Lumen* is often said to be the original work of the Scottish alchemist Alexander Seton, whom Sendivogius rescued from the dungeons of Christian II of Saxony. Seton had appeared suddenly and brilliantly on the alchemical landscape in 1602–3, performing a series of transmutations in Holland, Switzerland and Germany.[19] An account of his gold-making successes in Cologne, in the summer of 1603, was published in Ewald van Hoghelande's *Historiae Aliquot Transmutationis* in 1604. In July of that year, John Dee and Simon Forman were dining together and, Forman noted, Dee 'showed me a little book of one Hooger [i.e. Hoghelande] of transmutation of metals wrought by a noble Scot'.[20] We see how quickly European news and texts were picked up in English alchemical circles.

Sendivogius, whether or not his work was appropriated from the 'noble Scot', is typical of the alchemist flourishing in these first years of the seventeenth century, and he was highly esteemed by such powerful enthusiasts as Rudolf II and the Duke of Württemburg.[21] The *Novum*

Lumen is equally typical of this newly comprehensive alchemy. Its full title reads: 'The New Chemical Light, drawn from the Fountain of Nature and of Manual Experience.' Immediately we note the Paracelsist stress: illumination drawn equally from Nature and chemical 'Vulcanist' labour. Such Paracelsist terms as '*Archeus*, the servant of Nature'[22] confirm this allegiance. But, once again, this is fed back into the traditional structure of esoteric alchemy. The work is addressed to the 'sons of Hermes' in the classic terms of initiation, bidding them 'when the hidden has become manifest to them, and the inner gates to secret knowledge are flung open, not to reveal this mystery to any unworthy person'.[23] Sendivogius praises Lully and Geber, urges a return to the simplicity and profundity of the ancient adepts, and prefaces his text by quoting Arnald of Villanova's epigram of the *opus*: 'It has its birth in the earth, its strength it does acquire in the fire, and there becomes the true Stone of the ancient Sages.'[24] The division of the work into twelve treatises suggests the classic shape of the *opus*, concluding with 'the Stone and its Virtue'. Finally, this blend of Paracelsist and traditional themes is again channelled into the characteristic spiritual analogies: 'Man was created of the Earth and lives by vertue of the Air; for there is in the air a secret food of life which in the night we call Dew ... whose invisible congealed Spirit is better than the whole earth.'[25] The title suggests the irradiation of the *lumen naturae* perceived in chemical guise and channelled into the self by symbolic chemical procedure:

> Nature hath her proper light, which is not obvious to our eyes. The shadow of Nature is a body before our eyes. But if the light of Nature doth enlighten anyone, presently the cloud is taken away from before his eyes, and without any let he can behold the point of Our Loadstone, answering to each Center of the beams of the Sun and Earth, for so far doth the light of Nature penetrate and discovers inward things. . . . As therefore Mans body is covered with a garment, so also Mans Nature is covered with the Body.[26]

What links all the texts so far mentioned is not only their close chronological proximity – all of them between 1599 and 1605; it is also their characteristic way of blending the classic procedures of esoteric alchemy with the more contemporary interpolations of Paracelsist medicine and Hermetic mysticism. In this sense, one of the greatest texts of this 'alchemical renaissance' was the gigantic anthology, *Theatrum Chemicum*, published in three volumes by Lazarus Zetzner at Ober-Ursel in 1602. The blend of ancient and modern is suggested in the title-page promise of tracts 'concerning the antiquity, truth, justice, excellence and operations of Chemia and the Philosophic Stone' together with works of the 'devotees of the true Chemia and chemical medicine, who thence can reap a fruitful harvest of most potent remedies'.[27] Over eighty texts are contained in this densely printed miscellany, representing all shades of alchemical allegiance. In the first volume, a dozen tracts by Gerhard Dorn are flanked by classic works by Lully, Arnald and Nicolas Flamel; Ripley's *Compound* and Dee's *Monas* appear together in the second: it was this edition Tymme used for his lost translation of the *Monas*. Zetzner's *Theatrum Chemicum* crystallizes this quality of integration which is the keynote of early seventeenth-century alchemy. It is partly for this reason that I have gratefully purloined its title for my book.

*   *   *

Johann Thölde, editor and probable author of the Basilian *corpus*, is sometimes described as a 'secretary' of the Rosicrucian fraternity. There is no hard evidence for this, but when it is announced in the *Twelve Keys* that 'the old world passeth away, and the new is come in its place'[28] it seems to anticipate the clarion-call for a 'universal and general reformation of the whole wide world' which begins the first Rosicrucian manifesto, the *Fama Fraternitatis* of 1614.[29] And when Basil/Thölde describes man as the true recipient of the alchemical *corpus subtile* – 'when man shall be purified, he shall be like these Coelestial Spirits'[30] – he strikes the tone of the alchemy which plays so central a role in the Rosicrucians' 'new dawn' of magical enlightenment: the *Fama* asserts,

> The true philosophers are far of another mind, esteeming little the making of gold, which is but a *parergon*; for besides that they have a thousand better things. And we say with our loving father R.C.C.: *Phy: aurum nisi quantum aurum*, for unto them the whole Nature is detected. He doth not rejoice that he can make gold ... but is glad that he seeth the heavens open, and the angels of God ascending and descending, and his name written in the book of life.[31]

The 'loving father R.C.C.' is the legendary founder of the Order, Christian Rosencreutz, or Christian Rosy Cross, a mystic scholar-knight and physician who allegedly flourished in the fifteenth century.[32] Here too, the pseudo-medieval image of Basil Valentine chimes in with Rosicrucianism. The *Fama* announces: 'Under the name of Chymia many books and pictures are set forth in *Contumeliam gloriae Dei*, as we will name them in their due season, and will give to pure-hearted a Catalogue or register of them.'[33] Whatever the actual connections of Thölde with the Rosicrucian fraternity, it seems likely that this proposed catalogue of commendable alchemical texts would have included the *Twelve Keys*. We can hazard some other entries: Dee's *Monas*, for instance. Both its biblical epigram about the 'dew of heaven' and the actual symbol of the monad appear in original Rosicrucian publications, and Dr Yates has argued the *Monas* as a central ingredient of Rosicrucian thought.[34] Another text for the 'pure-hearted' would be the '*Vocabular* of *Theoph. Par. Ho.*' praised in the *Fama*,[35] probably Dorn's *Dictionarium Paracelsi*. Paracelsus was admired by the Rosicrucians not only as magician and alchemist, but as revolutionary healer: the first of six 'articles' in the *Fama* ordains that 'none of them should profess any other thing than to cure the sick, and that *gratis*'.[36] No doubt Basil's other great work, the Paracelsist *Triumphwagen Antimonii*, would also figure in the Rosicrucian *bibliotheca chemica*.

Thölde's Basil Valentine can be seen to foreshadow the kind of alchemy that figures in the Rosicrucian manifestos some fifteen years later. Similar connections could be argued for the *Triga Chemica*, published at Prague in 1599. This was the work of the Frenchman, Nicolas Barnaud, a prominent figure in the alchemical circles around Rudolf II. Like Thölde, Barnaud's name is vaguely associated with the early stirrings of Rosicrucianism and he is said to have travelled through France and Germany in the early 1590s seeking support for 'a secret convention of theosophical mystics, who were to devote themselves to a determined investigation of all Kabbalistic sciences'.[37] Also like Thölde, Barnaud presented himself as editor rather than original author. The most important of the three texts in the *Triga Chemica* is the *De Lapide Philosophico*, attributed to 'a noble ancient philosopher' called Lambspringk and now rendered into Latin verse by Barnaud. The engraving on the title-page shows Lambspringk wearing a tunic emblazoned with the eagle of Mercury and resting his hand on a miniature castle – no doubt Ripley's 'Philosophers Castle', for Barnaud's other collection of this year, *Quadriga Aurifera*, included a Latin version of Ripley's *Compound*. It also included, typically, a tract called *Elixir Solis Theophrasti Paracelsi*: again the mingling of diverse alchemical ingredients. Lambspringk's *De Lapide Philosophico* is a work of great poetic elegance and though there is nothing in the text to suggest that its antiquity is a fraud, it is absolutely typical of 1600: an adept revived, like Basil Valentine, within a context of contemporary alchemical interests, a context vaguely anticipatory of the Rosicrucians.

Historically, everything about the Rosicrucians is vague: the exact nature of their 'invisible college', its tangibility as an active secret society, remains unknown.[38] One alumnus, however, is certain: Johann Valentin Andreae, author of the *Chymische Hochzeit Christiani Rosencreutz* ('Chemical Wedding of Christian Rosycross') of 1616. This is an exotic allegorical romance fraught with alchemical meanings. It recounts the adventures of the heroic 'brother of the Red Rosie cross', who is invited to a royal wedding, and undergoes many ordeals, visions, ceremonies and labours there before he is dubbed a 'Knight of the Golden Stone' by the bridegroom-king. The raven and dove that haunt his journey (*nigredo* and *albedo*), the castle guarded by a 'Terrible Lyon', the 'mysterious and hidden wedding' of the King and Queen, the white and red roses which are Christian's chivalric emblem: all these provide the alchemical undertone, and certain episodes are quite lengthy allegorizations of alchemical processes.[39] And once again this work, the foremost expression of Rosicrucian alchemy, takes us back to the earliest years of the seventeenth century. For though it was first published in 1616, the *Chemical Wedding* is known to be a revised version of an earlier work. In his autobiography, Andreae recalls his first literary efforts as a teenaged student at Tübingen. Among them was a *'ludibrium'* – a light-hearted work, a *jeu d'esprit* – called the *Nuptiae Chymicae*, written some time in 1602–3.[40] The exact relationship between this lost early work and the *Chemical Wedding* of 1616 can only be conjectured, but it seems likely that, as Dr Yates suggests, the early version was 'a work of alchemical symbolism',[41] to which the specifically Rosicrucian elements were later added. It is fascinating to note that Andreae's other *juvenilia* from this time – certain comedies called *Esther* and *Hyacinth* – were written, he says, 'in emulation of the English actors'.[42] English touring companies were not uncommon in Germany, and Andreae may well have seen performances by the actors attached to the Garter embassy from James I to the Duke of Württemburg in late 1603, which visited Tübingen.[43] That early *Chemical Wedding* of c. 1603 might be seen as a contact between English theatrical influences and the German alchemical imagination. That contact, I hope to show, was two-way. The comings and goings of English players, 'strolling' between England and Europe, may well have been the channel by which 'Renaissance alchemy' found its way so swiftly into Shakespeare's plays.

* * *

We have seen how strong a presence alchemy had in England during the Shakespearean period, and those texts of the 1590s – revived authors like Ripley and Bacon, contemporaries like Hester and Forester – will continue to be of interest. To this must now be added the tremendous concentration of contemporary alchemical literature emanating from Europe in the years around 1600, works at once original and inclusive, blending various strains of alchemical practice and philosophy into an organic whole. A summarizing list shows how tightly they are clustered together in a specific period:

1599   Basil Valentine, *Zwölff Schlüssel* (Eisleben)
1599   Lambspringk, *De Lapide Philosophico* (Prague)
1602   Zetzner (ed.), *Theatrum Chemicum* (Ober-Ursel)
1602–3 Andreae, lost early version of *Chymische Hochzeit* (Tübingen)

1602–3   Tymme, lost translation of *Monas Hieroglyphica* (?)
1603   Quercetanus, *De Priscorum Philosophorum Verae Medicinae Materia* (St Gervais)
1604   Valentine, *Triumphwagen Antimonii* (Leipzig)
1604   Quercetanus, *Ad Veritatem Hermeticae Medicinae* (Paris)
1604   Sendivogius, *Novum Lumen Chemicum* (Prague)
1605   Tymme, *Practise of Chymicall and Hermeticall Physicke* (London)

I have also suggested that some of these texts can be seen as anticipating, by a decade and more, the kind of alchemy associated with the Rosicrucian manifestos: spiritual, healing, regenerating alchemy, part revival and part 'new dawn'. I have not intended by this to throw new light on these texts, or on Rosicrucianism, but to suggest a vital shift of dating. In her pioneering work, *Shakespeare's Last Plays*, Frances Yates has argued the presence of Rosicrucian themes and symbols in Shakespeare's so-called 'romances', i.e. the five last plays from *Pericles* (c. 1608) to *Henry VIII* (c. 1612). I am sure she is right, and my book is entirely indebted to the trail she has blazed. But there is one *caveat*: the chronology. The earliest Rosicrucian manifesto appeared in 1614,[44] two years after Shakespeare had packed his theatrical bags and retired to Stratford. The most typical works of Rosicrucian-influenced alchemy, those of Michael Maier, date from 1614 onwards, and his visit to England (where, typically, he sought out two medieval 'classics' of English alchemy, Norton's *Ordinall* and Abbot Cremer's *Testament*, in order to translate and publish them) appears to have been made in about 1616.[45] Similarly, Robert Fludd, the earliest Rosicrucian apologist in England, did not print until 1616.[46] It is all a bit too late, textually, and to call the alchemical and magical elements in Shakespeare's last plays 'Rosicrucian' means one is dealing with currents 'in the air', ideas disembodied of textual evidence.[47] Unless, that is, one stresses the whole alchemical element so integral to Rosicrucianism, and sees it flowering so luxuriantly in the first years of the seventeenth century. If we slip back to the year 1605, by which time all the texts mentioned above were in print, we are returning chronologically to the very heartland of Shakespeare's greatest work. It is there that the symbols and patterns of alchemy prospered.

The cloudy voice

'As it were in a riddle and cloudie voyce,' says the preface to the *Mirror of Alchimy*, 'they have left unto us a certaine and most excellent science.'[48] In a sense, the bizarre and nebulous language of alchemy is all that is left us today, the archaeological remains of a buried philosophy. But it is more than mere debris, for even when it lived and prospered alchemy was as much a literary phenomenon as a practised 'science'. The texts of this 'alchemical renaissance' fully exploited the rich dramatic and poetic qualities of alchemical language, and we can hear in them today the tones that might have excited a receptive ear in 1605. The *Twelve Keys* is typical. There is the poetry of colour, a kind of verbal pageantry:

> This is the Rose of our Masters, of a purple colour, and the Red Blood of the Dragon, whereof so many have written; it is that purple mantle, richly leaved in our Art, wherewith the Queen of Health is covered and wherewith all metals wanting heat may be revived.[49]

There is the exotic bestiary:

> If you adde to the Eagle the old Dragon, which hath a long time had his habitation among stones, and creepeth out of the caves; and put them both in the Infernal Pit; then will Pluto breathe upon them. . . .
>
> The Lion purifieth himself by the blood of the Wolf, and the tincture of his blood wonderfully rejoyceth in the tincture of the Lion. . . . When the Lion is satisfied, his spirit is made stronger than it was before and his eyes shine with great splendor.[50]

And there is the strong element of personification, symbolic human figures performing strange ritualistic functions:

> A double fiery man must be fed with a white Swan. . . . The rosted Swan will be food for the King, and the fiery King will exceedingly love the pleasant voice of the Queen, and out of his great love embraceth her, and satiateth himself with her, until both vanish and become one body.[51]

There is in all this an element of code, the 'cloudie voyce' as a veil against the impure and uninitiated. But there is also a tremendous quality of drama, an envisaging of chemical action as a series of internecine struggles, predatory feedings, phantasmagoric copulations, deaths, entombments and rebirths. All these took place within a vessel which could become a castle, a bed, a sea, a garden, an oven, or a grave according to the chemical events within it. In certain texts, these highly-charged readings were elaborated into alchemical 'parables', expounding the entire *opus* as a continuous allegorical narrative: we will look at some of these in a later chapter. But even as isolated emblems, one can see how integral the symbols of alchemy are to the whole enterprise. Through these bizarre acts of naming, the alchemist expressed, and to some extent created, the secret meanings of his chemistry. What he saw in the 'theatre' of the vessel became the death of a king, the blood of the dragon or the rose of the masters by virtue of a secretly repeated language. The symbol became a part of the substance or phenomenon it described: it expressed 'philosophical' properties actually resident in matter. And, like chemical matter itself, the symbol demanded to be probed and penetrated, for its meaning to be coaxed out like a quintessence: 'let me therefore admonish the gentle reader,' wrote Sendivogius in the *Novum Lumen*, 'that my meaning is to be apprehended not so much from the outward husk of my words, as from the inward spirit of Nature.'[52]

A fantastic range of alchemical symbols can be traced back to a single, though multiform, figure: the Mercury of the Philosophers, or Our Mercury. One could even suggest that Sendivogius' words above say that all alchemical language (the 'outward husk') refers to Our Mercury (the 'inward spirit'). This would be only a slight overstatement. Basil Valentine concludes the last of his *Twelve Keys* by saying: 'There is nothing, says the Philosopher, save a double Mercury. I say that no other matter has been named. Blessed is he who understands it: seek therein and be not weary.'[53] Our Mercury, we infer, comprehends the entirety of the *opus* and is therefore the only matter 'named' in the *Twelve Keys*. It has, in the words of Thomas Tymme, 'the Mastery in alchemy, & the α and ω in the worke'.[54] To understand this totality of Mercury so insisted upon, we must 'add up' the various symbols which describe it. Alchemical language is our only key to this 'inward spirit' of alchemy.

* * *

We have so far understood alchemical Mercury as the quintessential spirit concealed in the 'dark prison' of matter. It is 'a Soule, a substance bryght',[55] the *'spiritus et anima corporis'*,[56] a 'spiritual essence which is neither celestial nor infernal, but an aereal, pure and precious body, in the middle between the highest and lowest'.[57] Within the 'chemical theatre' of the vessel, this spirit manifested itself as volatile vapours, flying upwards out of disintegrating matter. 'Spirit of Mercury', writes Basil,

> is the principle to work Metals, being made a spiritual Essence, which is mere air, and flyeth to and fro without wings, and is a moving wind, which after its expulsion out of its habitation by Vulcan is driven into its Chaos, into which it entreth again.[58]

In this aspect, Our Mercury is described by symbols which stress its airy nature ('divine breath', 'wind of Hermes') or its wingedness ('eagle', 'white bird', 'snowy swan'). The properties of this spirit are only activated by, as Basil puts it, 'its expulsion out of its habitation'. After its expulsion and its flight to the 'heaven' of the vessel, it makes its reanimating return, entering again into the 'Chaos' from which it emerged. As the ascending spirit took the form of airy vapours, so the descending spirit is perceived as a water, a heavenly condensation falling back on the dead earth. Here Our Mercury is called 'dew of heaven', 'orient water', 'celestial water', 'milk of the virgin', 'our balm', 'our honey', 'spittle of the moon', 'moist pearl', 'water of Azoth', 'May dew', 'silver rain', etc.[59] It is 'our Pontic and Catholic water, which in its refluent course irrigates and fertilizes the whole earth, and is sweet, beautiful, clear, limpid, and brighter than gold, silver, carbuncles or diamonds.'[60]

As airy fugitive vapour and life-giving water, Our Mercury traces the circular journey of spirit, the 'oval orbit' of Dee's *Monas*. It comprehends *solve et coagula*: it is Mercury that is released when the solid is dissolved and Mercury that is recaptured when the spirit is congealed. Basil Valentine sums up:

> That very same spirit, say, is the Radix of the Life of our Bodies, and the Mercury of the Philosophers, from whence our liquid Water is prepared, which you must make again material. . . . Our beginning is a secret and palpable body, the middle is a fugitive spirit, and a golden water without any conversion. . . . The end is a most fixt medicine for humane and metalline bodies, which to know is rather granted to Angels than men.[61]

Fugitive spirit and golden water: in these shapes the spirit ascends and descends, drawing down the life of heaven to earth, imbuing matter with the 'power of the superiours and the inferiours'. Air and water are the elementary vehicles of Mercury's journey, the material roles assumed by spirit in the 'chemical theatre' of the vessel.

But Mercury as spirit is only one half of *Mercurius duplex*, the double Mercury. It has another and darker nature. For Mercury is also identified with the dragon that features so prominently in alchemical symbolism. In the *Aurelia Occulta*, the 'poison-dipping dragon' announces:

> My water and fire destroy and put together; from my body you may extract the green lion and the red. But if you do not have exact knowledge of me, you will destroy your five senses with my fire. From my snout there comes a spreading poison that has brought death to many. . . . By the philosophers I am named Mercurius; my spouse is the gold; I am the old dragon found everywhere on the globe of the earth.[62]

Mercury as dragon seems diametrically opposite to Mercury as spirit, for the dragon is fiery, poisonous and devouring, not animating and inspiring at all. Indeed the dragon, like the serpent and green lion which are part of the same nucleus of symbols, represents caustic, corrosive, vitriolic properties (the green lion is often identified as green vitriol, which contains in it the 'fiery poison' of sulphuric acid).[63] This vicious aspect of Our Mercury is further described by Philalethes in his masterly commentary on Ripley's *Vision*. Ripley writes:

> A Toade full rudde I saw did drink the juce of grapes so fast,
> Till overcharged with the broth, his bowells all to brast.[64]

The toad is Raw Stuff, the epithet 'ruddy' suggesting common gold.[65] The 'juice of grapes' is clearly a corrosive liquor penetrating the matter and causing, as Ripley so pungently puts it, 'his bowels all to burst'. This is how Philalethes glosses this caustic 'juice':

> This water they sometimes call *Aqua Ardens*, sometimes *Acetum Acerrimum*, but most commonly they call it their Mercury.... The Water soaks radically into our Body; being circulated upon it, according as the Philosopher saith. When its own Sweat is returned to the Body, it perforates it marvellously.... It pierceth radically to the very profundity of it; and makes it to alter its form. This is the Water which teareth the Bodies, and makes them no Bodies, but flying Spirits like a Smoak, Wind or Fume.[66]

Here again is the dragon Mercury: a 'burning water' and 'bitterest vinegar'; perforating, piercing, tearing. '*Acetum philosophorum*', the 'vinegar of the philosophers', is another expression of this; it is defined in Dorn's *Dictionarium* as 'a Mercurial Water, or otherwise called Virgin's Milk, wherein they say metals are dissolved'.[67] In terms of 'exoteric' chemical agents, then, the dragon Mercury is identifiable with vitriol, *aqua fortis, aqua regia*, etc. It is the liquor which is contained in the *balneum regis*, the caustic bath in which the King must be immersed and dissolved. More esoterically, it is identifiable as the alchemists' secret fire, the *ignis innaturalis* of which Ripley writes:

> Fire against Nature must doe thy bodies woe;
> This is Our Dragon as I thee tell,
> Fiercely burning as the fire of hell ....[68]

To understand the totality of Our Mercury, we must lock these opposite aspects together. The death-dealing dragon and the life-giving spirit form a whole which comprehends alchemical transformation. For as dragon, Mercury is the *agent* of dissolution; as spirit, Mercury is the *product* of dissolution. Death and birth, no less, are the twin poles of the double Mercury: 'Our living Western Quicksilver, which has placed itself above the gold and vanquished it, is that which kills and quickens.'[69] Or as Lambspringk puts it: 'This surely is a great miracle and without any deception: that in a venomous dragon there should be the great Medicine.'⁷⁰ After poisoning and devouring, Mercury nourishes and revives: after death, new life:

> He quickly consumes his venom,
> For he devours his poisonous tail.
> All this is performed on his own body,
> From which flows forth glorious Balm,
> With all its miraculous virtues.
> Hereat the Sages do loudly rejoice.⁷¹

Michael Sendivogius provides a beautiful panegyric for this double, indeed total, Mercury. He is called the son of Nature, and Nature herself descends to celebrate him with these words:

> He is all things, who was but one; he is nothing, and his number is entire; in him are the four Elements, and yet himself is no Element; he is a Spirit, and yet he hath a Body; he is a Man, and yet acts the part of a Woman; he is a Child and yet bears the Arms of a Man; he is a Beast, and yet hath the Wings of a Bird; he is Poison, yet cureth the Leprosie; he is Life, yet kills all things; he is a King, yet another possesseth his Kingdom; he flyeth from the Fire, yet Fire is made of him; he is Water, yet wets not; he is Earth, yet he is sowed; he is Air, yet lives in Water.⁷²

There are two directions in which alchemical Mercury leads us. On the one hand, it is a complex elaboration of chemical substance, its various qualities referring back to the properties of quicksilver, or 'common mercury'. The volatility of quicksilver and its aptitude for sublimation account for the fleeting, winged, vaporous nature of Our Mercury as airy spirit. The liquidity of quicksilver – the Greek name was ὕδωρ ἄργυρος, 'silver water' – is the springboard for Our Mercury as 'dew of heaven' or 'silver rain' of refluent spirit. And the corrosive properties of quicksilver – particularly its instant whitening of gold with which it forms an amalgam – lead to Our Mercury as *aqua ardens* or poisonous Dragon, mortifying the Raw Stuff. More generally, though still in the world of chemical performance, these aspects of mercury stand for all volatile vapours, all reviving liquors, all caustic acids. But there is another direction entirely, away from chemical matter. Mercury is not, finally, a substance, or even many substances: it is a process. Basil speaks of spirits and waters being elicited 'without any conversion'; Lambspringk says 'all this is performed on his own body'; Geber asserts that 'Our Stone is one: one medicine, to which we add nothing, from which we take nothing away.'⁷³ All these point to one crucial idea: that transformation is something intrinsic and contained *inside* matter. The dragon and the spirit (and their product, the Stone) are somehow there in the Raw Stuff: 'you have need only of one thing, which at any stage of our experiment can be changed into another nature.'[74] Mercury thus describes a whole potential for transformation: the capacity of matter to devour its corrupt body, to free its hidden spirit, and to be suffused and regenerated by that spirit when it returns to its body. Each stage of this self-devouring, self-generating process bears the name 'Mercury'. Mercury, in short, is alchemy itself. It is perhaps in this sense that 'no other matter has been named' in the *Twelve Keys*.

*   *   *

Eagle and dragon, philosophers' vinegar and virgin's milk, silver rain and secret fire: the terms which cluster round Our Mercury are typical of the redolence of alchemical imagery. They are also typical of the dangers of alchemical language: the single term, Our Mercury, has a whole retinue of synonyms which, in turn, provide a whole gamut of meanings. This is true of other alchemical names. The King, for instance, can be common gold, Our Sulphur or the Red Stone, according to context: the King undergoes changes of meaning just as the matter he represents undergoes changes of nature. And since the alchemists insist that true transformation is interior and self-contained — that the original matter (Raw Stuff), transforming energy (Our Mercury) and end product (Stone) are all 'one thing' — many of the key symbols of alchemy refer interchangeably to these apparently very different 'substances'. In *Bloomefields Blossoms* (1557), for example, the 'blessed Stone' is named, in quick succession, 'our greate Elixir, our Azot, our Basiliske, our Adrop, our Cocatrice, our Antimony, our Red Lead, our Crown of Glory and Diadem of our Head, the Lyon Greene, the Salamander, the Metalline Menstruall, a Substance Exuberate, Mercury of Metalline Essence, Limus Deserti, the Eagle flying from the North with great violence, a Toade, a Privy Quintessence'.[75] Here terms which suggest Raw Stuff (toad), volatility (eagle), corrosion (green lion) and spirit (essential Mercury) are all conflated into a single meaning: the Stone. The horizontality of alchemical language, its mobility among various meanings, must always be taken into account. Of course, any analysis of alchemy must continue to distinguish between the beginning, middle and end of the process: to call it all 'one thing' is worship rather than analysis. Sceptics suggested that this elusive lexicon had really no meaning at all — 'nothing but a masse of words,' wrote Bacon of Lully's teachings, 'much like a Fripper's or Broker's shoppe, that hath ends of everiething but nothing of worth.'[76] One can say, more charitably, that an appropriate meaning can be selected for a given alchemical symbol without denying the promise of other 'altered' meanings. The old injunction — *fac fixum volatile*, 'dissolve ye fixt' — is perhaps an aid to reading as well as performing alchemy.

'Of wit and of Alcumie'

Whatever its fabled secret meanings, the language of alchemy was, by the beginning of the seventeenth century, anything but secret. It was available to any who cared to read and listen. And for those whose business and pleasure was language – the brilliant assortment of poets, pamphleteers and playwrights astir in London, for instance – the rich exuberance of alchemical imagery was a gold-mine ripe for exploitation. It was, of course, those writers who wanted to make a particular satirical point about alchemical language who incorporated it most blatantly into their work. Here is Thomas Lodge, writing in 1595:

> Let us marke their misteries and spels,
> Their vaine *Ænigmata* and Problemes darke.
> First aske they where the flying Eagle dwels,
> Next of the dancing fooles, craft coyning clarke,
> Then of the Lyon Greene, and flying hart.
> Next of the Dragon, swallowing his tayle,
> Then of the swelling toade, they prattle art,
> Next of more blacke, then blacke, they chuse to rayle,
> Then of the crowes-head, tell they waighty things,
> And straight of Hermes seale, they sighing speake,
> Some of their *Lutum Sapientiae* sings,
> Thus on these toies, their bitter iests they breake.[77]

In *The Alchemist*, some fifteen years later, Ben Jonson's Surly resolves that 'Alchymie is a prettie kind of game, Somewhat like tricks o' the cards, to cheat a man With charming.' He challenges Subtle:

> What else are all your termes,
> Whereon no one o' your writers 'grees with other?
> Of your elixir, your *lac virginis*,
> Your Stone, your med'cine, and your chrysosperme,
> Your *sal*, your sulphur and your mercurie,
> Your oyle of height, your Tree of Life, your bloud,
> Your marchesite, your tutie, your magnesia,
> Your toade, your crow, your dragon, and your Panthar,
> Your sunne, your moone, your firmament, your adrop,
> Your *lato, azoch, chibrit, heautarit*,
> And then your red man, and your white woman. . . ?[78]

This cavalcade of alchemical terms – the repeated 'your' ironically echoing the alchemists' favoured possessive, 'our' – is further amplified in the course of the play, whether by esoteric synonyms for the Stone ('flower of the sun', 'perfect ruby', 'red ferment', '*sanguis agni*') or by technical jargon such as '*menstruum simplex*' (solvent), '*turris circulatorius*' (circulatory furnace), '*piger Henricus*' ('lazy Henry', nickname of the distillatory furnace), 'oile of Luna in kemia', 'reverberating in Athanor', and so on.

As Surly suggests, Dr Subtle cheats by 'charming'. His fantastical jargon works like a spell to transfix his dupes. As such, alchemical language is central to the play's comic purpose, rather than merely providing colour and verisimilitude. Jonson's infallible instinct told him that alchemy required nothing in the way of comic magnification. Its foibles and excesses were all there in the original: alchemy was its own caricature.

> 'Tis a stone, and not
> A stone; a spirit, a soule, and a body;
> Which if you doe dissolve, it is dissolv'd;
> If you coagulate, it is coagulated;
> If you make it to flye, it flieth.[79]

This is so nearly authentic, but given the lightest of pushes by Jonson it tumbles into empty syllogism. 'H. has his white shirt on?'[80] is somehow hilarious, though quite genuine as an alchemical figure, as in Bernard Trevisan's King who 'pulls off his robe and . . . receives a black shirt',[81] referring to the 'putrefaction' of gold *in balneo*. 'I sent you of his faeces there calcin'd,'[82] Subtle reminds Mammon: it sounds an unwelcome consignment, but again the humour plays upon an accepted chemical usage, 'faeces' meaning residue, sediment or calx. In these ways, Jonson found comedy inherent and ready-made in alchemical language. He did the same with other arcane terminologies: the sexy puns, for instance, as Subtle reads Dame Pliant's palm – '*stella* here *in monte Veneris*' and '*junctura anularis*'[83] (Jonson the classicist being mischievously aware that '*anulus*', the ring finger, was a diminutive of '*anus*'). Cabbalistic innuendoes, rabbinical rantings, addresses to the Queen of Fairies, even pidgin Spanish: *The Alchemist* is a cocktail of mystifications, with Subtle's alchemical jargon, in all its resonant meaninglessness, as its staple ingredient.

By virtue of the sheer volume of alchemical reference in the play, one can trace certain sources and thereby sketch out a very skeletal portrait of Jonson's alchemical reading. Often he turned to the old 'classics': Professor Duncan has shown that three passages from Act II are taken almost verbatim from the *Rosarium Philosophorum*, attributed to Arnald of Villanova.[84] Recent editions of this celebrated medieval text were published in 1585, 1593 and 1610. He also suggests that the long passage where Mammon expounds various classical myths as 'abstract riddles of our Stone' is indebted to Nicolas Flamel's *Hieroglyphicall Figures* (not translated into English until 1624, but long familiar in Latin and French versions). According to Mammon, the Golden Fleece sought by Jason was 'a booke of alchemie', Medea's potion 'the manner of our Worke' and the dragon guarding the Fleece 'our argent-vive'.[85] This may echo Flamel, who calls the dragons of Philosophical Sulphur and Mercury 'they upon whom Jason in his adventure for the Golden Fleece poured the broth or liquor prepared by the faire Medea'.[86] Mammon goes on to implicate 'th'Hesperian garden, Cadmus storie, Iove's shower, the boone of Midas, Argus eyes, Boccace his Demogorgon', etc., as alchemical myths. Here a sustained analogy has been noted[87] with Robert Vallensis' *De Veritate & Antiquitate Artis Chemicae* (1593): Jonson would have found this tract in the great *Theatrum Chemicum* anthology of 1602.[88]

Nearer to home, *The Mirror of Alchimy* may have provided a useful source for Subtle's lecture on the generation of metals. They derive, he tells us, from a prime matter 'commune to all mettalls and all stones', which turns

> to sulphur or to quick-silver,
> Who are the parents of all other mettalls.
> Of that ayrie
> And oily water, mercury is engendred;
> Sulphure o' the fat and earthy part.[89]

Though in itself a widely held theory, the particular expression of it is close enough to that in the *Mirror*, where it is said of mercury and sulphur that

> all mettalls and minerals, whereof there be sundry and divers kinds, are begotten of these two. . . . The grosnesse of water is so decocted and thickened, that in continuance of time it becometh argent-vive. And that of the fatness of the earth . . . sulphure is engendred.[90]

The extent of Jonson's debt to that other London text, George Ripley's *Compound of Alchymy*, has never to my knowledge been noted. Face reminds Mammon of the long hours he has put in at the furnace, how

> these bleard eyes
> Have waked to reade your severall colours, sir,
> Of the pale citron, the greene lyon, the crow,
> The peacocks taile, the plumed swan.

Compare this version of the colour sequence of the *opus* with that in Ripley's *Compound*:

> Pale & black wyth falce citrine. . . .
> The Peacocks feathers in colours gay. . . .
> The Spotted Panther, the lyon green, the Crowes byll. . . .
> These shall appear before the perfect White.[91]

This highly idiosyncratic rendition is clearly recognizable in Face's version, though (as the omission marks suggest) Jonson has pruned somewhat. He has also elaborated the 'perfect white' into the 'plumed swan'. One can trace the rapid, instinctive eye of the artist, plucking out the words at either end of Ripley's first line, skimming on down to the green lion and the crow, then back up to the eye-catching peacock. Nor is the exotic panther wasted: it turns up later in the scene, in Surly's catalogue quoted above. In that speech, the marchesite and tuttie, and the red man and the white woman, are also drawn from the *Compound*, and Ripley's 'Our Sunne, Our Moone, Our Ferment' becomes Surly's 'your sunne, your moone, your firmament'.[92] An extra syllable hardly matters when the words are just hot air. Ripley ends his verse on colours with the promise: 'Then hast thou a medicine of the thirde order, of his owne kinde multiplicable.' This surfaces in the next scene of *The Alchemist*, when Subtle triumphantly announces: 'The worke is done.... We have a med'cine of the triple soule.'[93] And the *Compound* was still open on Jonson's desk when, fifteen lines on, Subtle speaks of 'another Worke' which 'three dayes since past the Philosophers Wheele'. As announced on the title-page and referred to throughout the text, the Philosophers' Wheel was central to the *Compound*'s conception of the Great Work:

> to win to thy desire thou needst not be in doubt,
> For the Wheele of our Philosophie thou hast turned about.[94]

Ripley is also one of the alchemists mentioned by name in the play – 'A Lullianist? A Ripley? *Filius artis?*' – others being Paracelsus, Isaac Holland, John Dee and Edward Kelley.

Among the newer alchemical authors, Jonson had almost certainly read Sendivogius' *Novum Lumen Chemicum* of 1604. There is some evidence of this in *The Alchemist*. When Subtle asserts that 'we doe find Seedes of them [sc. mercury and sulphur] by our fire, and gold in them',[95] he is perhaps echoing, in *précis*, Sendivogius' discourse on 'the Eliciting of the Metallic Seed' and his belief that 'Fire doth act upon the Center of everything ... and maketh all seed to ripen'.[96] Much clearer illustration of Jonson's awareness of the *Novum Lumen* is provided by a later work, his chemical masque *Mercurie Vindicated from the Alchemists at Court* (*c.* 1615).[97] The scene is a 'Laboratory or Alchymists workehouse'; out of the furnace wriggles Mercury; he runs about the stage pursued by the desperate alchemist, Vulcan. The alchemist's constant desire to 'fix' Mercury, to congeal the vaporous spirit into body, is thus dramatized into a comic 'chase sequence'. Mercury takes advantage of his volatile freedom to deliver his complaint about the 'variety of torment' he has endured at the hands of Vulcan and his 'smoaky familie' of alchemists:

I am their crude and their sublimate; their praecipitate and their unctuous; their male and their female; sometimes their hermaphrodite; what they list to stile me. It is I that am corroded, and exalted, and sublim'd, and reduc'd, and fetch'd over, and filtred, and wash'd and wip'd; what betweene their salts and their sulphures, their oyles and their tartars, their brines and their vinegers, you might take me out now a sous'd Mercury, now a salted Mercury, now a smoak'd and dri'd Mercury, now a pouldred and pickl'd Mercury: never herring, oyster or coucumer past so many vexations. My whole life with 'hem hath bene an exercise of torture. One, two, three, foure, and five times an houre ha' they made mee dance the Philosophicall circle, like an ape through a hoope, or a dogge in a wheele.[98]

All this is extremely reminiscent of the *Dialogue* between Mercury, the Alchemist and Nature which forms an entertaining part of Sendivogius' *Novum Lumen*. The whole relationship between an arrogant, exasperated alchemist and an impish, fugitive Mercury is the same. Like Jonson's, Sendivogius' Mercury complains bitterly of his 'torment': the alchemist has 'done many evil things to me', dealt 'most sordidly with me' until 'I am tormented almost to death'. He pleads with the alchemist:

What wilt thou have me do more? Am I not tormented sufficiently by thee? Do I not obey thee? Do I not mix my self with those things thou wilt have me? Am I not sublimed? Am I not precipitated? Am I not made turbith, an Amalgama, a Past? Now what canst thou desire more of me? My body is so scourged, so spit upon, that the very stone would pity me. By vertue of me thou hast Milk, Flesh, Blood, Butter, Oyl, Water; and which of all the Mettals or Minerals can do that which I do alone? And is there no mercy to be had towards me? O what a wretch am I![99]

This surely is a primary source for the complaint of Jonson's Mercury: there is even the curious culinary touch – 'milk, flesh, blood, butter' – which might have inspired Jonson's mischievous business about Mercury being soused, salted and pickled like a herring. Once again, the alchemical literature of the day reveals to us the raw stuff on which a virtuoso imagination set to work. There are other parallels between the two works. The misguided attempts to put a spell on Mercury, for instance, Sendivogius' alchemist resolving to charm him 'as serpents used to be conjured',[100] Jonson's Vulcan bidding his 'threadbare Alchymists' to 'begin your charme, sound musique, circle him in and take him'.[101] There is also the whole perspective on Nature. In the *Dialogue*, the alchemist is heard to 'curse Mercury and revile Nature because she made him', while Jonson's Vulcan is accused of desiring the 'adultery and spoile of Nature' and 'her dishonour'.[102] The chemical tormenting of Mercury is defined as a crude assault on Nature and both pieces end suffused with redemptive *lumen naturae*. In the *Dialogue*, Nature is 'moved with anger, comes to the Alchymist and calls him, Ho Thou'; she admonishes him and delivers the beautiful panegyric of Mercury quoted in the previous section. In the masque, the scene changes to 'a glorious bowre, wherein Nature was placed'. The alchemists are commanded: 'Vanish, with thy insolence . . . and all mention of you melt before the Maiesty of this light.'[103]

Arnald's *Rosarium* and Flamel's *Figures*, the *Mirror* and the *Compound*, the *Theatrum Chemicum* and the *Novum Lumen*: add to these the Paracelsist translations of John Hester, which I suggested earlier may have been useful for the mountebank scene in *Volpone*, and one has just the beginnings of Jonson's alchemical bookshelf. He dug deeply into the literature of alchemy, almost exclusively for ammunition to use against it. When his library was destroyed by fire in 1623, a friend wrote that undoubtedly this was 'a mere act of retaliation on the part of Vulcan'.[104]

*   *   *

These traces of Jonson's alchemical reading help us to anchor the literary awareness of alchemy at the time. They are glimpses of his original data. As data, of course, alchemical language was also available to writers whose subjects had nothing to do with alchemy. Its popularity as a vocabulary, a stock of metaphorical resources, was enormous. One gets an inkling of alchemy's poetic promise as early as 1585, in John Lyly's *Gallathea*, a slight pastoral comedy whose sub-plot features an alchemist. Lyly's knowledge of alchemy is negligible: the terms and technicalities he mentions are lifted straight out of Chaucer's *Chanouns Yemannes Tale* (or, quite probably, out of Scot's *Discoverie of Witchcraft* (1584), which in turn imported wholesale from Chaucer). Lyly's attitude to alchemy is similarly unremarkable: his alchemist is a comic stereotype of deludedness, obsessed and cozened by his dreams of transmutation. And yet here, on the subject of transmutation, something flickers into life. Lyly seems to slip the leash of his hidebound conceptions to find his own delicately euphuistic versions of alchemy:

> Of his breath hee may make golden braselets, for often-times of smoke hee hath made silver drops. . . . With the fire of blood and the corosive of the ayre, he is able to make nothing infinit.
>
> Make of a dramme of winde a wedge of gold.
>
> Make the fire as it flames, gold; the winde as it blowes, silver; the water as it runnes, lead; the earth as it standes, yron; the skye, brasse; and men's thoughts, firme mettles.[105]

These have nothing to do with Chaucer, or with satirical stereotypes: they come straight out of Lyly's imagination. The alchemist is an incorrigible dreamer, but the dream is rich with poetic possibilities.

Through the decades either side of 1600, these possibilities were energetically exploited. In Robert Southwell's visionary poem, 'The Burning Babe', the infant Christ announces Himself as an entire alchemical process of redemption:

> My faultless breast the fournace is, the fewell wounding thornes,
> Love is the fire, and sighs the smoake, the ashes, shames and scornes;
> The fewell Justice layeth on, and Mercie blowes the coales,
> The mettall in this fournace wrought are mens defiled soules:
> For which, as now on fire I am to work them to their good,
> So will I melt into a bath, to wash them in my blood.[106]

In *Pierces Supererogation* (1593), Gabriel Harvey turns to alchemy to express a principle of literary criticism:

> In the best, I cannot commende the badd; and in the baddest, I reiect not the good: but precisely play the Alchimist, in seeking pure and sweet balmes in the rankest poisons. A pithy or filed sentence is to be embraced, whosoever is the Autor.... O Humanity, my Lullius, or O Divinitie, my Paracelsus, how should a man become that peece of Alchimy, that can turne the Rattes-bane of Villany into the Balme of honesty; or correct the Mandrake of scurrility with the myrrhe of curtesie, or the saffron of temperance.[107]

In his elegy, 'The Comparison', generally dated in the mid-1590s, John Donne draws a typically ingenious sexual analogy:

> Then like the Chymicks masculine equall fire,
> Which in the Lymbecks warm wombe doth inspire
> Into th'earths worthlesse durt a soule of gold,
> Such cherishing heat her best lov'd part doth hold.[108]

In a later poem – the 'Ecclogue' of 1613 – alchemical conceptions of the 'generation' of metals offer him a metaphor for social rather than sexual relations:

> The Earth doth in her inward bowels hold
>   Stuffe well-disposed which would faine be gold
> But never shall, except it chance to lye
>   So upward, that heaven gild it with his eye.
> As, for divine things, faith comes from above,
>   So, for best civil use, all tinctures move
> From higher powers; From God religion springs,
>   Wisdome and honour from the use of Kings.[109]

This alchemical cluster of associations – gold, sun, king – was also exploited by Shakespeare. It may be simple embellishment: the sun, in the *Sonnets* (*c.*1593–5),

> Kissing with golden face the meddowes greene,
> Guilding pale streames with heavenly alcumy.[110]

Similarly in *King John* (c. 1597):

> The glorious sunne
> Stayes in his course and playes the alchymist,
> Turning with splendor of his precious eye
> The meager cloddy earth to glittering gold.[111]

These embellish a visual effect, but equally Shakespeare uses alchemy to express less palpable enrichments:

> Shall I say mine eie saith true,
> And that your love taught it this Alcumie?
> To make of monsters, and things indigest,
> Such cherubines as your sweet selfe resemble,
> Creating every bad a perfect best
> As fast as objects to his beames assemble. . . .
>
> That which would appeare offence in us
> His countenance, like richest alchemie,
> Will change to vertue and to worthinesse. . . .
>
> How much unlike art thou Marke Anthony!
> Yet comming from him, that great Med'cine hath
> With his tinct gilded thee.[112]

Already these examples suggest the pliability, the range of applications, of alchemical metaphor. This is well illustrated by contrasting alchemical conceits in two works published in the same year, 1599. On the one hand, Sir John Davies's long meditation, *Nosce Teipsum*, personifies the soul as a kind of Lady Alchymia, vaporizing experienced phenomena into immaterial conceptual forms:

> Then what vast body must we make the mind,
> Wherin are men, beasts, trees, towns, seas & lands,
> And yet each thing a proper place doth find,
> And each thing in the true proportion stands?
> Doubtlesse this could not bee, but that she turnes
> Bodies to spirits by sublimation strange,
> As fire converts to fire the thing it burnes,
> As we our meats into our nature change.
> From their grosse matter she abstracts the formes

> And drawes a kind of quintessence from things
> Which to her proper nature she transformes,
> To beare them light on her celestiall wings.[113]

This elegant analogy contrasts sharply with the bizarre conceit dreamed up by Tom Nashe in his swan-song, *Lenten Stuffe*, written 'in praise of the Red Herring':

> Howe many bee there in the worlde that childishly deprave Alchumy, and cannot spell the first letter of it; in the black booke of which ignorant band of scorners, it may be that I am scorde up with the highest: if I be, I must intreate them to wipe me out, for the red herring hath lately beene my ghostly father to convert me to their fayth; the *probatum est* of whose transfiguration *ex Luna in Solem*, from his duskie tinne hew into a perfit golden blandishment, onely by the foggie smoake of the grossest kind of fire that is, illumines my speculative soule, what much more, not sophisticate or superficiall effects, but absolute essentiall alterations of mettalles, there may bee made by an artificiall repurified flame and diverse other helpes of nature added besides.[114]

Who other than Tom Nashe could see alchemy in the 'transfiguration' of herring into kipper? Perhaps this act of discernment is precisely the point. Alchemy was a way of perceiving how things happen. It offered – as well as an odd and colourful vocabulary – a description of processes and relationships that was portable into various contexts. It also provided, one finds, a self-reflexive metaphor for things happening in the mind of the writer. Marlowe's poets extracting a 'heavenly quintessence', Harvey's poet who is himself a 'quintessence of quicksilver', Shakespeare's sonnet which 'distils your truth', Nashe's tragedians as the 'Alcumists of eloquence', Middleton's poets that 'by divinest alchemy Did turn their ink to gold', Donne's patron who is 'that Alchemist which alwaies had Wit, whose one spark could make good things of bad'.[115] These suggest (though none of them very seriously) that the poet's creative performance was a type of alchemy. 'Only the poet', wrote Philip Sidney in his *Defence of Poesie* (1595),

> with the vigor of his own invention doth grow, in effect, into another nature, in making things either better then Nature bringeth foorth, or quite anew, formes such as never were in Nature.... Nature never set forth the earth in so rich tapestry as divers poets have done.... Her world is brasen, the poets only deliver a golden.[116]

Thus the alchemist's promises were reformulated, his secret 'art' annexed into a public art, his 'search and stirre to make gold' internalized as the search and stir of artistic creation. As Nashe happily asserts: 'It is almost impossible for anie to bee encumbred with ill spirites who is continually conversant in the excelent restorative distillations of wit and of Alcumie.'[117]

Isolated metaphors of the sort quoted in the last few pages display a delightful opportunism rather than any coherent 'attitude' to alchemy. Nor do they venture very far into the more esoteric formulations of alchemical language. For now, the point is that alchemy was very much *there* at the time: in the air, on the page, a mode of perception. We have seen how sharp a presence the alchemist was in the landscape of Shakespeare's England; how the twin guises of alchemy – esoteric-philosophical and medicinal-chymicall – were at the height of their influence; how this heyday was broadcast in a positive effusion of publications in the years around 1600; and, finally, how the exotic tones and promises of alchemy were scattered across the surface of contemporary English literature in the form of metaphors, conceits, embellishments, even bad Shakespearean jokes ('A philosopher with two stones moe than's artificial one',[118] punning on 'stone', testicle). All this is a context, the 'circumstantial evidence' from which the detection now proceeds. The rest of my book seeks to suggest ways in which a deeper and more systematic presence of alchemical symbol and pattern can be found in certain works by Jonson, Donne and, most strikingly, Shakespeare; to suggest, perhaps, that 'chemical theatre' was played out in spaces less obscure than the alchemist's smoke-stained vessel.

· 5 ·

Alchemical Patterns in Jonson and Donne

Eastward Hoe: a chemical morality

For all its racy urban wit, *Eastward Hoe* (1605) is anchored to a straightforward social morality – 'The gaine of honest paines is never base', 'Note but the reward of a thriftie course', 'Where ambition of place goes before fitnesse of birth, contempt and disgrace followe'.[1] The London settings – Goldsmith's Row, the Blue Anchor, Cuckold's Haven, the Counter prison – are the arena of a traditional contest between Vice and Virtue in socio-economic garb. The characters assemble on one side or the other, as Touchstone announces in the opening scene:

> As I have two Prentises, the one of a boundlesse prodigalitie, the other of a most hopefull industrie; so have I onely two daughters: the eldest, of a proud ambition and nice wantonnesse, the other of a modest humilitie and comely sobernesse.[2]

This moral symmetry – which no doubt helped to unify the assorted talents of Jonson, Marston and Chapman – is further integrated into the play by the recurrence of two *leitmotivs*. One, which gives the play its title, is topographical: 'Eastward Ho!' leads downriver to the royal court at Greenwich, 'Westward Ho!' to the gallows at Tyburn. The extremities of the city symbolize the polarities of social destiny within it. Quicksilver is warned, 'Eastward hoe will make you goe Westward hoe'[3] – vaunting ambition will lead to retribution. The other recurrent motif is drawn from alchemy.

The three principal characters in the play are the two goldsmith's apprentices, Frank Quicksilver and Master Golding, and their master, William Touchstone. Their characteristics are implicitly identified alchemically: the impish, mercurial Quicksilver; the virtuous, incorruptible Golding; the presiding, judicial Touchstone, against whom the purity and worth of his apprentices are measured. The play's moral grouping is underlined by an alchemical notation, drawing on pre-formed ideas and images associated with common mercury, gold and the touchstone. This gives an ingenious twist to a convention much favoured in the 'comedy of humours' genre (to which all three authors of *Eastward Hoe* had contributed) – the convention whereby a character's name identified his 'humour', i.e., his psychological type, his particular obsession, his 'hobby horse'. Jonson stuck to this convention throughout his career – Downright and Wellbred, Subtle and Surly, Sir Politic Would-be, Sir Epicure Mammon, Sir Moth Interest, Meercraft the projector, Eitherside the lawyer, Madrigal the poet. The transition from humour types to chemical types was no great leap: the four humours and the four elements derived from the same Aristotelean basis; we still use the word 'temper' in both a metallurgical and a psychological sense. But alchemy provides something more than a shorthand for character traits: it carries with it an implicit sense of *process*. The naming of Quicksilver and Golding is not only a means of defining what they are (an identity label like the 'humour-names'), but of suggesting what will happen to them. It also provides a formula for interaction, for as well as having individual chemical properties, mercury and gold had a specific relationship. By introducing a symbolic framework of chemical process and reaction, the authors of *Eastward Hoe* turn the essentially static humour-portrayal into an active participant in the play's unfolding.

This is immediately set to work in the play's opening scene, where Quicksilver urges Golding to rebel against the conformity of apprenticeship:

> God's my life! Sirrah Golding, wilt be ruled by a foole? Turne good fellow, turne swaggering gallant. . . . Looke not Westward to the fall of Don Phoebus, but to the East – Eastward Hoe!
>
> > 'Where radiant beames of lusty Sol appeare,
> > And bright Eous makes the welkin cleare.'
>
> We are both Gentlemen, and therefore should be no coxcombes; let's be no longer fooles to this flatcap, Touchstone. Eastward, Bully![4]

Straightaway, Quicksilver conforms to chemical type. He addresses himself to Golding just as crude mercury does to gold: he attempts to 'work on' him, to corrupt and corrode him. This was a familiar chemical confrontation – Paracelsus' description of it in his *De Compositione Metallorum* is one example of many:

> Common mercury . . . with its fume penetrates all other metals, and, as it were, breaks through them, calcines them, and disposes them to its own nature. . . . If gold be placed above mercury, so that the white fume of mercury touch and penetrate the body of the gold, the gold will be rendered fragile, and will melt with the greatest ease.[5]

But Golding refuses to be 'penetrated' or 'melted' or disposed to Quicksilver's own nature: he will not 'turne ... a drunken whorehunting rakehell like thy selfe.' Quicksilver 'offers to draw' but Golding 'trips up his heeles and holds him'.[6] The assault has been resisted; Quicksilver has temporarily been 'fixed'; Golding has, morally and chemically, maintained his integrity.

Quicksilver's restless ambition is chemically defined as his mercurial volatility. 'Whither art thou running?', 'mad Quicksilver', 'my nimble-spirited Quicksilver', 'my runagate Quicksilver' — all these characterize him according to chemical type. He is just as Jonson's Face describes Mercury in *The Alchemist*: 'a very fugitive, he will be gone'. In the second act, his continual drunkenness and insolence get him fired. He rejoices in his new-found freedom — 'free of my fetters', 'I am now loose'. Throwing off his apprentice's coat and cap, he cries: 'There lie, thou huske of my envassal'd state.'[7] In alchemical parlance, he 'flies out', evaporates: his volatility carries him up out of the solid, 'fixed' world of industrious respectability into the gaseous-chaotic[8] world of knavery, opportunism and get-rich-quick trickery. He falls in with bad company — Security the usurer, Sir Petronel Flash (husband of Touchstone's 'bad' daughter, Gertrude) — but his devious financial schemes only end him up, ship-wrecked and bare-headed, beneath the gallows at Wapping, there repentantly to moralize on his own fate:

> The drift of all unlawfull courses
> (What ever ende they dare propose themselves
> In frame of their licentious policyes)
> In the firme order of iust Destinie,
> They are the readie highwayes to our ruines.[9]

He is taken into captivity, as a 'masterless' vagabond, and as he is marched off to the Counter, Touchstone gloats: 'There is my Quickesilver fixt'.[10]

By a moral causality as relentless as a chemical process, Quicksilver has come to be 'fixed'. By a very precise chemical analogy, he has been *sublimated*: his liberation from social bonds was his *solve* or volatilization, his imprisonment his *coagula* or fixation. In this same year, 1605, Thomas Tymme describes the fixation of quicksilver thus:

> Chymists ... have seen it in the preparation of quicksilver, whose liquor and running nature, no exterior coldnesse, no Elementall frost, how great soever the same be, congeale or fixe. But if it be sublimed with Vitriol, it will come to passe that Mercury or quicksilver ... is made solid and firme, so as thou maiest easily handle it. Being brought into this forme, it is commonly called Subli-

mate. . . . He is now perfectly clensed, and is now no more common Mercurie or Hydragyre, but the Philosophers Mercury.[11]

The play's morality demands that Quicksilver's 'unlawfull courses' should be punished by 'iust Destinie', and this is underscored by an echoing sense of 'chemical destiny', whereby the 'liquor and running nature' of crude mercury is made 'solid and firme' and so is 'perfectly clensed'. For it is indeed a very cleansed and perfected Quicksilver that haunts the Counter, described as 'penitent . . . devout . . . singing of Psalmes and aedifying the whole prison',[12] and seen distributing clothes and money to his fellow-prisoners. One looks in vain for psychological motives for Quicksilver's penitence. It is an inevitable product of the play's moral-chemical unfolding: Vice is converted into Virtue, Crude into Sublimate. 'Inconstant Mercurie,' says Tymme, 'alwayes tendeth to his perfection.'

Golding's course through the play is similarly assured by an inevitability both moral and chemical. He is freed from his apprenticeship, granted the hand of Mildred (Touchstone's 'good' daughter) in marriage, admitted into the ranks of the select Goldsmiths' Company, and appointed deputy alderman. Golding's social elevation is metaphorically imaged as the exaltation of common gold into *aurum nostrum*. This is neatly encapsulated when Touchstone promises that he 'will weare scarlet shortly',[13] at once referring to the red velvet livery of the alderman and the donning of red robes which symbolizes the arrival at *rubedo*, the redness, the final stage of the *opus*, as in Grasshoff's *Tractatus Aureus* (1625): 'Now when that most precious scarlet garment had been finished, the great and mighty king appeared in great splendour.'[14] Or in Ripley's *Compound*:

> There shineth the Sunne . . . in rednes with glorie
> As king to raigne upon all mettals and Mercurie.[15]

The Red King 'reigns upon' crude mercury: this is exactly what Golding does. For when Quicksilver's 'dissolute and lewed courses' lead him and Sir Petronel into the clutches of the law, it is Golding that sits in judgment over them. Touchstone urges: 'Come over 'hem with some fine guird, as thus: "Knight, you shall be encountered", that is, had to the Counter; or, "Quicksilver, I will put you in a crucible." '[16]

One should not pursue the underlying alchemical pattern of *Eastward Hoe* any further than this, though there are many other chemical references scattered through the play. It is a pattern based – like the morality it reinforces – on simple contrasts. Quicksilver and gold suggest contrasted properties of social instability and stability. The volatility of quicksilver and the exaltation of gold represent alternative types of social mobility. The fixation of quicksilver is, neatly, both a punishment and a mode of conversion, a cleansing sublimation. At the end of the play, Quicksilver announces his intentions to become industrious and obedient – solid, true metal:

> I cast my coat and cap away,
> I went in silkes and sattens gay,
> False mettall of good maneres I
> Did dayly coine unlawfully....
> Now cry I, 'Touchstone, touch me stil
> And make me currant by thy skill'.[17]

*   *   *

It seems highly likely that Jonson was the main instigator of the chemical undertones of *Eastward Hoe*. They are his first tentative experiment with the possibilities of analogy between alchemical and social processes. In *The Alchemist*, written some five years after *Eastward Hoe*, Subtle reproves Face in terms which intensify the analogies suggested by the earlier play:

> Thou vermine, have I tane thee, out of dung,
> So poore, so wretched, when no living thing
> Would keepe thee companie, but a spider or worse?
> Rais'd thee from broomes, and dust, and watring pots?
> Sublim'd thee, and exalted thee, and fix'd thee
> I'the third region, call'd our state of grace?
> Wrought thee to spirit, to quintessence, with paines
> Would twise have won me the philosophers worke:
> Put thee in words, in fashion? Made thee fit
> For more than ordinarie fellowships? [18]

Subtle has (he claims) transmuted Face from a base state of servitude. He has drawn him forth from the broom-cupboard and liberated him into volatile independence, just as the alchemist freed the quintessence from the dark 'prison' of matter.[19] As arduously as two *magna opera* put together, Face has been rendered fit for decent company. Taking him 'out of dung' is not just Subtle's way of being rude about Face's former condition, but an echoing of the frequent alchemical claim that the Raw Stuff, the potential Stone, is to be found everywhere, even and especially in the dunghill: in the Hermetic *Tractatus Aureus* (a new edition of which appeared in 1610, the same year as *The Alchemist*) we find, 'Our most precious stone cast forth upon the dunghill, being most dear is made altogether vile';[20] and in Michael Maier's *Symbola Aureae Mensae* (1617), 'Take that which is ground in their dunghills'.[21] The dunghill partly refers to the use of dung to provide gentle heat in the preparatory stages of the *opus*, as well as to the general theme of extracting the precious from the vile. In this finely-wrought conceit (far surpassing anything in *Eastward Hoe*) the alchemist's work of perfection provides an apt metaphor for social ascent. A little later we hear of Drugger's gratitude to Subtle, whose 'skill does raise him in the world'.[22] The implications of change and movement inherent in alchemical terms have become a part of Jonson's depiction of a mobile society, scrabbling for status and money.

Aurum palpabile

> I, when I value gold, may think upon
> The ductilness, the application,
> The wholsomness, the ingenuitie,
> From rust, from soil, from fire ever free:
> But if I love it, 'tis because 'tis made
> By our new nature (Use) the soul of trade.'

John Donne – Elegie XVIII

The bait successfully laid for Dapper and Drugger, Face turns back to Subtle in triumph. 'Why now, you smoky persecuter of nature,' he says,

> Now doe you see, that something's to be done,
> Beside your beech-coale, and your cor'sive waters,
> Your crosselets, crucibles, and curcurbites?
> You must have stuffe brought home to you, to worke on?[23]

This is, perhaps, the central joke of *The Alchemist*. For the Raw Stuff ('stuffe') Face refers to is precisely the prosperous fools of London, the Dappers and the Druggers whom he talks into coming to consult the great Doctor. The true alchemy of the play is not Subtle's grandiose and empty chemical work with 'crosselets, crucibles, and cucurbites', but the alchemy of the swindle: transmuting the stuff of gullibility into the gold of profit. Their alchemy turns not lead, but fools, into gold. It is Face who (as he goes on to say) prospects down the shady by-ways of Jacobean London, 'searching out these veines, then following 'hem, Then trying 'hem out'. He is the miner, unearthing rich ores of credulity and open-handedness, the *prima materia* of their transmutations. Hence Surly's determination: 'I would not willingly be gull'd. Your Stone cannot transmute me.'[24]

This is a metaphor apt and intrinsic to a play so concerned with alchemy. But its very point is that this kind of alchemy, this financial extraction, is not confined to puffer and smoke-seller. Did you think, sneers Dol Common to Subtle, that 'you onely had The poulder to proiect with?'[25] Anyone with their wits about them can practise *this* kind of alchemy. So it is hardly surprising that much the same metaphor appears in *Volpone* (1605). At the outset of the play, Volpone discloses his strategy of attracting would-be heirs to his fortune:

> I have no wife, no parent, child, allie,
> To give my substance to; but whom I make
> Must be my heire, and this makes men observe me.
> This drawes new clients, daily, to my house,
> Women and men of everie sexe and age,
> That bring me presents, send me plate, coyne, iewels,
> With hope that when I die (which they expect
> Each greedy minute) it shall then returne
> Ten-fold upon them; whilst some, covetous
> Above the rest, seeke to engrosse me, whole ...
> All of which I suffer, playing with their hopes,
> And am content to coyne 'hem into profit.[26]

Volpone's game is an alchemy prefiguring Subtle's. His greedy devotees, his 'clients', are themselves eagerly pursuing an alchemical multiplication, hoping their 'plate, coyne, iewels' will 'returne Ten-fold upon them'. They bring their precious 'stuff' – Voltore plate, Corbaccio chequins, Corvino a pearl – just as Mammon and the Faithful Brethren bring their brass, pewter, etc., to Subtle in the hope of literal transmutation. The greediest seek to 'engrosse' Volpone whole, to dissolve him completely and turn him into gold. But he turns it all back on them: he can 'coyne 'hem into profit'. It is they, the battening clients, who are unwittingly the Raw Stuff. By a fine irony, they are the very materials of the fortune they hope to inherit. They that seek to devour him in fact enrich him. An alchemical paradox – as in the dragon's venom becoming the great medicine – wittily rehoused in an economic context.

It is reasonable to suggest that Jonson's brilliant anatomy of swindling in these plays is more than an object-lesson. It is a typification of the kind of society in which such predators flourish and such prey is plentiful. From its earliest days, Jonson's comic purpose was to 'shew an Image of the times, And sport with humane follies'.[27] The image he shows is of a predatory, exploitative, avaricious society. By the naming-convention of *Volpone*, Venice is seen as a kind of zoo: the fox plots cunningly, the fly (Mosca) buzzes infectiously, and the suitors are hovering scavengers – vulture (Voltore), raven (Corbaccio) and crow (Corvino). This characterizes the degraded social relations of a viciously acquisitive society.[28] In *Volpone* and *The Alchemist*, extortion acts as a caricature of existing and accepted economic relations: capitalism on the loose. Alchemy, so apt a metaphor for extraction and profiteering, thus plays its part in Jonson's wider critique of Jacobean social values. The Philosophers' Stone is, in fact, the very epitome of profit ('I will have A booke... shall prove a true Philosophers Stone to printers,' says Face.[29]) So one comes to an equation so obvious that it needed someone of Jonson's wit to make it: the equation between alchemy and capitalism. Transmutation is the dream business venture, where an initial outlay of Raw Stuff brings in a huge return of gold. The percentages are seductive:

> One part proiected on a hundred
> Of Mercurie, or Venus, or the Moone,
> Shall turne it to as many of the Sunne.[30]

Multiplication is even more essentially capitalistic: a stock of gold makes more gold, the motive principle of capital-ism. In these plays, Jonson diagnoses a body politic infected by a rampant dose of 'gold-fever'. He who can get gold, says Volpone, 'shall be noble, valient, honest, wise'. Mosca adds the refrain: 'Riches are in fortune A greater good than wisedome is in nature.'[31] This gold-fever – a hunger for money and status and a 'feverish pursuit of capitalistic enterprise', as Keynes typifies the period[32] – grips cozener and cozened alike. The extravagant gold-making claims of alchemy fit perfectly into this theme – an epitome of instant enrichment. To get one's hands on the Philosophers' Stone was to find *El Dorado* in one's own back-yard, and it was not just the Mammons of Jonson's imagination that lived in such hope: we remember Lord Burghley's begging letters to Kelley, requesting a small 'token' of the Tincture, just enough to make a 'sum reasonable' to refit the Queen's navy.

* * *

Sir Epicure Mammon – Mr Money himself, as his name tells us – ushers Surly in to the ante-room of Subtle's laboratory with the ringing promise:

> Now, you set your foot on shore
> In *Novo Orbe*; here's the rich Peru,
> And there within, sir, are the golden mines,
> Great Salomon's Ophir! He was sayling to 't
> Three yeeres, but we have reach'd it in ten months.
> This is the day wherein, to all my friends,
> I will pronounce the happy word, 'Be rich!'[33]

We are left in no doubt of Sir Epicure's conception of the Stone. Subtle's laboratory is a *Novus Orbis*, a 'rich Peru' – a source of gold as rich as the Americas which were, in economic fact as well as popular fiction, the material source of England's new and turbulent prosperity. Thus Seagull's vision of Virginia in *Eastward Hoe*:

> I tell thee, Golde is more plentifull there then Copper is with us; and for as much redde Copper as I can bring, Ile have thrice the waight in Golde. Why man all their dripping pans and their chamber pottes are pure Gold; and all the chaines with which they chaine up their streets are massie Golde ... and for Rubies and Diamonds, they goe forth on holy-dayes and gather 'hem by the seashore.[34]

For Sir Epicure, the Stone is a merchant venture, an argosy to the Americas which will return with untold wealth (Drake's *Golden Hind* returned with £600,000 worth of booty in 1580, a 4,700 per cent return on the original investment in the voyage[35]). The 'happy word', the crown of the venture, will be, 'Be rich!' Sir Epicure's delirium continues:

> this night I'll change
> All that is mettall in my house to gold.
> And, early in the morning, will I send
> To all the plumbers and the pewterers
> And buy their tin and lead up; and to Lothebury
> For all the copper ...
> Yes, and I'll purchase Devonshire and Cornwaile
> And make them perfect Indies![36]

He will turn England itself into a *Novus Orbis*, a Caribbean Indies. A 'brave new world' somewhat different from that conjured up in *The Tempest*, the contemporary and dramatic *doppelgänger* of *The Alchemist*: Sir Epicure's has only one requisite – plenty of precious metals to exploit.

Sir Epicure's gold is clearly money. But it is also 'alchemical gold'. To Sir Epicure they are one and the same thing, but as Jonson well knew from his chemical reading they are by no means the same thing. '*Aurum nostrum non est aurum vulgi*,' says the *Rosarium* unequivocally; 'The golde ingendred by this Art excelleth all naturall gold in all properties,' says the *Mirror*: (both of these texts on Jonson's 'alchemical bookshelf'). There are clearly two types of gold: the *aurum nostrum* of the alchemist and the *aurum vulgi* of the purse. And a vital element in Jonson's use of alchemy as a socio-economic metaphor is his way of manoeuvring these 'two golds' in relation to one another. Once again, *Volpone* anticipates *The Alchemist*, opening with its famous hymn to gold:

> Good morning to the day; and next my gold:
> Open the shrine, that I may see my saint.
> Haile the worlds soule and mine. More glad then is

> The teeming earth to see the long'd-for sunne
> Peepe through the hornes of the celestiall Ram,
> Am I, to view thy splendor darkening his:
> That, lying here, amongst thy other hoords,
> Shew'st like a flame, by night; or like the day
> Strooke out of chaos when all darkenesse fled
> Unto the center. O, thou sonne of Sol,
> (But brighter then thy father) let me kisse,
> With adoration, thee, and every relique
> Of sacred treasure in this blessed roome.[37]

This is a tremendous fanfare, and many of its resonances are drawn from alchemy. It is an alchemical celebration of gold set to blank verse. Gold is styled the *anima mundi* ('the worlds soule, and mine'); the 'son of Sol' whose brightness outshines the real sun; the *novum lumen* ('a flame by night'); the day 'struck out of chaos' (like the quintessence which 'restest invisible, till it be forced out of Chaos darke'[38]). Gold, Volpone continues, 'mak'st men doe all things'; it is 'vertue, fame, Honour, and all things else'.[39] Or, as the *Emerald Table* phrases the same belief, 'So shalt thou have the glorie of the whole worlde. All obscuritie shall flie away from thee. This is the mightie power of all power.'[40]

Volpone's apostrophe to gold differs in just that one respect from the hypothetical alchemical panegyric which it matches all down the line: it is addressed to a different kind of gold. Not the alchemists' *aurum nostrum* – that complex composite of chemical, medicinal and mystical aspirations, emblem of opposites harmonized, and metallic receptacle of supernatural virtues – but *aurum vulgi*, money, lucre. The spiritual fervour of Volpone's hymn – 'shrine ... saint ... adoration ... relique ... sacred ... blessed' – is directed to an unequivocally material idol. This is caught nicely later in the play, when Bonario threatens Volpone,

> Thou shouldst yet
> Be made the timely sacrifice of vengeance
> Before this altar, and this drosse, thy idoll.[41]

In alchemical terms, Volpone's gold is only dross, but he worships it none the less. We see quite clearly here what is implicit in Sir Epicure's speech: that the ecstatic reverence which the alchemist accorded to his 'symbolic' gold has been transferred wholesale to a merely financial gold. The sustained alchemical metaphor in these plays thus encompasses a whole perception of society as obsessively, worshipfully materialistic, and infiltrates that judgment into the linguistic texture of the plays. That acquisition is dangerous and nasty when it obliterates other values – i.e., when it becomes worshipped as a religion – was surely a comment

Jonson wished to make through these plays, an 'image' he wanted to show to the 'times'. It is this comment which is communicated in a constant undertone by the sustained misapplication of Volpone's and Mammon's alchemical rhetoric, and by playing off the two notions of gold – sacred and financial – which the alchemical metaphor brings with it. The critique cuts both ways – degrading alchemy by equating it with money-making, comically inflating money by making it the object of a florid, archaic and futile devotion. As well as being characterized as animals, Volpone and Mosca are a species of debased alchemist, suffused with the tarnished *lumen* of *aurum vulgi*. Their gold is in their souls – 'Hail the worlds soule and mine'. It has razed all other relationships from their lives – 'transcending All stile of ioy in children, parents, friends'. It has become the index of all sensations:

> a wench
> O' the first yeere! A beautie ripe as harvest!
> Whose skin is whiter then a swan, all over!
> Then silver, snow, or lillies! A soft lip,
> Would tempt you to eternitie of kissing!
> And flesh that melteth in the touch to bloud!
> Bright as your gold! And lovely as your gold![42]

Here gold provides a sexual climax. In Mosca's sinister ingratiation to Voltore, there is an even stranger gold-pathology at work:

> And, gentle sir,
> When you doe come to swim in golden lard,
> Up to the armes in honny, that your chin
> Is borne up stiffe with fatnesse of the floud,
> Thinke on your vassall; but remember me.[43]

Fetid tone apart, there is nothing in this ecstatic relationship with gold that differs from the alchemist's, except the debased nature of the gold. This double notation of gold is nicely caught in an exchange between Corbaccio and Mosca:

> *Corbaccio* ... See, Mosca, looke,
> Here, I have brought a bagge of bright *cecchines*,
> Will quite weigh downe his plate.
> *Mosca* Yea, mary, sir!
> This is true physick, this your sacred medicine.
> No talke of opiates to this great elixir.
> *Corbaccio* 'Tis *aurum palpabile*, if not *potabile*.[44]

Money is the cure-all, the universal medicine: gold to get your hands on, if not to drink.

* * *

This equation between alchemy and capitalism – this mystic pursuit of profit – continued to be of service to Jonson. In *The Divell is an Asse* (1616), the figure of Meercraft the projector is built around a pun on the word 'projector' in its topical economic sense (a kind of investment counsellor who dreamed up projects 'to enrich men or to make 'hem great'), and in its alchemical sense (one who projects the Stone on to base metals in order to transmute them.) His first speech brings the alchemical element into sharp focus:

> Sir, money's a whore, a bawd, a drudge
> Fit to runne out on errands; let her goe.
> *Via pecunia*! When she's runne and gone,
> And fled and dead, then will I fetch her againe,
> With *aqua-vitae*, out of an olde Hogs-head!
> While there are lees of wine, or dregs of beere,
> I'le never want her. Coyne her out of cobwebs,
> Dust, but I'll have her! Raise wooll upon egge-shells,
> Sir, and make grasse grow out o' marro-bones
> To make her come.[45]

Meercraft imagines investment as a kind of alchemical *circulatio*. The initial capital outlay – 'money' – is a financial version of alchemical Mercury. It must be volatilized, become, in Ripley's words, 'spirituall and flying'. Only when it is 'runne and gone, And fled and dead' can it be 'fetched again' – at a handsome profit. It will be quickened and resurrected, like *aqua vitae* (distilled alcohol) out of the dregs of old wine or beer. Money is the Mercury of the Profiteers, both the agent and the product of Meercraft's *opus*. He is a kind of financial adept, drawing money like a precious quintessence out of various vile substances – cobwebs, dust, eggshells.

The general analogy between social and alchemical processes initiated in *Eastward Hoe* appears here, some ten years later, honed down to a very precise conceit. Not only precise, but effective because so concrete. This was an essential attraction of alchemy as a metaphorical resource: it was so much a language of matter, its preoccupation with chemical substance being expressed through a richly substantive vocabulary. In Meercraft's brief discourse on the economics of investment there are so many *things*, bearing with them so many rough, sensory associations of touch, taste and smell: 'an olde Hogs-head', 'lees of wine ... dregs of beere', cobwebs, dust, wool, eggshells, grass, marrow-bones. None of these belong to alchemical language as such, but they are sucked into Jonson's 'sterling English diction'[46] by means of the presiding alchemical conceit of the lines. The heavy redolence of the nouns is precisely counterpointed by the fleeting momentum of the verbs – 'runne ... gone ... fled ... fetch ... raise ... grow ... come' – all words of physical movement through space, enforced by 'want ... have ... make' with their momentum of appetite and intent towards 'her' – *Mercurius-Pecunia*. This lightness and mobility is part of the speech's Mercury conceit (just as those properties were part of the Mercury characterization of Frank Quicksilver) and they are also part of Meercraft as a dramatic creation – this is his patter, agile and insinuating, full of volatile schemes, a refined Jacobean version of the mountebank's 'drug lecture' in *Volpone*. This language of momentum suggests another of alchemy's literary potentials: its dynamic quality, its constant interest in process and alteration. Time and again in these plays, alchemy fuels Jonson's expression of a restless, mobile, endlessly transacting society, a Snakes and Ladders of ascent and descent, volatility and fixation. Jonson saw Jacobean society as the alchemist saw matter: a ferment of constant change.

Love's alchemy

In his *Songs and Sonets*,[47] Donne returns again and again to a sense of fleetingness, a perception of life as – in the words of Montaigne – 'changing, motion and inconstancy'.[48] He cannot 'adde another houre, Nor a lost houre recall';[49] he cannot repair his 'unthrifty wast of breath and blood'.[50] The *Songs and Sonets* are so often poems of going, the many valedictions and obsequies, departures and deaths seeming to epitomize a world where everything is fugitive and ephemeral. Experience itself might be the fast and loose lady of Donne's 'Song':

> Though shee were true, when you met her,
> And last, till you write your letter,
> Yet shee
> Will bee
> False, ere I come, to two, or three.[51]

At times the poet revels in this perpetual motion. He is 'the Indifferent', for whom variety is 'love's sweetest part': he can 'love her, and her, and you and you ... any, so she be not true'. The only horror, in this mood of bravado, is fixity:

> Rob mee, but binde me not, and let me goe.
> Must I who came to travaile thorow you
> Grow your fixt subject, because you are true?[52]

But this is only one of a whole gamut of responses. The poems themselves

CONIVNCTIO SIVE
Coitus.

8 From anon., *Rosarium Philosophorum* (1550)

are a response – grabbing at a sensed moment, sheltering it in a poem ('We'll build in sonnets pretty roomes'[53]), philosophizing it with a rush of strange, elliptical associations that might just pinion it to some system of absolute values, were not that too 'all in peeces, all cohaerence gone'.

This perception of transience and inconstancy is, of course, typical of late Elizabethan poetry as a whole. 'Never-resting Time' was often invoked and railed at; '*carpe diem*' was common advice. But with Donne it seems to go deeper, invading his own psychology. Time and again, he asserts that he himself is temporary, a succession of births and deaths and fleeting identities:

> Since I die daily, daily mourne.

> When I dyed last, and, Deare, I dye
> As often as from thee I goe.

> But since that I
> Must dye at last, 'tis best,
> To use my selfe in jest
> Thus by fain'd deaths to dye.

> When a teare falls, that thou falst which it bore,
> So thou and I are nothing then, when on a divers shore.[54]

Daily deaths, feigned deaths, the self falling away like a tear: a sense of psychic vertigo prevails in these poems. So, in 'The Good-Morrow', all past lives are cancelled out by the poem's first lines:

> I wonder by my troth, what thou, and I
> Did, till we lov'd? Were we not wean'd till then?[55]

And, in the sarcastically titled 'Woman's Constancy', he addresses his lover:

> Now thou hast lov'd me one whole day,
> To morrow when thou leav'st, what wilt thou say?
> Wilt thou then antedate some new made vow?
>   Or say that now
> We are not just those persons, which we were?[56]

The *Songs and Sonets* perceive a world of instants, of situations on the brink of dissolution, a world in which we are never 'just those persons, which we were'. It is a perception that owes much to Montaigne, who wrote (in Florio's 1603 translation), 'I now and I anon are indeed two persons', and acclaimed himself 'at all hours ready to become anything I can be'.[57] And how Donne must have revelled in the psychology of Hamlet, who, like him, feeds 'of the Camelions dish' and eats 'the ayre Promis-cram'd'.[58] Hamlet's personality is as volatile as Donne's, as impossible to sustain, a prey to the 'thousand naturall shocks that flesh is heire to'.[59] Donne's wit and irony and agile intellectualizing – the hallmarks of his 'Metaphysical' style – are like Hamlet's 'anticke disposition',[60] a bizarre mental performance conjured up to confront an emergency. It is this aspect of emergency and existential crisis that makes Donne – and, for that matter, Hamlet – so 'modern': they were acutely conscious of the 'New Learning' and saw, centuries in advance, how far the questioning could go and how much it could destroy. They would understand Sartre when he wrote, 'At every point in history, man is forced to invent man', or Samuel Beckett, who wrote in his essay on Proust (1931):

> We are not merely more weary because of yesterday, we are other, no longer what we were before the calamity of yesterday. . . . The poisonous ingenuity of time results in an unceasing modification of [our] personality, whose permanent reality, if any, can only be apprehended as a retrospective hypothesis.[61]

But this shifting, fragmented world, constantly dissolving into new unexamined forms, was only *one* of the worlds that Donne and Hamlet inhabited. They were still intent – unlike Sartre or Beckett – on another world entirely. For Hamlet it remained an 'undiscovered country',[62] but in some of the *Songs and Sonets* that 'other' world is achieved. It contains all that the 'real' world lacks: permanence, harmony, stability. For Donne it was love, a particular passionate blending of sexual and spiritual love, which was this *novus orbis*: his mistress was 'my America! my new-found-land'.[63] Love offers an escape from the dislocations of the physical world: change and decay are the rule, 'but all such rules loves magique can undoe'.[64] Love transcends time:

> All other things, to their destruction draw,
>   Only our love hath no decay;
> This no tomorrow hath, nor yesterday,
> Running it never runs from us away,
> But truly keepes his first, last, everlasting day.
>
> Love, all alike, no season knowes, nor clyme,
> Nor houres, dayes, moneths, which are the rags of time.[65]

Love transcends space: it 'makes one little roome, an every where',[66] it 'makes my circle just, And makes me end, where I begunne'.[67] Love transcends vicissitude:

> Let sea-discoverers to new worlds have gone,
> Let maps to other, worlds on worlds have shown,
> Let us possesse one world, each hath one, and is one. . . .
> Where can we finde two better hemispheares,
> Without sharpe North, without declining West?[68]

These lovers are 'sea-discoverers' of another kind, embarked for a new world within themselves. The journey must begin at the here-and-now – "Tis true, 'tis day', 'So, so, breake off!', 'Marke but this flea'.[69] It is, at first, only the sudden, happening present that the poet knows: this moment, this self, the certainty of sexual contact. The poems accept nothing *a priori* but are always setting out on their search for transcendence, juggling and speculating their way towards the metaphysical. They are called 'Metaphysical Poems' in this sense: they search out strange avenues from the passing moment to 'eternal truths'. In certain *Songs and Sonets*, love is instanced as the supreme example of this transcendence. The lover is raised from chaos and corruption to harmony and perfection. Love is, in short, a transmutation.

This is a context in which to see the powerful presence of alchemical imagery in the *Songs and Sonets*. Lovers are they 'whom love's subliming fire invades':[70] the experience of love is like the alchemist's sublimatory furnace, in which matter (or self) is first reduced to vapour and then reconstituted into new, purer solidity. Similarly in 'The Extasie', 'so by love refin'd', and 'A Valediction: Forbidding Mourning', 'we by a love so much refin'd'[71], suggest purification, the exchange of gross for fine. And when Donne wishes to doubt the spiritual properties of love – transcendence being, like all else, a prey to doubt and irony – it is alchemy that provides the conceit:

> But if this medicine, love, which cures all sorrow
> With more, not onely bee no quintessence,
> But mixt of all stuffes, paining soule, or sense,
> And of the Sunne his working vigour borrow,
> Love's not so pure, and abstract, as they use
> To say, which have no Mistresse but their Muse,
> But as all else, being elemented too,
> Love sometimes would contemplate, sometimes do.[72]

The motif here is Paracelsist. Love is jestingly styled an *arcanum*, curing *similia similibus*, making one sorrow expel another. But is this medicine truly a quintessence – in the words of Paracelsus, 'a matter most subtly purged of all impurities and mortality'[73] – or is it merely a brew of raw impure 'stuffes', like the 'corporeal and grosse medicines' of the Galenists?[74] Is, perhaps, this 'pure and abstract' spiritual love ('abstract' nicely incorporating the chemical 'extract', both derivatives of *trahere*, to draw or drag) a mere poetic invention, and 'real' love a *massa confusa*, a mix of bodily elements and hidden spirit? Thus by devious dialectic – via an opposition of Paracelsist *arcanum* to Galenist concoction, and of quintessence to element – Donne arrives at a synthesis of spiritual love (which 'contemplates') and sexual love (which 'does').

These are instances of alchemical language providing an apt metaphor for the purifying, spiritualizing effects of love – love as subliming fire, refining process, quintessential medicine. Love is a kind of alchemy, redeeming the fragmentary 'base' nature of physical experience, exalting it to a condition of harmony and perfection, the 'spiritual solidity' of the Stone. So 'The Good Morrow', where love transforms 'one little roome' into 'an every where', closes with the words:

> Whatever dyes, was not mixt equally;
> If our two loves be one, or, thou and I
> Love so alike, that none doe slacken, none can die.[75]

A perfect equilibrium, 'none doe slacken', ensuring a perfect incorruptibility, 'none can die': precisely the condition sought in the form of *aurum nostrum* or Stone, described by Artephius as a '*corpus aequale*' (equal body) or by Thomas Norton with the injunction:

> Compound ye our Stone
> Equall, that in him repugnance be none;
> Neither division as ye proceede.[76]

The lovers are balanced, matching hemispheres of their microcosmic world, just as the elements are balanced without 'repugnance' in the Stone, itself called by Norton 'Microcosmos'.[77] The equation between harmony and incorruptibility was not, of course, peculiar to alchemy: Grierson cites Aquinas' *Summa Theologica*, 'non invenitur corruptio nisi ubi invenitur contrarietas'.[78] But where could one find more apt contemporary echoes of Donne's lines than in the alchemical literature of the day? In Sendivogius' *Novum Lumen* of 1604 we find: 'Whatsoever is conceived of two bodies is subject to the law of death, but the life of this fruit is the separation of all that is corruptible about it.'[79] And in Thomas Tymme's *Chymicall Physicke* (1605), gold is praised as a substance whose elements

> do so order themselves that the one doth not exceed the other, but being as it were equally ballanced and proportionated, they make gold to be incorruptible. . . . And the reason hereof is for that No equal hath any command or maisterie over his equal.[80]

*   *   *

Alchemy plays a different kind of role in the subtle, shifting argument of 'The Extasie'. It is generally accepted that the philosophical framework of the poem is strongly neo-Platonic: the *Enneads* of Plotinus (in Ficino's 1492 translation) and Ebreo's *Dialoghi d'Amore* (1535) have been cited as possible influences.[81] The ecstasy of the title is, certainly, the kind expressed by mystic neo-Platonists: a moving-out of the soul from the body; in Donne's own words, 'a departing, and secession, and suspension of the soul'.[82] That alchemy should be consistently present in the poem suggests Donne's awareness of its wider philosophical connections, and of its reference to psychic as well as material transformations.

'Where, like a pillow on a bed, A pregnant banke swel'd up.' The poem begins, characteristically, with a neat *mise-en-scène*: a grassy slope with flowers. But it is more: a poem about mystical ecstasy whose first two lines introduce 'bed' and 'pregnant', a very sexual causality. Thus the poem's conjugation of physical and metaphysical, sexual and spiritual, begins at the very outset. The lovers – 'one anothers best', again the image of harmony and reciprocation – sit on the bank, their hands 'cimented With a fast balme, which thence did spring'. This 'balm' appears often in Donne's poems. In the 'Nocturnall', winter is depicted with the bleak line, 'The generall balme th'hydroptique earth hath drunke':[83] there the balm is a kind of 'world sap', a liquor of growth and fertility drained and drunk up by the dropsical earth. More specific is the balm which Donne writes of in a verse-letter to Lucy, Countess of Bedford:

> In every thing there naturally growes
> A *Balsamum* to keepe it fresh, and new,
> If 'twere not injur'd by extrinsique blowes.[84]

Similarly, in the 'Anatomie of the World', the world is 'dead, yea putrified' because 'shee Thy intrinsique balme, and thy preservative, Can never be renew'd'.[85] This animating, preserving balm or balsam was drawn straight out of current alchemical, particularly Paracelsist, terminology. In his *Chymicall Physicke*, Tymme writes of 'our naturall Balsam' as 'the only meane to conserve our life ... the only immediate putter away of sicknesses, and of all corporeal infirmities'. It is the 'burning lamp of the fire of our nature' and sickness occurs whenever it is 'diminished or hurt by any occurrent outwardly' (or, as Donne phrases it, 'injur'd by extrinsique blowes').[86] Still weightier claims are pressed in Martin Ruland's *Lexicon Alchemiae* (1612), where *balsamum* is praised not only for 'preserving its body from corruption' but also as the 'liquor of external Mercury ... the firmamental essence of existences, the Quintessence'.[87] By pursuing these alchemical overtones we see how Donne's opening lines prefigure in miniature the poem's whole enquiry into the relation between sexual and spiritual love. Explicitly, the balm that springs from the lovers' hands is the sweat and moisture of sheerly physical touch. Implicitly, however, it is the animating *balsamum*, the quintessential spirit of life itself. A covert pun – sticky palms or firmamental essence – ushers in the poem's dialectic. Donne may have had in mind a similar juxtaposition in Shakespeare's *Venus and Adonis*:

> she ceazeth on his sweating palme,
> The president of pith and livelyhood,
> And trembling in her passion, calls it balme,
> Earth soveraigne salve to do a goddesse good.[88]

Now the ecstasy begins, the secession of the soul from the body:

> As 'twixt two equall Armies, Fate
>   Suspends uncertaine victorie,
> Our soules, (which to advance their state)
>   Were gone out,) hung 'twixt her, and mee.
> And whil'st our soules negotiate there,
>   Wee like sepulchrall statues lay. ....
> This Extasie doth unperplex
>   (we said) and tell us what we love,
> Wee see by this it was not sexe,
>   Wee see, we saw not what did move:
> But as all severall soules containe
>   Mixture of things, they know not what,
> Love, these mixt soules, doth mixe againe,
>   And makes both one, each this and that.

Again love is an alchemy, or rather an alchemist. The soul is extracted ('gone out') like the Mercurial *anima* from matter, consigning the body to mortification and death ('sepulchrall statues'). This ecstasy 'doth unperplex': it is an unravelling of soul from body, an alchemical separation. Thus Ripley in the *Compound*:

> Separation doth each part from other divide,
> The subtile from the gross, the thick from the thinn....
> Till earth remaine beneathe in colours bloe,
> That earth is fixed to abide all woe:
> The other parte is spirituall and flying.[89]

Thus extracted, the lovers' souls are purged of all division by the alchemist – 'love these mixt soules doth mixe againe, And makes both one' – and conjoined into one indissoluble unity:

> When love, with one another so
>   Interanimates two soules,
> That abler soule, which thence doth flow,
>   Defects of lonelinesse countroules.
> Wee then, who are this new soule, know
>   Of what we are compos'd, and made,
> For th'Atomies of which we grow,
>   Are soules, whom no change can invade.

This 'interanimation' of souls into a single incorruptible 'abler soule' sounds much like the alchemical *coniunctio*. After solution and separation, the next 'gate' into Ripley's castle is conjunction:

> For untill the time the Soule be separate
> And cleansed from his originall sinne ...
> The true Coniunction maist thou never begin.[90]

A slightly pruned passage from Siebmacher's *Hydrolithus Sophicus* sums up the course of Donne's alchemical ecstasy so far:

> Let it be your first object to solve this substance, or first entity.... Then it must be purged of its watery and earthy nature.... Thus, by a final sublimation, the heart and inner soul contained in it may be separated and reduced to a precious essence.... Then the extracted heart, soul and spirit must once more be distilled and condensed into one.[91]

At this point, the poem reaches its crucial *volte-face*, the turning-point of the soul's circular journey:

> But O alas, so long, so farre
>   Our bodies why doe wee forbeare?

> They are ours, though they are not wee, wee are
>   The intelligences, they the spheare.
> We owe them thankes, because they thus,
>   Did us, to us, at first convay,
> Yielded their forces, sense, to us,
>   Nor are drosse to us, but allay.

After *solve* must come *coagula*, 'For the Dissolution on the one side corporall Causeth Congelation on the other side spirituall.'[92] The extracted soul, the now 'abler' *anima*, must return to re-animate the body. This pre-ordained circle, traced out by vapours and droplets in the alchemist's chemical theatre, becomes the framework of Donne's purposefully circular argument. It becomes clear that 'The Extasie' is not just a celebration of love as a mystic union of souls, but rather a highly philosophical incitement to love as a sexual union of bodies. A sharp chemical distinction brings home the point: our bodies are not 'drosse' but 'allay', not mere refuse and residual earthy 'faeces' left behind by the soul, but an alloy, an essential constituent to which the soul must be recompounded.

> So must pure lovers soules descend
>   T'affections, and to faculties,
> Which sense may reach and apprehend,
>   Else a great Prince in prison lies.

Without the *descensus* of spirit back into matter the *opus* remains incomplete; the secret 'entrance into the shut palace of the king' remains unopened. Ingeniously, it is sex that consummates the mystic ecstasy:

> To our bodies turne wee then, that so
>   Weake men on love reveal'd may looke;
> Love's mysteries in soules do grow,
>   But yet the body is his booke.

This is the pay-off, the clinching of Donne's alchemical seduction. The argument he uses here is found, for example, in Petrus Bonus' *Pretiosa Margarita Novella*:

> It is the body which retains the soul, and the soul can show its power only when it is united to the body. Therefore when the artist sees the white soul arise, he should join it to its body in the same instant, for no soul can be retained without its body. . . . The body is stronger than the soul or spirit, and if they are to be retained it must be by means of the body. The body is the form, and the ferment, and the Tincture of which the sages are in search.[93]

None of this is intended to displace in any way the neo-Platonic structure of 'The Extasie'. It might be said, however, that the persistent alchemical undertone works as a way of re-aligning the neo-Platonic convention so characteristic of Renaissance love poetry. Donne does not drain love of sexuality by asserting that all bodily beauty is merely a cipher of inner, spiritual purity. Rather, he draws down spiritual meanings into the sphere of sexuality, just as the alchemist breathed spiritual powers into matter: he creates a 'metaphysical' body as the alchemist a 'philosophical' stone. The alchemical terminology is not obtruded, but is always there – balm, refined, concoction, mixture, multiplies, atomies, dross, allay – always providing that anchorage to the world of substance and action which proves to be the goal of the poem's argument.

This same kind of alchemically-tinged Platonism is the staple of Donne's 'Anatomie of the World', written and published in 1611 in memory of Elizabeth Drury, daughter of the poet's patron at the time.[94] The poem's presiding conceit is Platonic: the dead girl is idealized as the Soul of the World and so the world has, by her death, become a 'carcasse', since 'shee which did inanimate and fill the world be gone'.[95] The idea of the world's 'death' allows much sombre lucubration on the decadence and parlousness of the age – "tis all in peeces, all cohaerence gone' being a well-known example. Once again, alchemical language is a channel for the poem's discourse, indeed the first image Donne uses to depict the world's exanimate plight is a chemical one:

> This World, in that great earthquake languished;
> For in a common bath of teares it bled,
> Which drew the strongest vitall spirits out.[96]

The world is metaphorically immersed in a chemical *balneum* of sorrow, a caustic 'bathe of tears' in which it is dissolved and decomposed, freeing its quintessential 'vitall spirits'. The world becomes dross, dead exanimate earth, an alchemical type of 'carcass'. It undergoes the *nigredo* of putrefaction:

> it be too late to succour thee,
> Sicke world, yea, dead, yea putrified, since shee
> Thy intrinsique balme, and thy preservative,
> Can never be renew'd, thou never live.[97]

Alongside the Platonic image of the idealized She as the *anima mundi*, Donne cultivates the image of her as quintessential spirit, intrinsic balm, Our Mercury. As such, she hovers ready to return, the promise of *coagula* after the disintegrations of *solve*:

> The twilight of her memory doth stay;
> Which from the carcasse of the old world, free,
> Creates a new world, and new creatures bee

> Produc'd: the matter and the stuffe of this,
> Her vertue, and the forme our practice is.⁹⁸

Thus the possibility of resurrection is dangled before the world. Her 'vertue' – in the operative, Paracelsist sense – will redeem the dead earth and quicken it into a 'new world'. As Basil Valentine promised in his *Twelve Keys*, 'After that burning, a new Heaven and a new Earth shall be formed and the new man shall more gloriously shine forth . . . for he shall be purified.'⁹⁹ She, as Mercury, will be the 'matter and the stuffe' of this transformation; we, by imitating her virtue and honouring her memory, will perform this global *opus* ('our practice' will be 'the forme' of it). In a highly-wrought eulogy, Donne brings the alchemical motif to a climax:

> Shee in whom vertue was so much refin'd;
> That for Allay unto so pure a minde
> Shee tooke the weaker Sex; shee that could drive
> The poysonous tincture, and the staine of Eve,
> Out of her thoughts and deeds; and purifie
> All, by a true religious Alchymie.¹⁰⁰

Her virtue is 'so refin'd' – therefore so quintessential and potent – that it could join itself to (take for 'allay') the earthiest of elemental matter ('the weaker sex') and transmute it, cleansing its intrinsic corruption, the alchemist's 'Filth originall' here laid at the door of Eve. This virtue is no pale moral purity, but an energy, a potential for transformation. Donne knew well the claims of 'true religious Alchymie' so eloquently advanced in the first years of the seventeenth century: that man could be transmuted as well as metal, and that the true province of the Mercurial spirit was within the self.

In the 'Anatomie', alchemy provides a quality of momentum and process, as it did, in a very different way, for Ben Jonson, whose *Alchemist* was on the London stage when Elizabeth Drury died. The Platonic framework of the poem – the fled *anima mundi* and ailing *corpus mundi* – is made dynamic. This dynamism is expressed both through a Paracelsist idea of spirit as operative and therapeutic –

> She, for whose losse we have lamented thus,
> Would worke more fully, and pow'rfully on us,
> Since herbes, and roots, by dying lose not all,
> But they, yea Ashes too, are medicinall,
> Death could not quench her vertue so.¹⁰¹

– as well as through the classic alchemical figure of transmutation:

> She, from whose influence all Impressions came,
> But, by Receivers impotencies, lame,

> Who, though she could not transubstantiate
> All states, yet guilded every state. . . .
> She that did thus much, and much more could doe,
> But that our age was Iron, and rustie too.[102]

Working on, flowing in, transubstantiating all to gold. To infuse his conception of *anima* with vigour, presence, perhaps even meaning, Donne calls it *arcanum* and Our Mercury.

*   *   *

Another way of looking at 'The Extasie' and 'The Anatomie' is to say that in these poems alchemy is involved in a particular 'reading' of death. Death often figures in the *Songs and Sonets*, I suggested earlier, as a perception of temporariness: it describes the inevitable dissolving of all situations and selves. These 'feign'd deaths' lead nowhere: they suggest a sort of 'picaresque' perception of experience as a series of interrupted scenarios and discarded masks. 'The Extasie' offers a quite different reading of death. The ecstasy of the title is precisely a death – 'Wee like sepulchrall statues lay' – but it is death as an essential phase in a process of transformation. Only by that 'mortification' of the body can the lovers' souls 'advance their state'. Death is no longer an amputation of life, but a channel to new, transcendent life. 'No generation without corruption' was the alchemist's phrase for this death, the *nigredo* of putrefaction through which matter must journey, Ripley's Gate of Blackness. 'Nothing in the world can be naturally born or animated,' asserts Raymond Lully, 'except after its corruption and putrefaction and mortification, because it is then that nature is changed into nature.'[103] This idea of 'through death to life' is, of course, universal, a feature common to devotional and magical systems of all sorts. But time and place throws up a language, a specific instrument for a universal purpose: the searching mind of *c.* 1610 found that language in alchemy.

'The Extasie' and 'The Anatomie' are not the only examples of alchemical metaphor as part of Donne's conception of 'mystic death'. In his 'Elegie' on Lady Markham (1609), the grave is imaged as an alchemical vessel, an arena of transmutation:

> So at this grave, her limbecke, which refines
>   The Diamonds, Rubies, Sapphires, Pearles, and Mines,
> Of which this flesh was, her soul shall inspire
>   Flesh of such stuffe, as God, when his last fire
> Annuls this world, to recompence it, shall
>   Make and name then, th'Elixar of this All.[104]

Here again death is an exaltation, a liberation of the soul to perform its office as Our Mercury. A much more extended alchemical conceit, similarly designed to conjure death into affirmation, is found in 'The Dissolution'.[105] It begins:

> Shee is dead; And all which die
> To their first Elements resolve.

Death is a dissolution into 'first Elements', a transition from the four bodily elements to the pre-existent spiritual (or formal) elements. The distinction is explained by Thomas Tymme in his *Chymicall Physicke*:

> The invisibles are the elements simple, formall, the astral seedes, and spirituall beginnings.... The visibles are all one and the same, but yet covered with a materiall body. The which two bodyes, spiritual and material, visible and invisible, are contained in every Individuall.[106]

Her death is, then, a return to the 'elements simple, formall', the spiritual beginnings of matter. Tymme goes on to say how 'Chymestry is to be extolled' for separating 'Elements and their beginnings', and how a 'workeman' can dissolve the bodily elements and 'bring them to nothing'. Clearly, in 'The Dissolution', death is that 'workeman', or alchemist. Having suggested, in the first two lines, what the effect of death has been on her, Donne then turns to complain of its effect on him, the lover left alive:

> And wee were mutuall Elements to us,
> And made of one another.
> My body then doth hers involve,
> And those things whereof I consist, hereby
> In me abundant grow, and burdenous,
> And nourish not, but smother.
> My fire of Passion, sighes of ayre,
> Water of teares, and earthly sad despaire,
> Which my materialls bee,
> But neere worne out by loves securitie,
> Shee, to my losse, doth by her death repaire.

She has abandoned her bodily elements and, because the two lovers were 'made of one another' and mutually 'involved', he is left with them. Her flight into ethereal form is contrasted with his stasis, weighed down with this 'double helping' of materiality. His body grows 'burdenous'; it 'smothers'; his 'materialls' are, by this discomfiting paradox, repaired and fortified by her death. But then comes the argument's crucial gymnastic twist:

> I might live long wretched so
> But that my fire doth with my fuell grow.

> Now as those Active Kings
> Whose foraine conquest treasure brings,
> Receive more, spend more, and soonest breake:
> This (which I am amaz'd that I can speake)
> This death, hath with my store
> My use encreas'd.
> And so my soul more earnestly releas'd,
> Will outstrip hers.

The pivot is fire, and here Donne draws on a rather specialist item of alchemical theory. According to the traditional Aristotelean anatomy of matter, fire was simply one of the four elements, and so Donne treats it in the first part of the poem: 'fire of Passion' is one of the elements with which he is, by her death, over-endowed. Certain contemporary Paracelsians, however, questioned the elemental status of fire, reducing the complement of elements to three, matching the *tria prima* ('three beginnings') of mercury, sulphur and salt postulated by Paracelsus. Thomas Tymme's *Chymicall Physicke* provides the new theory in its form most accessible to Donne:

> Aristotle did more than was needefull to appoynte a quaternarie number of Elements, out of the quaternarie number of the fower qualities, Hote, Colde, Drie, Moyst.... Forasmuch as Moses in the first Chapter of his *Genesis* (wherein he sheweth the creation of all things) maketh no mention of Fier, it is more convenient that we leave it rather to the opinion of the divine prophet, then to the reasons of an Ethnicke Philosopher. And therefore wee acknowledge no other Fier than Heaven, & the fiery Region which is so called of burning. Therefore it ought to be called the fourth formall Heaven and essential Element, or rather, the fourth essence, extracted out of the other elements, because it is indued with far more noble vertues then the most simple elements.[107]

Fire, then, is promoted. It is no longer an element, but a 'fourth essence' akin to – or, correctly, replacing – the quintessence. This is a celebration of fire as process rather than constituent. Fire – the Paracelsist Vulcan, the alchemist's Athanor, the dragon Mercury – was, in Tymme's words, 'the author of all formes, powers, and actions, in all the inferior things of nature'. That Donne was aware of this modification of elemental theory is apparent from a famous couplet in 'The Anatomie':

> And new Philosophy calls all in doubt,
> The Element of fire is quite put out.[108]

That this is the lynch-pin of 'The Dissolution' has never, to my knowledge, been pointed out. For Donne argues that, if all his bodily elements are increased to a burdensome abundance, this must entail an increase in his fire – 'my fire doth with my fuell grow'. And, just as the enrichment of kings by 'foraine conquest' leads to their downfall (they 'spend more, and soonest breake'), so his elemental enrichment will paradoxically hasten his dissolution, because his fire, the 'author of all formes', will dissolve and transform him. Her death 'hath with my store My use encreas'd': there the two readings of fire are juxtaposed, fire as bodily element ('store') and fire as active principle ('use' – using him up). The latter will ensure the desired dissolution: his freed *anima*, being more violently expelled ('earnestly releas'd') will soon overtake hers on the way to the ethereal regions of the first elements.

In 'The Dissolution', death is juggled by alchemical argument into liberation. As in 'The Extasie', 'The Anatomie' and the 'Elegie' on Lady Markham,[109] Donne quarries an idea of death as transformation, and makes it confront that other idea of death as curtailment. We might find in alchemy a vital channel between the love lyrics of the *Songs and Sonets* and the divine poems of Donne's later years, where Christ's resurrection is a *magnum opus* –

> Hee was all gold when he lay downe, but rose
> All tincture, and doth not alone dispose
> Leaden and iron wills to good, but is
> Of power to make even sinfull flesh like his.[110]

– and where God, like the alchemist's fire, dissolves to transform:

> Burne me, O Lorde, with a fiery zeale
> Of thee and thy house, which doth in eating heale.
>
> Recreate me, now growne ruinous:
> My heart is by dejection, clay,
> And by selfe-murder, red.
> From this red earth, O Father, purge away
> All vicious tinctures, that new fashioned
> I may rise up from death, before I am dead.[111]

*   *   *

The poem which deliberately turns all this upside down is the great 'Nocturnall upon S. Lucies Day'[112] where alchemy's metaphoric association with love and death is twisted into a strange negation. It begins, like 'The Anatomie', with an evocation of a dead world, the *corpus mundi* as carcass. It is the depth of winter, the 'yeares midnight':

> The Sunne is spent, and now his flasks
> Send forth light squibs, no constant rayes:
>   The world's whole sap is sunke:
> The generall balme th'hydroptique earth hath drunk.

The landscape image — baleful wintry sun and parched frozen land — is underscored by alchemical hints: the mortified Sol sputtering in his retorts, the draining away of quintessential *balsamum*. Seasonal and alchemical deaths bear with them a promise of renewal, but this is nowhere to be felt in the lines. The poet announces himself as even worse off than the 'shrunke, Dead and enterr'd' world:

> For I am every dead thing,
> In whom love wrought new Alchimie.
> For his art did expresse
> A quintessence even from nothingnesse,
> From dull privations, and leane emptinesse:
> He ruin'd mee, and I am rebegot
> Of absence, darknesse, death, things which are not.
>
> All others, from all things, draw all that's good,
> Life, soule, forme, spirit, whence they beeing have;
> I, by loves limbecke, am the grave
> Of all, that's nothing.

This is another of Donne's images of love's alchemy — the working on a corrupt Raw Stuff ('dead thing'), the drawing out ('expresse') of quintessence, the purifying distillation in the alembic ('limbecke') of love. But the effects of this transmutation are drastically inverted. So irredeemably dead is the Raw Stuff that this alchemy can only fashion its elemental constituents — 'nothingnesse, dull privations, leane emptinesse' — into their concentrated spiritual equivalents, 'absence, darknesse, death'. Ingeniously, and fittingly for the poem's 'gothic' nocturnal setting, Donne has created a kind of horror-story. The 'Nocturnall' is, perhaps, a prototype of what might be called the Mad Scientist genre, where something goes 'dreadfully wrong' with the experiment, and some mutation or destructive force is unleashed.[113] In the 'Elegie' on Lady Markham, the idea of alchemical death — the grave as limbecke — produced the 'Elixar of this All'. In the 'Nocturnall' it produces precisely the opposite: an Elixir of nothing ('I am ... Of the first nothing, the Elixer grown'). The lover becomes that Nihilixir which Siebmacher warns against in the *Hydrolithus Sophicus:* 'If thou strivest unduly to shorten the time thou wilt produce an abortion. Many persons have, through their ignorance, or self-opinionated haste, obtained a Nihilixir instead of the hoped-for Elixir.'[114] The alchemy of the 'Nocturnall' is an aborted *opus*, a transmutation producing death instead of new life, a quintessence of nothing instead of the seed of something. So the last stanza begins: 'But I am none, nor will my Sunne renew.' The mortified gold, the stricken self, remains dead.

This dark antic is a kind of coda; it is the exception which proves the rule of Donne's conception of alchemy. He found in alchemy, as did Ben Jonson, a valuable and pliant language. Through it, he expressed a recurrent idea of transcendence, alteration and renewal. In various of the *Songs and Sonets*, love is imagined as a form of alchemical change, a flight from the chaos and vertigo of physical experience to a world of serene harmonies where 'none doe slacken, none can die'. Also, by a blending of alchemical and Platonic terminologies, one idea of death – as an annulment brooding over all moments – is confronted by another 'mystical' death, an alchemical separation or Platonic ecstasy which liberates the quintessential *anima* from the confines of the body. Alchemy is only one of Donne's rich metaphorical resources, only one of the languages he ransacked for images, arguments, precedents and general metaphysical scaffolding. But the motif it expresses is a central one – transcendence, perhaps the one constant mental movement to which all the jagged, elliptical patterns of his poetry finally conform.

· 6 ·

Alchemical Bearings on *King Lear*

The *longissima via*

'Nonulli perierunt in opere nostro.'

Rosarium Philosophorum

In one of the earliest extant alchemical texts – the third-century treatise Περι ἀρετή ('Concerning the Art') by Zosimos of Panoplis – the alchemical process is narrated in a series of dream-visions. In the first of these, the narrator sees a sacrificial priest (ἰεουργός) standing high upon an altar shaped like a bowl or dome (φιαλη). The priest, Ion, describes the ordeal of his consecration:

> I have performed the act of descending the fifteen steps into the darkness, and of ascending the steps into the light. And he who renews me is the sacrificer, by casting away the grossness of the body; and by compelling necessity I am sanctified as a priest and now stand in perfection as a spirit. . . . I am Ion, the priest of the inner sanctuaries, and I submit myself to an unendurable torment. For there came one in haste at early morning, who overpowered me, and pierced me through with the sword, which he wielded with strength, and mingled the bones with the pieces of flesh, and caused them to be burned upon the fire of the art, till I perceived by the transformation of the body that I had become spirit. And that is my unendurable torment.[1]

This is probably the earliest example of the rich narrative and allegoric vein in alchemical literature. The priest personifies the Stone, or *corpus subtile* – matter which has 'cast away the grossness of the body' to 'stand in perfection as a spirit'. The bowl-shaped altar on which he stands is an emblem of the alchemical vessel.[2] The 'unendurable torment' of dismemberment and burning suggests chemical processes of dissolution and separation. The descent into darkness is the alchemical death, the *nigredo* or blackness. In the vision of Zosimos, the alchemical *opus* is presented as a story of sacrifice and redemption, torment and transformation.

9 Killing the king. From Daniel Stolcius, *Viridarium Chymicum* (1624)

Another early example of alchemical narrative is the *Visio Arislei* ('Vision of Arisleus'), which forms part of the *Turba Philosophorum* ('Convention of Philosophers'), a tenth-century Islamic text probably derived from Greek sources. Arisleus and his companion-alchemists journey to 'the end of the world', to a kingdom where 'nothing multiplies'. Only like mates with like. Learning that there are no philosophers in this barren land, the alchemists offer to teach the king the secrets of generation. 'If there were a philosopher among you, your sons could multiply, your trees would not die, your seed would grow.'[3] They decree that the king's son and daughter, Thabritius and Beya, must couple. At the moment of sexual union, Thabritius dies – in one version[4] he is swallowed up into his sister's body: 'she embraced Thabritius with so much love that she absorbed him completely into her own nature.' The alchemists are cast into prison, a 'triple glass house', together with the corpse of Thabritius. They promise to restore him to life if Beya can also remain in prison with them. There they languish 'in the shadows of the waves, the intense heat of summer, and the turbulence of the sea'. After eighty days, Thabritius is reborn. The alchemists are released, the king rejoices, the kingdom prospers.

As a narrative, the *Visio Arislei* clearly has qualities we do not find in Ion's description of his sanctifying ordeal. There is the element of journey, a redeeming arrival in the 'waste land'. There is a choreography of characters – alchemists, king, king's children – whose fates are intertwined. The alchemists themselves participate in the ordeal within the glass prison of the vessel, in contrast to Zosimos, whose alchemist is an implacable 'sacrificer'. The alchemical process itself is more precisely allegorized in the *Visio Arislei*. The emphasis is shifted to the central act of *coniunctio*. Thabritius is sulphur (Arabic *kibrit*), Beya mercury (Arabic *al-baida*, the white one): the incestuous coupling of prince and princess is a version of the 'Chymical Wedding' of King and Queen. Nevertheless, though different in type and detail, these two narratives recount the same basic 'story': a sacrificial submission to some mysterious, fatal and finally redeeming process. The death of Thabritius, like the torture of Ion, represents 'casting away the grossness of the body': Thabritius, as sulphur, is the bodily and earthy aspect of matter. Both must pass through the *nigredo* – Ion descending the fifteen steps to darkness, Thabritius putrefying for eighty days in the glass prison. Both stories end with reanimation – Ion sanctified as priest-spirit, the king's son reborn, matter regenerated as *corpus subtile*. The basic alchemical rhythm – *solve et coagula* – is recounted as painful loss and sanctifying recovery.

This genre of 'alchemical parable' – the depiction of the *opus* in allegorical narrative guise – is to be found in all periods of alchemical literature. Within the medieval English tradition, the best example is undoubtedly George Ripley's *Cantilena* (Song), a 38-stanza Latin poem composed *c.* 1450–70, and translated into Elizabethan English verse sometime before 1581.[5] It shares elements with the *Visio Arislei* – a 'certaine Barren King' bewailing his lack of issue; an incestuous union followed by a death, 'Ranke Poison issuing from the Dying Man'; a passage through the 'grievous Plight' of putrefaction followed by a glorious regeneration, a 'Second Birth' ensuring 'Firtile and Sweete Fruits'. We shall return to look at Ripley's *Cantilena* in greater detail later. Another, very different parable is found in Thomas Charnock's *Breviary of Philosophy* (1555). It tells of 'two poore Men', who present themselve to the alchemist with a plea:

> Master, for the love of God and our Lady,
> Give us your Charity whatsoever you please,
> For we have not one peny to do us ease;
> And we are now ready to the Sea prest,
> Where we must abide three moneths at the least;
> All which tyme to Land we shall not passe,
> No although our Ship be made but of Glasse,
> But all tempest of the Aire we must abide,
> And in dangerous roades many tymes to ride;
> Bread we shall have none, nor yet other foode,
> But only faire water descending from a Cloude:
> The Moone shall us burne so in processe of tyme,
> That we shalbe as black as men of Inde:
> But shortly we shall passe into another Clymate,
> Where we shall receive a more purer estate;
> For this our Sinns we make our Purgatory,
> For the which we shall receive a Spirituall body.[6]

In this beautiful passage of sixteenth-century alchemical poetry, the *opus* is depicted as a perilous journey at sea. The vessel becomes a glass ship, the activity of chemical elements within it become climatic elements surrounding it: the sea, the 'tempest of the Aire', the burning moon. Luna (silver) here suggests Our Mercury ('living silver', 'moon water'): this burning moon is the fiery, caustic aspect of *Mercurius duplex*. The travellers' weatherbeaten complexion – 'black as men of Inde' – ingeniously expresses the *nigredo* through which they pass. Charnock presents a purgatorial journey which, through hardship, starvation and exposure, arrives at exaltation. Their 'Sinns' are cleansed, they feed on Mercurial dew – 'faire water descending' – and finally 'receive a Spirituall body', a *corpus subtile*. Once again, the narrative vehicle has its own peculiarities, but the pattern it expresses is familiar.

Having touched on examples of alchemical allegory from Alexandrian, Islamic and medieval English texts, we might conclude with an example from the literature of the 'alchemical renaissance'. 'Once upon a time, when I was walking abroad in a wood. . . .' So begins the 'Parable in which the Mystery of the whole Matter is declared' that concludes Grasshoff's *Tractatus Aureus* (1625).[7] Although this date places the work outside the scope of the Shakespearean period, it is still absolutely typical of the type of alchemical literature that flowered in the years immediately after 1600. The wandering alchemist of this *parabola* finds himself on a path 'rough, untrodden, and . . . beset with briars'. He is forced on down it – 'I . . . strove to retrace my steps, but it was not in my power to do so; for so violent a tempest blew upon me from behind' – until he emerges in a beautiful sunlit orchard, which he learns is called the 'Meadow of Happiness'. Many symbolic vignettes follow – a discussion with aged philosophers, a struggle with a Lion, an arduous entry into a garden surrounded by immense walls and locked gates. But, as in the *Visio Arislei* and Ripley's *Cantilena*, the central action concerns a marriage and a death.

> Now when the bridegroom, in his bright scarlet robe, with his bride, whose silk dress gave out shining rays, reached the old men, they were straightway joined together. And I marvelled that the maiden, who was said to be the mother of her bridegroom, was of so youthful an appearance, that she might have seemed his daughter. But I know not what sin they had committed, except that brother and sister had been drawn to each other by much passionate love that they could no more be separated; and, being charged with incest, they were shut up forever in a close prison, which, however, was as pellucid and transparent as glass. . . . Here they were to do penance for their sins with ever-flowing tears and true sorrow.

The incestuous *coniunctio* is, as with Arisleus and Ripley, a dissolution of the male, a 'casting away' of the bodily and sulphurous, for the pair 'embraced so passionately that the husband's heart was melted with the excessive ardour of love, and he fell down broken in many pieces'. His bride-mother-sister wept over his corpse 'until he was quite flooded and concealed from view'. Then, 'weary with exceeding sorrow, she at length destroyed herself.' The *nigredo* follows: the corpses, 'black as coals', putrefy and emit a 'grievous smell'. This in turn is succeeded by re-animation:

> Towards evening I noticed that many vapours rose from the earth through the heat of the sun. . . . Afterwards, when night fell, they watered the earth as fertilizing dew, and washed our bodies, which became more beautiful and white the oftener this sprinkling took place.

The return of the Mercurial spirit produces the *albedo*, and 'the clarified body of the Queen . . . was straightaway restored to life'. She addresses the alchemist with these words:

> The most High God is one God, who has power to set up and pull down kings. He makes rich and poor as He wills. He has killed, and raised again. I was great, and was brought low; but now, having been humbled, I have been made Queen of many more kingdoms.

Finally, wrapped in a 'most precious scarlet garment', the reborn King emerges with 'indescribable magnificence' as the Red Stone. 'I do not remember', concludes the alchemist, 'to have seen a more glorious man, or more glorious deeds.'

We see here the syn-chronology of alchemical literature, the continuity which underlies all these examples of alchemical allegory, ranging through third, tenth, fifteenth, sixteenth and seventeenth centuries. That continuity is, of course, the *magnum opus*, the unfolding chemical-spiritual process to which all these allegories refer. The *opus* is their 'deeper meaning'. I have tried fitfully to illuminate that meaning beneath, but what really interests me here is what is *on the surface*. For on the surface are people. Symbolic, two-dimensional people, moving across an emblematic landscape, but 'characters' nevertheless: kings and queens, princes and princesses, a priest, two poor wayfarers and, of course, the ubiquitous alchemist, part onlooker and part hero. They are 'characters' in both senses: persons and ciphers. These alchemical parables depict alchemy in emblematic but *human* terms, and the pattern which emerges in those terms is this. The beginning is grossness, infertility, ailment, poverty, incompleteness. The middle, or 'action' of the narrative, entails a submission to some drastic, mysterious and overwhelming process: the torment of dismemberment, the perils of a sea journey, the incestuous marriage which leads to imprisonment and death. The end, as a result of this purgatorial process, is exultant. The qualities which prevailed at the beginning are each healed, transformed into spirituality, abundance, health, riches and wholeness. This dangerous but ultimately healing journey through darkness and ruin is (the texts themselves tell us) the human pattern which corresponds to the alchemical process. And this journey or pattern is one we know well, in a form more familiar to our ears than alchemical allegory. It is the basic pattern of tragic drama.

*   *   *

Tragedy is at root sacrificial and therapeutic. The fall of the tragic hero is cathartic – literally, purgative. He falls, we are cleansed. Thus Aristotle's classic definition of catharsis as the desired effect of tragedy:

'by means of fear and pity bringing about the purgation of such emotions.'[8] Or, as a modern American critic puts it: 'By watching tragedy we can wish, and thereby create, our own psychic health. It is the miracle of tragedy that failure belongs to fictitious characters and health to its creators, the artist and the audience.'[9] The submission of the tragic hero to the 'will of the Gods', or to fate, or destiny, is a sacrificial necessity. Racine described Phèdre as: 'Ni tout à fait coupable, ni tout à fait innocente. Elle est engagée par sa destinée, et par la colère des Dieux dans une passion illégitime dont elle a horreur.'[10] It is not a question of guilt and innocence – even the classic 'flaw' of *hubris* (ὔβρις) is inextricable from the necessary greatness of the hero (ἥρως). It is a question of destiny, the 'anger of the gods'. The machinery of some supernatural process requires the hero. His journey through darkness and ruin is indeed a healing journey. *He* is not healed: he is destroyed. But by his sacrifice the community he represents – a kingdom within the drama, an audience outside it – is redeemed, and inasfar as he is a part of that community he is a recipient of that redemption. The king must die so that the king may prosper.

Tragedy is, then, a drama of redemption. Its unremitting depiction of loss and doom is placed in a wider context of renewal. This is as true of *Oedipus* as *Hamlet*. Thebes is polluted by 'an unclean thing ... which must be driven away'. Oedipus must 'bring everything into the light':[11] make the occult manifest. The pollution turns out to be Oedipus himself. His humiliation, blindness and banishment achieve the cleansing of Thebes. Similarly, 'something is rotten in the state of Denmark.' Hamlet reluctantly accepts his healing mission: 'O cursed spight, That ever I was borne to set it right.'[12] At the end, the stage is littered with corpses – including that of the 'sweete prince' – but 'th'election lights on Fortinbrasse'. A new order grows into the emptiness left by the uprooting of old corruption. Tragedy is 'unnatural' in that it inverts the 'survival of the fittest' into the sacrifice of the fittest. Rather, it is *super-natural*: it appeals and submits to a higher Nature – the workings of fate or 'the Gods', a secret Nature which transforms and purifies.

This is the underlying pattern, the classic model, of tragedy: purgatorial and redemptive. It is also the underlying pattern of alchemy. Indeed, those alchemical parables quoted earlier could be said not only to allegorize the alchemical process, but also to *dramatize* it – to present the *magnum opus* as a symbolic drama of purgation and redemption. As such, we find in them stringent echoes of tragic heroism and destiny. Ion resigns himself to the 'compelling necessity' of the alchemical process just as the tragic hero is ineluctably *'engagé par sa destinée'*. Prince Thabritius undergoes dissolution in order to heal the sickness of the kingdom. If he had a mind and a voice he might well complain, like that other prince did, 'O cursed spight, That ever I was borne to set it right.'

In the *Tractatus Aureus*, the revived Queen speaks of the God (in tragedy, usually 'the Gods') who 'has power to set up and pull down kings', who 'makes rich and poor as He wills'. We can think of many tragic heroes and heroines who could say, with her: 'I was great, and was brought low.' In alchemy, matter is the hero. It submits itself to torment and blackness so that the kingdom of which it is a part – the entirety of material nature – can be redeemed.

Both alchemy and tragedy define a journey: a road to wholeness that goes by way of dismemberment and dissolution. We are close here to the Jungian interpretation of alchemy. Jung drew deep and detailed analogies between the alchemical process leading to the Stone, and the psycho-therapeutic 'individuation process' leading to an integrated, whole personality.[13] He saw the succession of *solve et coagula*, separation and conjunction, as a 'separation and synthesis of psychic opposites' within the self.[14] That quality of difficulty and danger which all alchemists ascribe to the *opus* – 'Not a few have perished in our work,' warns the *Rosarium*[15] – reflects the essentially psychic nature of the journey. Jung writes:

> The right way to wholeness is made up, unfortunately, of fateful detours and wrong turnings. It is a *longissima via*, not straight but snake-like, a path that unites the opposites in the manner of the guiding caduceus, a path whose labyrinthine twists and turns are not lacking in terrors.[16]

For Jung this *longissima via* is both the alchemist's *opus* and the patient's path to psychic integration. In its classic form, tragedy is precisely therapeutic. The tragic hero also travels down this 'longest road', on a journey through terrors to wholeness.

I have, of course, one particular tragic hero in mind: King Lear. He has been lurking at the edge of this chapter from the beginning. He is one who endures the torment of inner dismemberment; an ailing king who must learn the secrets of generation; one who was great – 'every inch a king' – but is brought low – a 'poore naked wretch'.[17] In speaking of the 'underlying pattern' of tragedy, one must not forget the variety of drama that pattern supports. Shakespeare's tragedies do not always insist on their redemptive, healing purpose. There is no word of a new order or fertilized kingdom at the end of *Othello*, *Coriolanus* or *Timon*. The tragedy of *King Lear* is, however, profoundly redemptive. Indeed, the Victorian critic A. C. Bradley was moved to ask:

> Should we not be at least near the truth if we called this poem *The Redemption of King Lear*, and declared that the business of 'the gods' with him was neither to torment him, nor to teach him a 'noble anger', but to lead him to attain through apparently hopeless failure the very end and aim of life?[18]

Many would question Bradley's reading of the play, feeling that it compromises *Lear*'s unquestionable status as tragedy – making, in other words, a faulty equation between tragedy and 'pessimism'. I think Bradley is quite right. In the midst of the storm, Lear utters the words which prophesy the ultimately enriching nature of his ruin:

> The Art of our necessities is strange,
> And can make vilde things precious.[19]

They are alchemical words. Hardship, he says, is a secret art: an alchemy which transforms the vile into the precious. It is the central assertion of this book that the whole story, the whole unfolding process, of *King Lear* is deeply and intentionally alchemical. That *King Lear* is a masterpiece of 'chemical theatre'.

The Wheel of Fire

'The Wheele of our Philosophie thou hast turned about.'

George Ripley – *The Compound of Alchymy*

10 From anon., *Speculum Veritatis* (MS, seventeenth century)

In *King Lear*, the tragic *longissima via* is defined not as a linear journey but a circular revolution. When the vexed king is reunited with Cordelia, his first words on awakening are:

> You do me wrong to take me out o' th' grave
> Thou art a soule in blisse but I am bound
> Upon a wheele of fire, that mine owne teares
> Do scald like molten lead.[20]

The Wheel of Fire is the image we take away from the play, an icon of Lear's purgatorial torment. But the Wheel takes many forms throughout the play. Kent settles down to a night in the stocks with the words, 'Fortune, goodnight; smile once more; turne thy wheele'.[21] Next morning, the Fool counsels him: 'Let go thy hold when a great wheele runs downe a hill, least it breake thy necke with following.'[22] These both invoke the Wheel of Fortune, the traditional image of mutability. Fortune is 'painted also with a wheele', says Fluellen in *Henry V*, to signify that 'shee is turning and inconstant, and mutabilitie and variation'.[23] In *Lear*, the Wheel is also a wheel of generations, a genetic cycle. Its turning bears Lear and Gloucester downwards and their vicious offspring upwards, for, as Edmund says, 'the yonger rises when the old doth fall'.[24] But it goes on turning, to complete its revolution: Edmund dies the stateless bastard he began as. He says: 'the wheele is come full circle; I am heere.'[25]

The purgatorial Wheel of Fire, the fickle Wheel of Fortune, the cyclic Wheel of Generations: each of these contributes its meaning to the central motif of the Wheel in *Lear*. It is literally a 'motif': a moving force, an impulsion at the heart of the play. The Wheel is the 'journey' undertaken by all the central characters. Lear, Gloucester and Edgar all go through a revolution in that literal, circular sense. All fall from their status at the 'top' of the Wheel to enact a role diametrically opposite. The king becomes an outcast, poor and naked; the earl becomes a blind wayfarer; the first-born becomes a beggar possessed by fiends. All climb back up from this, the 'bottom' of the Wheel, to achieve some kind of reinstatement, an altered nobility, a supremacy quite different from their original, purely social altitude. The revolution transforms them. The Wheel becomes, in Shakespeare's hands, a process of spiritual growth. Its message for the characters is that in order to attain knowledge and love, they must first travel through negation and no love. The road to the zenith goes by way of the nadir, which is the destitution of their former state and the dissolution of their former self. As Eliot wrote in the *Four Quartets*:

> In order to arrive at what you are not
> You must go through the way in which you are not.[26]

It is only when madness and the storm have reduced Lear to nothing that he can be redeemed. He was wrong when he told Cordelia, 'Nothing will come of nothing'.[27] Lear himself will grow out of nothing. As Eliot also wrote: 'In my end is my beginning.'[28]

Lear is king, furthermore: representative hero. The play is patterned according to the old harmony it so violently tests. The kingdom, the material world, the supernatural are ever-widening circles concentric on the king. They too revolve on the axis of Lear's fortunes. In the first chaos of his madness, a fantastic courtroom is set up in a farmhouse shed. The lunatic king, the naked beggar and the dying fool pass sentence on an empty chair which they presume to contain Queen Goneril. It is pure hallucination, a lurid fantasia. The next scene takes us to Gloucester Castle, where the 'real' Queen Regan holds the chair while her husband gouges out Gloucester's eyes. That is no fantasia: it is happening. The reign of Goneril and Regan *is* the madness of the kingdom. 'A dogg's obey'd in office.' 'Humanity must perforce prey on itself, Like monsters of the deep.'[29] The degradation of the king is inextricable from that of the kingdom. So too, as the Wheel passes through its nadir, Nature is sealed up in sterility at the king's command. His earlier curse on Goneril –

> Heare, Nature, heare deere Goddesse, heare:
> Suspend thy purpose, if thou didst intend
> To make this creature fruitfull!
> Into her wombe convey stirrility!
> Drie up in her the organs of increase.

– becomes a universal blight:

> Strike flat the thicke rotundity o' th' world!
> Cracke Natures moulds, all germaines spill at once.[30]

The king's discontent is indeed a winter. Finally, the gods themselves revolve on the Wheel. At the beginning, Lear swears by Apollo and Jupiter, but these are eclipsed by the 'Prince of Darkenesse'[31] and a score of lesser fiends with names like Smulkin, Hoppedance and Flibbertigibbet. The fall of the king is a social, natural and cosmological disaster: the entire 'frame of Nature' is 'wrencht . . . From the fixt place'.[32] But, because this tragedy is a drama of redemption, each of these breaches will be healed by the end of the play. Edgar will 'rule in this realme, and the gor'd state sustaine'. Cordelia will make the 'vertues of the earth Spring with my teares'. Lear himself will become one of 'gods' spies' (or 'God's spies': the Folio punctuation gives no clue), a looker into 'the mystery of things'.[33]

This telescopic pattern of concentric revolutions extends inwards too. Inside the king is a man. The turning of the Wheel that makes the king a beggar makes the man a madman. Lear's mind is itself a microcosmic kingdom undergoing a revolution. 'Sovereign reason', as the Elizabethans often called it,[34] is deposed by the clawing ascendancy of his madness:

> Downe, thou climing sorrow!
> Thy element's below . . .
> O me my heart! My rising heart! But downe.[35]

It is the same Wheel, now inside the self. Something dark and monstrous being borne upwards, reason and refinement being dragged downwards. We see that in this inner arena, Lear's circular *via* leads through the depths and recesses of his self. To place alongside this mental *abaissement*, there is throughout the play a bizarre anatomical topography in which reversal occurs. There is the Fool's 'if a man's braines were in's heeles'.[36] There is Lear's outburst of sexual disgust which begins: 'Behold yond simpring dame, whose face betweene her forkes presages snow'[37] – her face between her legs, the most elevated anatomical feature in the place of grossest (as Lear sees it). The image catches an echo of a line of Kent's: 'let . . . the forke invade the region of my heart.'[38] The 'fork' he means is an arrow-head, but the covert image is of a rising from beneath: the sex invading the heart. In *Lear*, the body is as much an arena of convulsive reversal as all the larger worlds around it. Its duality of 'above' and 'below' mimics the cosmological one – thus Lear's anatomy of womankind:

> Downe from the waste they are Centaures,
> Though women all above:
> But to the girdle do the Gods inherit,
> Beneath is all the Fiends: there's hell, there's darkenes,
> There is the sulphurous pit.[39]

The Wheel is, in short, the pattern of *King Lear*. At every level of the play's unfolding, the process proves finally to be circular. A shattering fall is followed by an arduous ascent – from king to beggar, from Edgar to Poor Tom, from reason to madness, from Lear's kingdom to Goneril and Regan's kingdom, from generation to sterility, from gods to fiends, from face to fork. And back again: not back to the same king or self or kingdom, but to these transformed. A precarious, flawed, corrupt stability is upended and dissolved because only through that dissolution can true strength and wholeness ensue. The final reinstatement is transcendent. The revolving wheel, the *longissima via*, is a process of transformation.

* * *

Alchemy is just such a Wheel, a circular process of transformation. The classic dicta – *solve et coagula* and *fac fixum volatile et volatile fixum*

– describe a chemical circle. Solid resolved into spirit, spirit coagulated into solid. Thus the *Emerald Table*, translated in 1597: 'It ascendeth from the Earth into Heaven, and again it descendeth into the earth, and receiveth the power of the superiours and inferiours.'[40] Chemically speaking, these describe sublimation. The other prototypical alchemical operation – distillation – also tended to circularity. Among the alchemist's array of stills, alembics and cucurbites, a significant vessel was the 'pelican'. Its delivery spouts leading back into the bowl of the vessel enabled a continuous process of evaporation and condensation, the distillate flowing back into the residue to be re-distilled, a chemical *perpetuum mobile*. This process, known in modern chemistry as 'refluxing', is referred to in European and English alchemy as 'circulation'.[41] In the *Tractatus Aureus Hermetis Trismegisti* (1600) – not to be confused with the *Tractatus Aureus* of 1625, quoted in the previous section – we find the following description:

> Circulation of spirits, or circular distillation, that is, the outside to the inside, the inside to the outside, likewise the lower to the upper; and when they meet together in one circle, you could no longer recognise what was outside or inside, or lower or upper; but all would be one thing in one circle or vessel. For this vessel is the true philosophical Pelican, and there is no other to be sought for in all the world.[42]

The medieval English alchemists also placed great emphasis on circulation. Thomas Charnock claimed he had performed over 600, each one lasting a week, before he attained the Red Stone.[43] Their concept of circulation went far beyond the merely distillatory – 'such Circulation,' wrote Thomas Norton dismissively, 'is but only a rectification, better serving . . . for correction than for transmutation'. Norton spoke of the more profound alteration effected by 'Circulations of Elements' –

> Our Circulation is from Fier on high,
> Which endeth with Water his most contrary.
> Another circulation beginneth with Ayre,
> Ending with his contrary cleane Earth and faier.

There are, he continues, circulations for which 'the Red worke hath desire', others that 'be better for the White'. Some take 'full thirtie Weeks', others shall 'oft time have lesse'. The circles within the microcosmic vessel are keyed to celestial planetary circles:

> Every Circulation hath her proper season,
> As her lightnesse accordeth with reason,
> For as one Planet is more ponderous
> Then is another and slower, in his course.[44]

As in *King Lear*, there are circles within circles, wheeling revolutions concentric on an emblematic substance – the Raw Stuff, the ailing king.

These circular operations – sublimation, distillation, and the more complex 'circulation of elements' – are all included in the *magnum opus*, the 'great work' of transmutation. They can be seen both as actual chemical phases of a cumulative *opus*, and as emblematic operations reflecting the overall process and purpose of the *opus*. *Solve et coagula* iterates in miniature the circular route of the 'great work' in which matter – the *massa confusa*, 'our chaos' – is redeemed by its journey through the depths of *nigredo*. As the Elizabethan alchemist, William Bloomfield, has it:

> Bring them first to Hell, and afterwards to Heaven,
> Betwixt lyfe and death thou must then discusse,
> Therefore I councell thee that thou worke thus.
> Dissolve and separate them, sublime, fix and congeale,
> Then has thou all.[45]

We have already seen certain correspondences between the alchemical process and the pattern of tragic drama, correspondences that emerge most clearly when one looks at those alchemical allegories in which the *opus* is, to some extent, 'dramatised'. We are now beginning to see that Shakespeare's particular rendition of that pattern in *King Lear* makes those correspondences still stronger. His insistence on the tragic pattern as circular, revolutionary, an inexorable Wheel upending and reinstating at every level of the play, is clearly comparable to the alchemists' emphasis on circulation as the pattern of the *opus*. Shakespeare's treatment of the characters in *Lear* could be aptly epitomized in the words of Bloomfield above: 'Bring them first to Hell, and afterwards to Heaven.'

It is the very image of the Wheel which provides the concrete link between *King Lear* and contemporary alchemy. For, considering the compulsively symbol-forming mind of the alchemist, it is not surprising that the circular nature of the *opus* was given its emblem, and that one of those emblems was the Wheel. Thus the title-page of Ripley's *Compound of Alchymy* promises to reveal the workings of 'his Wheele', and this 'Wheele of Philosophie' is repeatedly invoked throughout the text. In one form it describes that 'circulation of elements' which Norton also spoke of. In the first 'Gate' of the *Compound*, Ripley counsels:

> First of these elements make thou rotacion,
> And into water thine earth turne first of all;
> Then of thy water make ayre by levigacion;
> And ayre make fier; then Maister I will thee call
> Of all our secrets great and small:
> The wheele of Elements then canst thou turne about,
> Truely conceiving our writings without doubt.[46]

But, more importantly, Ripley's Wheel is a depiction of the whole *opus*, the alchemical process *per se*. Here is the crucial passage in full:

> Each bodie . . . hath dimencyons three:
> Altitude, Latitude, and also Profunditie,
> By which allgates turne we must our Wheele;
> Knowing that thine entrance in the West shall be;
> Thy passage forth to the North if thou doo weele,
> And there thy lights lose their lights each deele,
> For there thou must abide by ninetie nights
> In darknes of purgatorie withouten lights.
> Then take thy course up to the East anone,
> By colours passing variable in manifold wise,
> And then be winter and vere nigh overgone.
> To the East therefore thine ascending devise,
> For there the Sunne with daylight doth uprise
> In Sommer, and there disporte thee with delight,
> For there thy Worke shall become perfect white.
> Foorth from the East into the South ascend,
> And set thee downe there in the chaire of fire,
> For there is harvest, that is to say an end
> Of all this Worke after thine owne desire:
> There shineth the Sunne up in his Hemisphere,
> After the Eclipses in rednes with glorie,
> As King to raigne upon all mettals and Mercurie.[47]

To understand the iconography of this circulation, we must first make the equation: Raw Stuff = Common Gold = Sol. The circular journey begins in the West: the setting sun suggests gold in its unredeemed state, resplendent but decadent. The North is thus the nadir of this solar Wheel – night, winter, the blackness of putrefaction: 'thy lights lose their lights . . . in darknes of purgatorie'. The ascent into the East represents the first stage of resurrection – sunrise, summer, delight, *albedo* – culminating in the 'midday' triumph of the Stone. The Red King, in his chair of fire, reigns over all metals. In the 'Recapitulation' which concludes the *Compound*, Ripley sums up the turning of his Wheel:

> The West was the beginning of thy practise,
> And the North the perfect meane of profound alteration,
> So in the East after them the beginning of speculation is;
> But of this course up in the South the sun maketh consumation.[48]

For the beginning to arrive at the consummation, the journey must pass through 'profound alteration' and 'darkness in purgatory'. The Wheel in Ripley's *Compound* and the Wheel in Shakespeare's *Lear* trace the same circle and perform the same transformation.

This alchemical Wheel appears in other contemporary publications. In Heinrich Khunrath's *Von Hylealische Chaos* of 1597, we find: 'Through circumrotation or a circular philosophical revolving of the Quaternarius, it is brought back to the highest and purest simplicity of the plusquamperfect Catholic monad.'[49] And in one of Gerhard Dorn's tracts in the *Theatrum Chemicum* of 1602: 'The wheel of generation takes its rise from the *materia prima*, whence it passes to the simple elements.'[50] Again the Wheel summarises the overall process of the *opus*: from the Quaternarius (the elemental 'fourness') to the Monad (perfect 'oneness') by way of dissolution (matter being broken down into its 'simple elements'). Similarly, in the seventeenth-century *Speculum Veritatis* ('Mirror of Truth'), the *opus* is pictorially represented as a Wheel (fig. 10) divided into eight segments marked with colours and symbols. An armoured man, apparently representing Philosophical Sulphur, turns the Wheel, watched by Vulcan at his furnace.[51] The supervision of Vulcan reminds us that the alchemical Wheel is indeed a Wheel of Fire. Fire is the agent of transformation – both the outer fire of the furnace and the secret *ignis innaturalis* (or Mercurial dragon). As the *Mirror* says, 'His perfection and proceeding consisteth in the fire, which is the cause of his life and death.'[52] It is fire which turns the transforming Wheel.

The Wheel that moves through *King Lear* is, we have seen, a composite of images – a Wheel of Fortune, of Generations, of Fire. As such, it clearly contains traditional elements. The Wheel of Fortune was an ancient and popular image; the Wheel of Fire, we are told, was 'traditional in the medieval legends and visions of Hell and Purgatory' drawn from the New Testament *Apocrypha*.[53] Equally clearly, the Wheel in *Lear* alters those traditional readings which are part of it. It is not just the inconstancy of Fortune, because its turning has a purpose and end; it is not just an instrument of apocalyptic torture, because the moment that Lear announces that he is 'bound upon a wheele of fire' is also the moment of his awakening and rebirth in the arms of Cordelia. Shakespeare's Wheel is more than the sum of its metaphoric parts, and I suggest that in defining the inexorable tragic process in *Lear* as a Wheel, he drew not only on traditional associations, but also (and primarily) on a very contemporary symbolism: the Wheel as alchemical process. Ripley's Wheel of Philosophy, passing through eclipse and profound alteration; Dorn's Wheel of Generation, which generates by dissolving: these are the prototypes of Shakespeare's Wheel of Fire, which carries king and kingdom through death to new life. The vast upheavals of *King Lear* are the pains of birth, and, in the words of the alchemical mystic Jakob Boehme, 'the form of the birth is as a turning wheel.'[54]

An alchemical reconstruction

The first known performance of *King Lear* was on the night of 26 December 1606, when it was played before King James at Whitehall. The play was almost certainly written in 1605. In that year, the *True Chronicle History of King Leir* – the old chronicle-play which provided the skeleton of Shakespeare's *Lear* – was first published. When Shakespeare's Gloucester speaks of 'these late eclipses in the sun and moone', and relates them to 'discord ... treason ... machinations',[55] we again sense 1605, a year that saw an eclipse of the moon in September, an eclipse of the sun in October, and the Gunpowder Plot in November. These are circumstantial evidence, but it has been generally accepted that 1605 is the year of *Lear*, if only as shorthand for a more scrupulous '*c.* 1604–6'.

The year 1605 has cropped up often during the course of this book. I began by quoting extracts from Bacon's *Advancement of Learning* and Tymme's *Practise of Chymicall Physicke*, both published in that year. We have seen that the first wave of the 'alchemical renaissance' reached a high-water mark around that year – the major works of Valentine, Sendivogius and Quercetanus; the influential anthology, *Theatrum Chemicum*; the early version of Andreae's *Chemical Wedding*: all these are clustered in the opening few years of the century. Also 1605 was the year of Ben Jonson's first excursion into his own version of 'chemical theatre': *Eastward Hoe* and (probably) *Volpone* were both performed that year.[56] Donne's *Songs and Sonets*, though individually undateable, certainly belong to the decade of which 1605 is the centre. If Shakespeare were to embrace alchemy into his poetry, as these writers did, then 1605 might be the year, *King Lear* the play.

With these chronological pointers in mind, I offer in the following chapter an 'alchemical reading of *King Lear*'. It will concentrate almost exclusively on what happens to Lear himself. The previous pages – arguing a correspondence between alchemy and tragedy intensified in *Lear* by the shared symbol of the Wheel – have, I hope, provided the background for this more focused scrutiny. What Lear undergoes is something very like an alchemical transmutation. I shall try to show this by a series of parallels between the plot and imagery of the play and the patterns and symbols of alchemy, drawn almost exclusively from contemporary alchemical texts which Shakespeare could have read.

I can only say he 'could have' read them. A writer's relationship to ideas is more elusive and complex than any reconstruction four centuries later can encompass. The alchemical publications of the period are the only remaining clue to what alchemy meant to someone in 1605; the text of *King Lear* is, for that matter, the only remaining clue to what was going on in Shakespeare's mind when he wrote it. If one places these publications and this play together, the results are extraordinary. But what is their exact relationship – what late-night conversations, philosophical arguments, chance encounters or skimmed pages lie between them – I cannot say. I suspect, on this subject, that the flourishing alchemical illustration of the period – beautiful and suggestive emblems – may have been a more amenable channel of alchemical meanings to Shakespeare than the literature *per se*, which tended to hide its gems amid much prolixity. But the exact provenance of alchemical meanings in *King Lear* must remain a mystery, part of the deeper mystery of Shakespeare's capacity to imbibe everything around him and extract everything inside him. Keats was referring to this inclusiveness of Shakespeare's when he wrote: 'What shocks the virtuous philosopher, delights the chameleon poet.'[57] It has been the effect of a smugly virtuous modernism to consign alchemy to a role of buffoonery and quaintness. The chameleon poet of 1605 made no such rejection.

I hope, therefore, that this 'alchemical reading' will also be an alchemical reconstruction: that it will restore to *King Lear* certain meanings which have been lost in intervening centuries. Those meanings in no way displace other interpretations of the play; there are no final solutions to *King Lear*. They are, however, meanings more immediately visible to contemporary eyes than to our own. If John Dee or Robert Fludd, James Forester or Francis Anthonie had watched *King Lear* (they were all living in London at the time); if Sir Walter Raleigh and the Wizard Earl had been at the Globe instead of the Tower; if, for that matter, Francis Bacon or Ben Jonson or John Donne had seen the play; these are, I believe, some of the meanings they would have drawn from it.

So let us take our seats at some early and mighty performance of *King Lear*, having borrowed for the occasion the outer casing and inner reflexes of some judicious Jacobean playgoer.[58] He is well-versed in the scientific and magical speculation of the day, and particularly in the alchemical aspects of it. He is familiar with the English writings of George Ripley and Thomas Tymme, and with the classic European texts of the early seventeenth century – Valentine's *Twelve Keys*, Lambspringk's *De Lapide Philosophico* and Sendivogius' *Novum Lumen*. What he sees on the stage and what he has read in those texts mingle and react in his mind, each echoing and illuminating the other. He understands, perhaps for the first time, the terrors and exaltations of which the alchemists write. He sees a drama which he privately calls 'The Transmutation of King Lear'.

· 7 ·

The Transmutation of King Lear

By transmutation I meane, when any thing so forgoeth his outward forme, and is so changed, that it is utterly unlike to his former substance and woonted forme, but hath put on another forme, and hath assumed another essence, another colour, another vertue, and another nature and property.

Thomas Tymme – *The Practise of Chymicall Physicke*

11 From Lambspringk, *De Lapide Philosophorum Emblemata* (1678)

The King

> *Kent* I thought the King had more affected the Duke of Albany then Cornwall.
> *Gloucester* It did alwayes seeme so to us; but now in the division of the kingdome, it appeares not which of the Dukes hee valewes most.[1]

King Lear begins with this innocuous exchange between two noblemen. The focus is instantly on Lear, and with it a faint note of danger. The division of the kingdom, the halving of his affections. Our first information about Lear speaks of an imminent partition, something about to be broken apart. With a flourish of trumpets and a bustle of retinue, Lear himself enters to announce his 'darker purpose' to the court:

> Give me the map there. Know that we have divided
> In three our kingdome; and 'tis our fast intent
> To shake all cares and businesse from our age,
> Conferring them on yonger strengths, while we
> Unburthen'd crawle toward death.[2]

So begins Lear's speech of abdication. Two things strike us as we watch him: his regal bearing and his extreme old age. It is the conflict of these two which enforces his abdication, conferring the 'cares and businesse' of kingship on 'yonger strengths'. His power as king is compromised by his infirmity as a man. This is emphasised, perhaps more darkly than Lear intends, by the flourish of his pronouncement breaking like a wave against the abjection of 'crawle toward death'. He continues:

> Our son of Cornwall,
> And you, our no lesse loving sonne of Albany.

But they are not his sons. Lear has no son. Dynastically he is barren. The smooth mechanism of succession and primogeniture cannot function. Instead there is the rigmarole of this first scene, with its bartering, competitions and power-sharing. As the King's Son, Cornwall and Albany are travesties. The kingdom must undergo its revolution in order to find a King's Son, the tried and tested Edgar. Even in these opening moments of the play, the first cracks appear in Lear's status as king. Infirmity and barrenness are only half-concealed by pomp and circumstance.

Lear's edict rapidly degenerates into charade. 'Tell me, my daughters,' he says,

> Which of you shall we say doth love us most?
> That we our largest bountie may extend
> Where Nature doth with merit challenge.

Goneril and Regan hasten to humour this senile whim. Their florid protestations of love are rewarded with lands, wealth and power. Only Cordelia refuses to abuse love by bartering with it: she offers only to 'love and be silent'.[3] Lear's demand for a public show of love is clearly a symptom of the impaired vision he displays throughout the scene. He is ensnared in the world of public display and courtly ritual which has lain so long at his kingly disposal that he can recognize no other. When he pointed to the map and awarded Goneril lands 'even from this line to this',[4] he saw only the map and his own theatrical largesse. He did not recognize the real space, the actual earth, he was giving away. He could not foresee that the castle doors will soon close behind him, that he will be one more 'houselesse head' among the 'poore naked wretches' of his kingdom. So too when he hands Cornwall and Albany the crown – 'which to confirme, This coronet part betweene you' – the ritual gesture conceals from him the true nature of his surrender. He has forgotten the meaning of the map and the coronet: he sees only the sign. In the *realpolitik* of his vicious daughters' world, there are no symbols – no maps and coronets, only lands and power. Lear's fond belief that he can surrender 'the sway, revennew, execution' of kingship and still retain 'the name and all th'addition to a king'[5] has no place in their world. The name is nothing without the sway. Lear will later come to see that even 'a dogg's obey'd in office', but for now he is pampered with delusions of anointedness. He has, in the words of one of Shakespeare's sonnets, 'drunk up the monarch's plague, this flattery'.[6]

This first scene is like an inventory of Lear's incapacities as king. He is aged, infirm, issueless. He is corrupted and deluded by his own kingship, craving the glitter of a self-inflating symbolism while unaware that the pattern of social and moral allegiances which underpins that symbolism is doomed by his own actions. Lear's kingship is theatrical, two dimensional, a brittle surface concealing unexamined tensions and violences. For, at the first hint of obstruction, his charade shatters. It is broken by a single word: 'Nothing'. Cordelia refuses to speak: she absents herself from Lear's world of comforting signs. Within moments, Lear's composure erupts. The royal edict, the daughters' love rhetoric, the whole strained regal pantomime, degenerate into the barbarism of Lear's famous curse on Cordelia, beginning with arcane invocations to the 'mysteries of Hecate' and plunging to a gruesome imagery of dismemberment and cannibalism – 'He that makes his generation messes To gorge his appetite.'[7] Beneath the kingly surface, something monstrous is glimpsed.

THE TRANSMUTATION OF KING LEAR

* * *

The figure of an ailing, barren king is one we encounter often in alchemical writings and emblems. Such a king is the protagonist, or 'hero', of George Ripley's *Cantilena*:

> There was a certaine barren King by birth,
> Composed of the purest, noblest Earth,
> By Nature sanguine (which is faire) yet hee
> Sadly bewailed his Authoritee.
>
> Wherefore am I a King, and Head of all
> Those Men and things that be Corporeall?
> I have no Issue, yet (I'le not deny)
> 'Tis mee both Heaven and Earth are ruled by.
>
> Yet there is either a Cause Naturall
> Or some Defect in the Originall. . . .
> My Nature is so much restrain'd
> No Tincture from my Body can be gain'd. . . .
>
> I cannot generate: my Blood growes cold:
> I am amaz'd to think I am so Old.[8]

The King here symbolizes gold. 'There are seven bodies,' says the Hermetic *Tractatus Aureus*, 'of which gold is the first, the most perfect, the king of them and their head.'[9] More particularly, the King represents gold as Raw Stuff – common gold (*aurum vulgi*) ripe for transmutation into philosophical gold (*aurum nostrum*). Ripley's King is barren and issueless because, in the words of Sendivogius' *Novum Lumen Chemicum*, 'vulgar gold is like an Herb without seed; when it is ripe, it brings forth seed; so gold when it is ripe yields seed, or Tincture'.[10] Thus Ripley's King is unripe, or raw: 'No Tincture from my Body can be gain'd.' He is a king in outward show only: his 'authoritee' is hollow and incomplete, reigning only over 'Men and things that be Corporeall'. The resplendence of common gold is inwardly flawed and corrupt: there is some 'defect in the Originall' causing Nature to be 'so much restrain'd'. (Sendivogius explains that common gold cannot 'produce seed' because of the 'crudity of the ore'.[11]) The King is crude and dark beneath the glittering royal surface. For all his material 'authoritee', he is ailing and feeble. His 'blood growes cold': he is 'amazed' to think he is so old.

The alchemical King in his symbolic status as common gold or Raw Stuff is a curious mixture of nobility and corruption. We must not forget that the King is also frequently a symbol for the Stone itself, the triumphant Red King reborn. This is to say that (again in the words of Sendivogius), 'the Philosophers Stone or Tincture is nothing else but gold digested to the highest degree'.[12] The Raw Stuff contains the Stone *in potentia*. Hidden within the ailing, degenerate, corporal king are the seeds of the transcendent, tingeing Red King: in order to discover them and bring them to fruition, the King must undergo the dissolutions and darkness of the *opus*. A flawed superficial kingship must be resigned for the true one to be won.

The King as Raw Stuff, purest of the impure race of metals, is met with in other contemporary texts. The first of Basil Valentine's *Twelve Keys* insists on gold as the only fit subject for transmutation:

> Our Masters require a pure and undefiled body, which is not adulterated with any spot or strange mixture: for the addition of another thing is a Leprosie to our metals. The King's Diadem is made of pure gold, and a chast Bride must be married unto him.[13]

In Salamon Trismosin's *Splendor Solis* (1598), the Raw Stuff entering the *nigredo* of putrefaction is likewise 'dramatized' as a king:

> The old Philosophers declared they saw a Fog rise, and pass over the whole face of the earth, they also saw the impetuosity of the Sea, and the streams over the face of the earth, and how these same became foul and stinking in the darkness. They further saw the King of the Earth sink, and heard him cry out with eager voice, 'Whoever saves me shall live and reign with me for ever in my brightness on my royal throne.'[14]

We find a similar passage in Michael Maier's *Symbola Aureae Mensae* (1617):

> Although that King of the Philosophers seems dead, yet he lives, and cries out from the deep: 'He who shall deliver me from the waters, and bring me back to dry land, him will I bless with riches everlasting.'[15]

The terrors of *nigredo* are yet to come in *Lear* – the King sinking into darkness, drowning in stormy waters – but already this alchemical characterization of the King as Raw Stuff strikes chords in the play. Lear is a sick King – ailing, barren, corrupt and dark beneath the outward splendour of royal authority. He clings to 'the name and all th'addition to a king', yet has not the power or vision to enforce that title. Only when every vestige of that false kingship has been dissolved away can Lear truly call himself 'every inch a king'. The treatment of kingship in the play is, of course, many-faceted. It can clearly be read politically: in the first uneasy years of James's reign, succession and disunion (between England and Scotland) were much in the mind. But among the many possible and interdependent readings of Lear as King, there is, I believe, an alchemical one: an emblematic reading of the King as Raw Stuff, matter that must undergo the pains of transformation, the 'hero' of the

12 From Petrus Bonus, *Pretiosa Margarita Novella* (1546)

redemptive drama of the *magnum opus*. This becomes much stronger and clearer as the play unfolds, but I think our Jacobean playgoer-cum-chymical-enthusiast would have readily identified this symbolic alchemical theme from the outset. For him, the King that features so frequently in the alchemical texts of the day lends resonances and meanings to the King he sees on the stage.

One last quotation, a little alchemical vignette, seems to sum up these hidden connections between matter at the outset of the *opus* and Lear at the outset of the play. It is from the *Pretiosa Margarita Novella*, a fourteenth-century text first published in 1546 and often thereafter (a copy of the 1554 edition was in the Wizard Earl's library).[16] One of the engravings in this work (fig. 12) depicts the King seated on his throne: the word '*oro*' stamped upon him makes clear his identity as gold. Kneeling before him are his 'son' (Mercury) and his five 'servants' (silver, copper, iron, tin and lead). The accompanying text reads:

> You may enter into the Palace which has fifteen mansions, where the King crowned with the diadem will be on his lofty throne, holding in his hand the sceptre of the whole world: before whose majesty, his son, together with five servants variously clad, entreat the King on bended knees that the kingdom should be divided [*ut regnum impertiri dignaretur*] equally among the son and the servants. The King himself makes no reply to their pleas.[17]

The king on his throne; the division of the kingdom; hints of a mutiny of children and subjects which the King, soon to be deposed and broken, does not heed. The world of *King Lear* and alchemical allegory are strangely close.

The Dragon

We left Lear poised at the brink of that first great disintegration: his curse on Cordelia. We had been half-prepared for some such collapse – growing uneasy about Lear's behaviour, sensing the flawed, feeble and deluded nature behind the ceremonial gestures. But nothing could have prepared us for the savagery that bursts from Lear when his ceremony is punctured, his gestures annulled, by that single word, 'Nothing'. His gory incanting curse seems to rise up out of the hidden unreclaimed depths of his self. Later he will cry: 'Downe, thou climing sorrow! Thy element's below.' But by then the relentless process will be under way. It begins here, with the curse on Cordelia. As the kingly surface shatters, another Lear entirely, someone dark and monstrous, rises up.

The effects of his curse are immediate and drastic. The first and

13 From Lambspringk, *De Lapide Philosophorum Emblemata* (1678)

deepest is the expulsion of Cordelia. The curse ends with the words, 'Thou my sometime daughter'. Cordelia is, suddenly, a thing of the past: she is thrust out of Lear's narrow self-indulgent field of vision – 'Hence and avoid my sight.' 'Better thou hadst not beene borne then not t'have pleased me better.' 'We have no such daughter, nor shall ever see that face of hers againe.'[18] It is more than an expulsion: it is an extinction. Cordelia is deleted from Lear's landscape. Much the same happens to Kent, who dares to question the 'hideous rashnesse' of Lear's treatment of Cordelia. Lear screams at him the same command as to Cordelia: 'Out of my sight!' Kent too must be removed. Henceforth, if 'thy banisht trunke be found in our dominions, The moment is thy death.'[19] The words of this monstrous Lear are full of mutilation and dismemberment – making 'his generation messes To gorge his appetite'; turning the noble Kent into a 'trunk', a stateless, limbless torso.

This is what Lear's outburst achieves: he curses his most loving daughter and banishes his most faithful servant. And, by a precisely compensatory movement, he embraces his cruel daughters and exalts his malevolent sons-in-law:

> Cornwall and Albanie,
> With my two Daughters dowres digest the third;
> Let pride, which she cals plainnesse, marry her.
> I doe invest you iointly with my power,
> Preheminence, and all the large effects
> That troope with maiesty.[20]

It is, in short, the first turn of the Wheel, the first drastic and fatal reversal of hierarchies, the extinction of truth and love in favour of flattery and hatred. What you have just done, says Kent to Lear, is to 'kill thy physition, and thy fee bestow Upon the foule disease'. And he takes his leave of Lear with another image of inversion: 'Sith thus thou wilt appeare, Freedome lives hence and banishment is here.'[21] Later the Fool will play mercilessly on this sense of Lear's upside-down world: 'Truth's a dog must to kennell; hee must be whipt out, when the Lady Brach may stand by th' fire and stinke.'[22] Lear has whipped out truth in the person of Cordelia and instated falsehood, the 'brach' (bitch) Goneril. But the opposite of truth is not only falsehood: it is also folly, or madness. Lear's vituperation of Cordelia is his first act of madness. Kent recognizes this instantly:

> be Kent unmannerly,
> When Lear is mad. What wouldest thou do, old man?
> Think'st thou that dutie shall have dread to speake
> When power to flattery bowes? To plainnesse honour's bound
> When Maiesty falls to folly.[23]

The King is 'mad'; he 'bows'; he 'falls to folly'. The words of Kent – uttered, he says, in an emergency which makes 'unmannerly' truths necessary – bring home to us the full significance of Lear's outburst. It is here, at the moment of his estrangement from Cordelia, that Lear's fall begins. All that happens to him follows on from this: in resigning the kingdom and whipping out truth, he resigns himself to beggary and madness. He himself becomes the true object of his devouring curse.

In the 'hideous rashnesse' of his curse, that 'other' Lear rises up from the depths, the 'element below'. Hidden within or beneath the King is the opposite, or shadow, or negation of the King. By this first turning of the Wheel, it ascends to displace the King. And, if we look back to the end of Lear's curse on Cordelia, we find that we are told exactly what this hidden creature, this 'id-King', is:

> *Lear* . . . He that makes his generation messes
> To gorge his appetite, shall to my bosome
> Be as well-neighbour'd, pittied, and releev'd,

> As thou my sometime daughter.
> Kent Good my Liege –
> Lear Peace, Kent!
> Come not betweene the Dragon and his wrath.[24]

At this precise first moment of fall, this first turning of the Wheel, Lear becomes the Dragon. The Dragon rises up to devour the King.

*   *   *

Alchemically, the Dragon is a complex figure. In one sense, it is a further characterization of the Raw Stuff. While the sick King represents matter (typically, gold) as corrupt and infertile beneath its resplendent metallic form, the Dragon suggests the green, raw and primordial heart of matter. It is – as in *Lear* – the monster that lurks 'beneath' the King: the chaos of unredeemed nature. As such, it is related to other reptilian or bestial emblems for the Raw Stuff – serpent, toad, salamander, green lion. Always the imagery suggests lurking hiddenness: Ripley's 'serpent within a well'; 'The Dragon who watches the crevices (and) shuns the sunbeams' in the *Tractatus Aureus*; Valentine's 'Old dragon which . . . creepeth out of the caves'.[25] The mythological slaying of the Dragon thus becomes the alchemist's conquest of unregenerate material nature.

Yet, paradoxically, the Dragon also represents the Secret Fire itself, the agent of dissolution and purification. It is the Mercurial dragon, breathing fire and destruction. Thus Ripley in the *Compound*:

> Fire against Nature must doe thy bodies woe;
> This is our Dragon as I thee tell,
> Fiercely burning as the fire of hell.[26]

It is the 'poison-dipping dragon' of the *Aurelia Occulta*:

> My water and my fire destroy and put together. . . . If you do not have exact knowledge of me, you will destroy your five senses with my fire. From my snout there comes a spreading poison that has brought death to many. . . . I am the old dragon found everywhere on the globe of the earth.[27]

It is the green lion as dissolving corrosive, whose properties Abraham Andrewes describes in *The Hunting of the Greene Lyon*:

> full quickly can he run,
> And soone can overtake the Sun:
> And suddainely can hym devoure . . .
> And hym eclipse that was so bryght.[28]

The sun here is, once again, gold as Raw Stuff, dissolved – 'devoured' and 'eclipsed' – by the vitriolic green lion. Pictorial representations of this process – the green lion eating the sun – are found in the *Rosarium Philosophorum* (1550) and the *Philosophia Reformata* of Johann Mylius (1622).²⁹

This is the twin aspect of the Dragon. It is at once the primordial chaos within the Raw Stuff and the Secret Fire (fiery Mercury, *aqua ardens*, vitriol) which attacks and disintegrates the Raw Stuff. The Dragon is that which is dissolved and it is that which dissolves: it represents, in other words, that 'mercurial' capacity within the Raw Stuff to devour *itself*. The Stone, we are always told, is 'one thing'; transmutation is an intrinsic process, a self-inflicted death and a self-generated resurrection. As so often in alchemy, alternative readings of a symbol lock together into a 'total' reading.

The Dragon is thus the self-devourer. This certainly is suggested by its most ancient and familiar guise: the Dragon eating its own tail. Originally a Gnostic symbol, this figure is found in the earliest surviving alchemical manuscripts. It is the serpent Ourobouros, the 'tail-eater', which features in the second century *Gold-making of Cleopatra*: within the circle of its self-devouring body appear the words, ἕν τὸ πᾶν ('the One, the All').³⁰ Jung calls this emblem 'the basic mandala of alchemy', and connects the Dragon with Mercury as symbolic of the entire circle of transformation:

> Time and again the alchemists reiterate that the *opus* proceeds from the one and leads back to the one, that it is a sort of circle like a dragon biting its own tail.... Mercurius stands at the beginning and end of the work: he is the *prima materia*, the *caput corvi*, the *nigredo*; as dragon he devours himself and as dragon he dies, to rise again as the *lapis*.³¹

The Dragon becomes another manifestation of the circular *opus*, the death-dealing and life-giving Wheel.

The description and illustration (see fig. 13) of the Dragon in Lambspringk's *De Lapide Philosophico* sums up its paradoxical and circular qualities:

> A savage Dragon lives in the forest,
> Most venomous he is, yet lacking nothing:
> When he sees the rays of the Sun and its bright fire,
> He scatters abroad his poison,
> And flies upward so fiercely
> That no living creature can stand before him,
> Nor is even the Basilisk equal to him.
> He who hath skill to slay him, wisely
> Hath escaped from all dangers.
> Yet all venom and colours are multiplied
> In the hour of his death.

> His venom becomes the great Medicine.
> He quickly consumes his venom,
> For he devours his poisonous tail.
> All this is performed on his own body,
> From which flows forth glorious Balm,
> With all its miraculous virtues.[32]

This portrait of the alchemical Dragon, first published in 1599, brings us close again to the world of *Lear*. It lurks in the 'forest' – the hidden unreclaimed depths of matter or the self. It 'flies upward so fiercely that no living creature can stand before him', just as the Dragon inside Lear does. Lear too 'scatters abroad his poison' – his 'vitriolic' curses, his dismembering banishments. And just as the alchemical Dragon is the true object of its own poison – 'he quickly consumes his venom, for he devours his poisonous tail' – so too Lear is the true recipient of his own curses: it is he who is banished and dismembered, 'made nothing' by the storm and his madness.[33] As Lambspringk prophesies: 'All this is performed on his own body.' When Lear cries to Kent, 'Come not betweene the Dragon and his wrath', we see that the Dragon and the true object of its wrath are one thing: Lear himself. How could Kent interpose between them? And beyond the wrath, beyond the total eclipse of Lear now beginning, lies the promise of transformation. 'His venom becomes the great Medicine': the Wheel that begins with the expulsion of Cordelia will end with her return, and the flowing forth of 'glorious balm with all its miraculous virtues'.

This is the alchemical reading of this vital moment in the play. Lear's first and drastic act of madness is signalled as his 'becoming the Dragon'. This transition from one emblem of the Raw Stuff to another marks the moment of activation. The sick King announces himself as the Dragon and begins to devour himself. The transmutation of King Lear is under way:

> A weakling babe, a greybeard old
> Surnamed the Dragon: me they hold
> In darkest dungeon languishing
> That I may be reborn a King.[34]

The daughter

Madness and beggary lie ahead for Lear, the nadir of the circular journey he is now embarked upon. They are the twin aspects of his fall – of loss of reason and loss of kingship: inner and outer reversals: wheel within wheel. The agent of these reversals has also two forms, inner and outer. The Wheel of Lear's fortunes is turned both by forces inside him and by

14 From Michael Maier, *Atalanta Fugiens* (1618)

people outside him. What has so far happened suggests that Lear's fall is self-generated; it springs from that single fatal blunder, the 'hideous rashnesse' against Cordelia which expels the good and instates the evil. That this blunder is 'necessary', that it activates a process, is a sense of destiny at once psychological (the madness is in Lear before he actually 'goes' mad) and tragic (rashness as the 'tragic flaw' which releases the hero into his role as sacrificial victim). In this sense, the momentum which bears Lear downwards is provided by forces inside him – the 'Dragon force'. But it has too its outward aspect – characters that cluster around Lear, performing their roles destructive or healing, contributing to the process of his transformation. The patterning of these characters into opposed camps embodies the conflict of opposed forces inside Lear's self. This is most clearly seen in the characters tightest around him: his daughters. The struggle between madness and truth inside Lear is precisely embodied in the struggle, or opposition, between Goneril and Cordelia 'around' Lear. Regan is, for the purpose of this argument, simply a satellite of Goneril – as she herself says, 'I am made of the selfe-same metall as my sister'.[35] So diametric is the opposition between Cordelia and Goneril that they seem two halves of a single figure: Lear's daughter. Let us see, through the words of the play, how these two women – the positive and negative of the daughter – are defined.

Planted in Lear's wrathful attack on Cordelia is a reminiscence of her true – or, at least, her previous – meaning to him. 'I loved her most,' he says, 'and thought to set my rest On her kind nursery.'[36] The words of France tell the same story – 'even but now' Cordelia was

> your best obiect,
> The argument of your praise, balme of your age,
> The best, the deerest.[37]

Cordelia was something precious ('best object'), something soothing ('kind nursery'), something healing ('balm of your age'). This is Cordelia's true nature, as her later role as Lear's redeemer will show. Her healing properties, her status as precious balm – at present reviled by Lear – will triumph. Goneril's true nature, as hidden from Lear at this moment as Cordelia's is, will also be revealed by the play's unfolding. Lear will soon have cause to describe her cruelty and violence:

> She hath tied
> Sharpe-tooth'd unkindnesse, like a vulture, heere. . . .
> Look'd blacke upon me; strooke me with her tongue,
> Most Serpent-like, upon the very heart.[38]

Here she is a vulture and a serpent. Elsewhere she is called a 'detested kite', and has a 'wolvish visage'.[39] She has all the qualities of the child that the enraged Lear wishes upon her:

> Create her childe of spleene, that it may live
> And be a thwart disnatur'd torment to her!
> Let it stampe wrinkles in her brow of youth,
> With cadent teares fret channels in her cheekes,
> Turne all her Mother's paines and benefits
> To laughter and contempt; that she may feele
> How sharper then a Serpent's tooth it is
> To have a thanklesse childe![40]

The prevailing image of Goneril is as a vicious predatory creature – serpent, wolf – and as a bird of death – vulture, kite. Her properties are all attacks – 'sharp-toothed', 'struck me', 'torment . . . stamp wrinkles . . . fret channels'. Her particular violence is a biting, corroding, caustic type: also harsh and burning, as when Lear says to Regan:

> Thy tender-hefted Nature shall not give
> Thee o're to harshnesse: her eyes are fierce, but thine
> Do comfort and not burne.[41]

He says it, of course, to the wrong daughter: it is not Regan who is gentle and comforting. She is 'the selfe-same metall' as Goneril. Metal indeed, for the true opposition between Cordelia and Goneril is expressed by Shakespeare in implicitly *chemical* imagery. Cordelia is a precious balm which soothes and heals, Goneril a fierce caustic which burns and corrodes. The daughter has two natures, healer and destroyer. We see now a more precise application of Kent's words about Lear's expulsion of Cordelia being to 'kill thy physition and thy fee bestow upon the foule disease'. Lear will later call Goneril 'a disease that's in my flesh'[42] and Cordelia will return to heal the disease with 'balm'.

*   *   *

Basil Valentine writes in the *Twelve Keys*: 'There is nothing, says the Philosopher, save a double Mercury.'[43] The daughter in *Lear* – caustic and balm, torment and nurse, disease and physician – has precisely the double nature of alchemical Mercury. In an earlier chapter (see above, pp. 92f) I attempted some incursion into the mysteries of Our Mercury. We saw that it described the total process of the *opus*, that it embodied both *solve* and *coagula*. As the secret fire or Dragon, Mercury is the corrosive agent of dissolution. But, as Lambspringk says, the 'venomous Dragon' becomes the 'great Medicine': Mercury's other nature is as quintessence or *anima*, released as fugitive vapour and returning as 'dew of heaven' or 'divine breath' to reanimate the blackened Stuff. The double Mercury is, like Lear's daughter, destroyer and healer. So Paracelsus in the *Opus Paramirum*: 'Mercury cures the holes it has provoked.'[44] And so Cordelia in *Lear*, returning to 'repair those violent harmes that my two sisters Have in thy reverence made'.[45]

That most magical moment of the play, Cordelia's awakening of Lear, is still far ahead. It is the goal of Lear's *longissima via*, and we are still only at the beginning. In that scene of reunion, Shakespeare forces home with every available image his identification of Cordelia as quickening spirit, *anima Mercurii*. It is not until that scene that the tragic process in *Lear* becomes fully clear: Lear's journey can only be revealed as circular at the moment when the circle joins up: the occult purpose of his suffering becomes manifest at the moment of redemption. At the beginning of the play that purpose is there, but occult – hidden. The Wheel is felt only as fall. The final return of Cordelia is hidden beneath its opposite, the expulsion of Cordelia. So too the alchemical meanings that radiate from the reunion scene are present but hidden at the outset of the play. I must invoke once more the alchemical playgoer of *c.* 1605. He saw the sick King as an emblem of corrupt matter, and heard the Dragon as a signal for that matter to embark on its purgatorial journey. He would also discern, in the images describing Cordelia and Goneril, certain echoes from alchemical literature which suggest to him that Lear's 'daughter' is that twofold transforming principle the alchemists call 'Our Mercury'.

First, the casting-out of Cordelia. The dominant imagery is one of *devaluation*. She that was precious – 'your best obiect' – is accounted worthless:

> When she was deare to us we did hold her so,
> But now her price is fallen.[46]

She has become a 'little-seeming substance'. Lear's present estimation of her fallen value is, of course, a hollowly financial one: he thinks of the dowry he has withdrawn, the 'opulent' third of the kingdom; Cordelia is now only 'dow'rd with our curse'. But his words tell of the deeper degradation as yet unseen by him: the casting-out of 'the best, the dearest', the whipping-out of truth, the killing of the physician. In taking Cordelia's hand, however, the King of France is aware of her 'true worth'. She is 'herself a dowrie', he says; his rival cannot 'buy this unpriz'd precious maid of me'. He woos her with these beautiful words:

> Fairest Cordelia, that art most rich, being poore,
> Most choise forsaken; and most lov'd, despis'd!
> Thee and thy vertues here I seize upon:
> Be it lawfull I take up what's cast away.[47]

This depiction of Cordelia as a precious thing reviled is immediately suggestive of a frequent alchemical formulation. The Hermetic *Tractatus Aureus* of 1600 has: 'Our most precious stone cast forth upon the dung-hill, being most dear is made altogether vile.'[48] Lambspringk's *De Lapide Philosophico* has:

> There is only one substance
> In which all the rest lie hidden. . . .
> In this one vile thing
> You will discover and bring to perfection the
> whole work of Philosophy.[49]

Grasshoff's *Tractatus Aureus* of 1625 has:

> It is clearer than day that the substance of our Blessed Stone is one (although different sages call it by different names), and that Nature has made it ready to the hand of the adept. . . . This Matter lies before the eyes of all; everybody sees it, touches it, loves it, but knows it not. It is glorious and vile, precious and of small account.[50]

The alchemists' description of this precious substance is extraordinarily close to the descriptions of Cordelia at the moment of her expulsion. It is 'cast forth', Cordelia is 'cast away'; it is 'most dear... made altogether vile', Cordelia is 'most lov'd despis'd'; it is 'precious and of small account', Cordelia is 'unpriz'd precious'.

What, then, is this mysterious substance the alchemists speak of? Its quality of preciousness reviled is clearly related to a quality of ubiquity: it is 'before the eyes of all', yet everybody 'knows it not', it is present and yet occult. It is, in other words, spirit, the *anima* indwelling in all things, ubiquitous but invisible: it must be liberated and revealed by the alchemist, and so become 'the substance of our Blessed Stone'. Ripley speaks of this 'mervelous thing', which is everywhere –

> Foules and fishes to us doth it bring,
> Every man it hath, and it is in every place,
> In thee, in me, and in each thing, time and space.

– and goes on to identify it: 'Mercurie it is I wis, But not the common called Quicksilver'.[51] So the precious thing accounted vile is none other than 'Our Mercury', the spirit hidden within matter. This 'vileness' not only expresses its ignored ubiquity: it also conveys the whole idea of being 'cast forth', i.e. it describes that *expulsion* of the Mercurial spirit which signals the beginning of matter's mortification. Cordelia is the *anima mercurii* cast out from the Raw Stuff, the sick King Lear. As the Dragon within the King begins to rage and devour, the life-giving spirit flies up. She becomes, in the words of Basil Valentine, 'mere air' which 'flyeth to and fro without wings ... after its expulsion out of its habitation'.[52] If we look again at the words of Cordelia's wooer, France, we find a beautiful and covert image of her volatility: 'I *take up* what's cast away', the ascent of the liberated *anima*. This fleeing Mercury is also the Paracelsist *arcanum*, full of healing 'virtues': so France says, 'thee and thy vertues here I seize upon'. In a fine passage from the *Fons Chymicae Veritatis*, Philalethes expresses this 'precious-vile' aspect of Our Mercury in tones which seem to pick up echoes from Cordelia, Lear and France:

> Our Mercury ... is a most pure virgin, and is loved of many, but she meets all her wooers in foul garments, in order that she may be able to distinguish the worthy from the unworthy. Our beautiful maiden abounds in inward hidden graces; unlike the immodest woman who meets her lovers in splendid garments. To those who do not despise her foul exterior, she appears in all her beauty, and brings them an infinite dower of riches and health. Our Queen is pure above measure, and ... the Sages ... style her their quintessence.[53]

It seems that France, who takes Cordelia to be his Queen and knows her graces to be a dower, is displaying a wisdom alchemical as well as emotional.

As the casting-out of Cordelia is the expulsion of the spirit Mercury, so the instatement of Goneril is the complementary empowering of the other aspect of Mercury: the venomous, caustic, dissolving Mercury, 'the Water which teareth the Bodies, and makes them no Bodies'.[54] The biting, pecking, striking, fretting, burning properties which we have seen associated with Goneril identify her harsh and corrosive treatment of Lear, her tearing the body of the King and making it nobody. We shall look at this dissolving process, the mortification of the King, in the next section. Goneril is also described with animal images: how do these express her identity as destroying Mercury? She strikes Lear 'most serpent-like', is 'sharper than a serpent's tooth'. Here the emblem is exact: the serpent is precisely the fiery nature of Mercurius. It is 'the *serpens mercurii*' which 'creates and destroys'.[55] Mercury, says Jakob Boehme, is 'the fiery wheel of the essence in the form of a serpent'.[56] Goneril as serpent indeed plays her part in turning Lear's Wheel of Fire. The description of the alchemical serpent (from the *Tractatus Aristotelis*) sounds not unlike the slinky Goneril: 'The serpent is more cunning than all the beasts of the earth; under the beauty of her skin, she shows a harmless face.... She causes the nature wherewith she is united to vanish.'[57] This annihilating serpent is, of course, another aspect of the devouring dragon. The Dragon in Lear and the serpent Goneril are the inner and outer agents of Lear's dissolution.

Goneril is also the wolf ('thy wolvish visage'), and here too she figures as the mortifier of the King. In the *Twelve Keys* Basil Valentine, having identified the Raw Stuff as as a King crowned with gold, orders this treatment for him: 'Take the most ravenous grey Wolf.... He is very hungry. Cast to him the King's body, that he may be nourished by it; and when he hath devoured the King, make a great fire.'[58] The same instruction is found in Daniel Stolcius' *Viridarium Chymicum* (1624): 'hand over the King to be eaten by the ravening wolf.'[59] This potent image appears as the 24th emblem of Michael Maier's *Atalanta Fugiens* (fig. 15): we see in the background the reborn King stepping forth from the fire, for like the dragon and the serpent, the wolf's ravages promise transmutation. Basil Valentine's 'grey wolf' is particularly identifiable as antimony, one of the many 'exoteric' substances (quicksilver, *aqua fortis*, vitriol, *et al.*) which performed the role of Mercury-as-destroyer. Professor Read glosses this passage from Valentine thus:

> The alchemical wolf in general represents a corrosive or 'biting' agent, sometimes an acid. In this case, the grey wolf is clearly antimony, which was known to the alchemists as *lupus metallorum*, or 'wolf of the metals'.... On account of its use in purifying molten gold – the impurities being removed in the form of a scum – antimony was also called *balneum regis*.[60]

As serpent and wolf, dissolving mercury and biting antimony, the role of Goneril is to devour, to reduce and so finally to cleanse the corrupted body of the King, to eat him away as in a bath of corrosive acid. This she proceeds to do with hideous efficiency.

Harshness and balm, serpent-wolf and fleeing preciousness: with these touches Shakespeare identifies Goneril and Cordelia as the polarities of Our Mercury. Lear's daughter is like the Mercury of Sendivogius' *Novum Lumen*, a beast with the wings of a bird and a poison that ultimately cures.[61] The essential identifications have now been made for *King Lear* as alchemical process to be set in motion. The sick King becomes the Dragon: the Raw Stuff embarks on the purgatorial *opus*. At that moment, the two natures of Mercury separate: the life-giving spirit flees and the death-dealing dissolution begins. This double Mercury has an inner and outer form: truth and madness, Cordelia and Goneril. 'Thus our Body has been rendered fit for the first stage of our Work.'[62] The mortification of the King begins.

The mortification

At the end of the first scene Goneril and Regan, now joint Queens of the realm, are left alone on stage. They now speak harshly of the father for whom they recently professed such love. 'How full of changes his age is', his 'poore iudgement ... appeares too grossely', he shows 'the infirmity of his age', 'the imperfections of long ingraffed condition', 'unruly waywardnesse', 'infirme ... cholericke ... unconstant'.[63] We have no reason to disagree with their assessment: it is a precise inventory of Lear's corruption. Gross, ailing, imperfect, changeable: matter fit for transmutation. The scene ends on a note of undefined menace:

> *Regan* We shall further thinke of it.
> *Goneril* We must do something, and i'th'heate.[64]

The wayward King is to be melted down in the 'fire of the treatment'.

The mortification of Lear takes outwardly the form of a relentless reduction of status. Goneril commands Oswald to treat Lear with 'what weary negligence you please'. The father, bereft of authority, becomes no more than a troublesome child: 'Old fools are babes again, and must be us'd With checks as flatteries.'[65] At first, it is the *name* of King which must be demolished. Oswald's first snub begins this process:

> *Lear* Oh you Sir, you, come you hither, Sir.
> Who am I, Sir?
> *Oswald* My Ladies father.[66]

15 From Michael Maier, *Atalanta Fugiens* (1618)

From King to 'my Lady's father': the next peg down is provided by the Fool (whose paradoxical allegiances we must soon examine):

> *Fool* That lord that counsell'd thee
>   To give away thy land,
> Come place him here by me,
>   Do thou for him stand:
> The sweet and bitter fool
>   Will presently appear;
> The one in motley here,
>   The other found out there.
> *Lear* Dost thou call me fool, boy?
> *Fool* All thy other titles thou hast given
>   away; that thou wast born with.[67]

From King to father to bitter fool: the re-naming of the King now reaches its nadir, again in the words of the Fool: 'Thou wast a pretty fellow when thou hadst no need to care for her frowning; now thou art an O without a figure. I am better then thou art now; I am a Foole, thou art nothing.'[68] So, in name, Lear becomes nothing. This is a symbolic stripping-down preparatory to the real one, the King becoming a naked beggar in the storm. The Fool gives another definition of Lear-as-nothing:

 Lear Who is it that can tell me who I am?
 Fool Lear's shadow.[69]

This is what Lear must become: his own shadow, the negative of the King, *sol niger* (the black sun).

This nominal extinction of Lear is accompanied by another reduction, one accomplished by Goneril and Regan with a positively mathematical precision. Goneril's particular displeasure is aimed at the 'disorder'd' and 'debosh'd' behaviour of Lear's retinue of one hundred knights. To Lear, still clinging to the remnants of his symbolic world, the hundred knights *are* his kingdom: they shelter him from the truth of his dispossession. They must be stripped away. Goneril's steely request begins this reduction:

>   be then desir'd
> By her, that else will take the thing she begges,
> A little to disquantity your traine.[70]

Lear rages and curses; he and his ragged 'kingdom' troop off to Gloucester Castle, already colonized by Regan and Cornwall. The 'disquantitying' of his retinue now accelerates. Regan:

> I pray you, Father, being weake, seeme so.
> If, till the expiration of your moneth,
> You will returne and soiourne with my sister,
> Dismissing halfe your traine, come then to me.[71]

He is down to fifty now, but not for long: Regan again –

>   what! fifty followers?
> Is it not well? What should you need of more?
> Yea, or so many? . . .
>   If you will come to me,
> (For now I spie a danger) I entreate you
> To bring but five-and-twentie.[72]

Believing Goneril will still tolerate fifty, Lear turns once more to her: 'Ile go with thee: Thy fifty yet doth double five-and-twenty, And thou art twice her love.' It is the same cracked old man, counting up the points, making love a currency to barter in. Only this time he has no power to bargain, no 'sway, revennew, execution'. This time his bluff is called:

> *Goneril* Heare me, my lord.
> What need you five-and-twenty? ten? or five . . . ?
> *Regan* What need one?[73]

In grand theatrical manner, Lear had divided up his kingdom in the first scene. This now is the real division, the unrecognized meaning behind that symbolic partition of the map. His retinue relentlessly halved away – 100, 50, 25, 10, 5, 1 . . . 0. 'Now thou art an O without a figure', a shadow King ruling over a people-less kingdom. The Wheel that began with the word 'Nothing' carries Lear down towards nothingness.

* * *

The alchemical concept of mortification entails exactly this breaking down of the Raw Stuff. Our Mercury, writes Philalethes, 'is the Water which teareth the Bodies, and makes them no Bodies'.[74] Ripley compares the dissolution of 'our Bodies' to a casting down of mountains into a corrosive sea:

> Into the deepnes therefore of Mercurie
> Turne them. . . .
> Then hath the bodies their first forme lost.[75]

Goneril disintegrates Lear's former substance, decomposes his flawed solidity, removes his 'first forme'. She is the wolf antimony, devouring the King; the serpent Mercury, who 'causes the nature wherewith she is united to vanish'; the 'mercuriall sharpish liquor' which is, in the words of Thomas Tymme, 'so pearcing that it is able to open and unlock the most strong and hard gates of Sol and Luna'.[76] In disintegrating Lear's protective substance as King, Goneril penetrates into the heart of Lear, breaks down the 'gate' enclosing his true self. (Lear himself uses the word 'gate' in this sense: striking his head, he cries, 'O Lear, Lear, Lear! Beate at this gate, that let thy folly in, And thy deere iudgement out!'[77])

As a removal of previous metallic form, the mortification is a process of stripping down. Philalethes writes of the 'regimen of Mercury' as the operation which 'despoils the King of his golden garments'.[78] The 'regimen' of Goneril (a word apt to describe her regimental rule) has just this effect. Lear is first stripped of the name of King and then of the vestiges of his kingdom (the hundred knights). This is a prelude to the actual nakedness of Lear in the Storm – 'unbonneted he runs', 'Off, off you lendings'.[79] He himself tears off the 'golden garments' of the King, becomes a 'poore naked wretch'. Yet this denuding, begun by Goneril and made actual by Lear himself, is also a stripping-away of illusions, a removal of shelters: in nakedness he discovers the true and essential nature of 'unaccommodated man', in nothingness he finds 'the thing it selfe'.[80] Once again we see that Goneril's harsh and caustic treatment of Lear is a necessary penetration, a breaking down of 'gates' to free something hidden and essential in Lear. She is 'vitriolic' in the full esoteric reading of vitriol expressed in the acrostic, '*Visita Interiora*

Terrae Rectificando Invenies Occultum Lapidem'.[81] Visit the interior of the earth and by purifying you will find the hidden stone. Vitriol is a vehicle for that journey inwards, a key to unlock the secret 'entrance into the shut palace of the King'. Goneril breaks Lear down, but it is Lear himself that must make the journey into the dark interior of his self.

The Fool

Goneril and Regan lead Lear to the brink of nothingness, then cast him out into the storm which will complete the extinction of his former kingship. Similarly, they send him mad and then leave his madness to dissolve his former self. They are the agents of his dissolution, but they do not work alone. The Fool plays a vital role in all this. He constantly needles Lear with images of reversal: 'Why this fellow has banish'd two on's daughters, and did the third a blessing', 'thou boar'st thine asse on thy backe o're the durt', 'thou mad'st thy daughters thy mothers'.[82] These insist on Lear's own fatal reversal of values. The Fool forces Lear to relive his follies: 'Can you make no use of nothing, Nunckle?' he asks: what else can Lear do but reply, 'Why, no boy; nothing can be

16 From Daniel Stolcius, *Viridarium Chymicum* (1624)

made of nothing,'[83] and so twist the knife of recollection in his own wound? The Fool makes Lear see that he has injured not only Cordelia but also himself, that he is the object to be devoured:

> For you know, Nunckle,
> The hedge-sparrow fed the cuckoo so long,
> That it's had its head bit off by its young.[84]

Finally, as we have seen, he calls the King the Fool – 'that such a King should play bo-peepe and goe the Foole among', 'Thou wouldst make a good Foole'. 'Fool', he says, is Lear's natural 'title': 'all thy other titles thou hast given away; that thou wast born with.' Later, in the height of his madness, Lear will style himself 'the Naturall Foole of Fortune' and say, 'When we are borne, we cry that we are to come To this great stage of Fooles.'[85]

All this guides Lear towards the brink of madness and collapse as actively as Goneril and Regan do. In being named 'Fool', Lear is diagnosed as mad (the word 'fool' derives immediately from Old French '*fol*': in modern French, '*le fou*' means both madman and jester.) The Fool is undoubtedly exercising his professional function as irritant of pomp, eroder of delusions. His derision of the King is that 'burlesque of sacred things and persons' which the anthropologist Julian Steward defines as one of the clown's universal functions.[86] The Fool's action makes him 'in league' with Goneril: it is solvent, corrosive: it breaks Lear down and brings him face to face with nothingness. 'Now thou art an O without a figure ... thou art nothing.' Yet clearly the Fool detests Goneril. She is the stinking 'Lady Brach', the cuckoo that bites off its foster-parent's head. Goneril returns the feelings, complains of the 'all-lycenc'd Foole' and sends him packing with the words, 'You, sir, more knave then foole, after your master'.[87] The Fool and Goneril work together yet stand quite opposed.

The Fool and Cordelia have a strange and beautiful kinship. An unnamed Knight tells Lear: 'Since my young Lady's going into France, Sir, the Foole hath much pined away.'[88] With a deft stroke, Shakespeare etches an emotional bond between them – a bond which, in the purely human terms of the drama, seems just right. For all his barbed and biting wit, the Fool is a delicate, affectionate and loyal creature. 'A fragile, hectic, beautiful-faced, half-idiot-looking boy,' was how Macready envisaged him, 'of a light delicate frame, every feature expressive of sensibility, even to pain.'[89] This is undoubtedly a Romantic perception – a touch of the *pierrots* – but it carries a certain authority: it was Macready's 1838 production of *Lear* that restored the Fool to the play, after Nahum Tate's Fool-less adaptation (or rather, lobotomy) of the play had held the stage for a century and a half.[90] 'What a noble heart, a gentle and a loving one, lies beneath that parti-coloured jerkin,' said

Macready. It could almost be Cordelia beneath the motley, the way he describes it, and he did indeed give the part to a 19-year-old actress. It has even been suggested (by Quiller-Couch among others) that the parts of Cordelia and the Fool were originally taken by the same boy-actor. They are never on stage together. Cordelia departs at I, i and returns at IV, iv; the Fool appears at I, iv and utters his last riddling line – 'And Ile go to bed at noone' – at III, vi. In a production during the First World War, with male actors scarce, both parts were played by Sybil Thorndike. All of these are reactions to the Fool which seem to seek some theatrical expression for an elusive but potent impression of kinship between Cordelia and the Fool. Within the play, this impression is most powerfully generated by that haunting line that begins Lear's last speech. Looking down at the face of the dead Cordelia, he says: 'And my poore Foole is hang'd.'[91]

Against all this, the patterning of the play would seem to suggest that the Fool is the precise opposite of Cordelia. She is the truth that Lear has 'whippt out', he is the folly that now reigns. She heals, he dissolves. This is the paradox of the Fool: he is in league with Goneril yet far removed from her, he is close to Cordelia yet the opposite of her. Paradox is, of course, the universal Fool's *metier*. The jester professionally inhabits two worlds: the ordered and meaningful world in which he performs, the chaotic nonsense world which he conjures safely within the circle of his clowning. He is a go-between, the scout sent out by society to reconnoitre anarchy. He wears his double nature as a costume – the parti-coloured motley. This encompassing of opposites confers on him an unexpected completeness (cf. representations of the Harlequin as hermaphroditic[92]), and this is perhaps how to read the paradoxical nature of Shakespeare's Fool.

Though the Fool partners Goneril in performing the office of solvent, the vital difference between them can be expressed in simple terms: the Fool *loves* Lear. In a touchingly frail way, he watches over Lear with a loyalty at once protective and dependent. He calls Lear 'Nuncle', Lear calls him 'my boy': they relate as irascible father to impertinent son. In all his derision of Lear, all his caustic renaming of the King as Fool and nothing, is contained a suspended, potential, half-expressed kind of love. He thus provides a continuity with Cordelia: a reminder and promise of her, a hovering presence of that fled *anima*. When Cordelia is whipped out, the Fool takes over – her opposite, Cordelia in motley, but her ally in the provision of love to the broken King. It is the Fool, and he alone, who accompanies Lear out into the storm, while Goneril simply wipes the blood off her hands and turns to size up the next victim. Only the Fool, we see, can perform two roles at once. Cordelia cannot hurt Lear, Goneril cannot love Lear: the Fool does both. When madness must prevail, Cordelia is helpless. She can only speak truth. The Fool, however, is familar with madness, it is his stock-in-trade: he alone can express truth *through* nonsense. He is the archetypal 'wise fool', and so provides a continuous presence of disguised truth as well as love. While Cordelia and Goneril are each confined to a single role – redeemer and destroyer – the motley Fool wears two colours and has two natures. And while Lear must fall on the Wheel from zenith to nadir, the Fool frequents both places like an acrobat. In the strange world of *King Lear*, the Fool alone is complete.

This is the secret presiding role of the Fool. We see a noble and generous Fool – one that (in his own words) 'breaks his neck with following', one that journeys out with Lear into certain destruction because, like the *chevalier* of romance, only he in all the kingdom can make that journey. And if we place this Fool inside the impish, irritating, nonsense-mongering Fool that skips across the darkened dramatic surface of the play, we begin to see why the Fool is among the most haunting of all Shakespearean characters. The question now is: how is this newly 'elevated' Fool to be depicted in alchemical terms? The answer seems clear, if strange. If the Fool includes both Goneril and Cordelia in his double nature, then alchemically he includes both aspects of the double-natured Mercury. The Fool *is* Mercury. And in making this identification, it is just possible that we are for the first time finding alchemy in the actual sphere of visual, theatrical representation.

*   *   *

Like all the metals most familiar to the early chemist, Mercury bore the name of a god. The metal was invested with properties associated with the god. Mercury's liquidity as chemical corresponded with the nimbleness of the god, just as iron was strong like Mars and lead old (decadent) like Saturn. The more esoteric application of Mercury – as volatile spirit, ascending and descending *anima* – related to Mercury as winged messenger between gods and men. The dissolving aspect of alchemical Mercury links with the mischievous character of the god Mercury, patron of rogues, vagabonds and pickpockets. The gods also gave their names to the planets, and poured down their particular qualities on any person, moment or talisman 'ruled' by their planet. By a combination of these routes – mythological, chemical, astrological – the gods became descriptive of certain human qualities. We no longer acknowledge Mars, Saturn and Jupiter as gods or metals (only planets) but we still retain their associations when we describe someone's behaviour or appearance as 'martial', 'saturnine' or 'jovial'. So too with 'mercurial'. Someone who is mercurial is (according to the Oxford dictionary) 'sprightly, ready-witted and volatile', and (according to Roget) changeable, mobile, quick, excitable, elusive. If we called the Fool in *Lear* 'mercurial', we would, looking back through those synonyms, be describing him to a T. Not only might this word define him within an invisible alchemical pattern; it would describe his character, behaviour, gestures, his very presence on the stage of the Globe theatre.

With their natural tendency towards dramatization, alchemical writers often presented Our Mercury in the guise of a 'mercurial' character. He is called 'inconstant' by Gerhard Dorn, *'versipellis'* (skin- or form-changing) by Aegidius.[93] He is the 'pissing manikin' of alchemical illustration (the mercurial water being sometimes styled *urina puerorum*).[94] Philalethes warns, 'You must be very wary how you lead him, for if he can find an opportunity he will give you the slip, and leave you to a world of misfortune.'[95] These characterize Mercury as elusive, baffling, malicious, potentially faithless. They could once again describe the Fool as he appears – the tormenting Fool whom Lear calls 'a pestilent gall to me', the apparently disloyal Fool who counsels Kent to 'let go thy hold when a great wheele runs downe a hill'.[96] And just as the alchemists' exasperation must be seen in the context of their reverence of Mercury as the secret and agent of transformation, so the Fool's impishness belies his true love; he ignores his own common sense, clinging stubbornly to the runaway wheel of the King's misfortune. Jung sums up the paradoxical nature of Mercury-as-character in alchemical writings:

> All through the middle ages, he was the object of puzzled speculation on the part of the natural philosophers: sometimes he was a ministering and helpful spirit, a πάρεδρος (literally 'assistant, comrade') or *familiaris*; and sometimes the *servus* or *cervus fugitivus* [fugitive slave or stag], an elusive, deceptive, teasing goblin who drove the alchemists to despair.[97]

Again we hear the Fool in all this, the contradictory aspects of his relationship to Lear. Comrade and fellow-spirit in suffering, or teasing goblin: which one is the Fool? He is, like Mercury, both: *Mercurius versipellis*, the motley fool.

The fullest personification of Mercury in alchemical literature is the *Dialogue* between Alchemist and Mercury in Michael Sendivogius' *Novum Lumen Chemicum*. It dramatizes the difficulties and pitfalls of the *opus* as the attempts of a deluded, overbearing alchemist to regulate an impish, taunting Mercury. It sounds, at times, quite like the relationship between Lear and his Fool. The exasperated threats are the same:

> *Alchemist* Now go to, be now therefore obedient, or else it shall be the worse for thee. . . . O thou art a Devil, and not a good Mercury . . . truly I will go to work with thee again. . . . Wife, bring hither the Hogs-dung, I will handle that Mercury some new ways![98]

Lear A pestilent gall to me. . . . Take heed Sirrah, the whip. . . . And you lie, Sirrah, wee'l have you whipt.[99]

The stung reaction to insolence is the same:

Alchemist What, dost thou still deride me?[100]

Lear Dost thou call me fool, boy?[101]

Mercury does indeed call the alchemist a fool, someone blind and deluded:

Mercury What wilt thou have of me, thou fool? Why has thou thus accused me?
Alchemist Art thou he that I have longed to see?
Mercury I am, but no Man that is blind can see me.
Alchemist I am not blind.
Mercury Thou art very blind, for thou canst not see thyself. How then can you see me?
Alchemist O now . . . thou contemnest me: thou dost not know perhaps that I have worked with many Princes. . . .?
Mercury Fools flock to Princes' courts, for there they are honoured and fare better than others.[102]

Is this not the substance of the Fool's message to Lear? That Lear is blind to himself and his situation: we have eyes, he tells Lear, so that 'what a man cannot smell out, he may spy into',[103] and Lear has failed to smell or see his danger. That his royal pedigree is no guarantee against folly – 'All thy other titles thou hast given away; that thou wast born with.' That the court is indeed the headquarters of folly – 'No, faith, lords and great men will not let me . . . have all the fool to myself; they'll be snatching.'[104] If we allow for different contexts, and for a certain disparity of literary talents, a correspondence emerges. The insolence and piquancy of Sendivogius' Mercury; the tyrannous yet desperate bearing of his alchemist: these seem like two-dimensional versions of Shakespeare's Fool and Lear.

To identify the King and the daughter in *King Lear* as Raw Stuff and Mercury – i.e., transformed and transformer – is to suggest that these characters in the play and these substances in alchemy embody the same hidden meaning. Such correspondences belong in the sphere of meaning alone: they are communicated by language, by *Lear* as poetry. They do not necessarily penetrate to the surface of the dramatic action, except in as far as certain eyes in the audience might see certain groupings on stage and 'freeze' them into emblems of the sort reproduced throughout this chapter. The Fool as Mercury is, however, an identification which works on a visual and gestural plane as well as a symbolic one. He is experienced by the audience as 'mercurial' – a swift, garrulous, boyish, elusive, apparently malicious figure, full of jibes and songs. This reminds us of another character on the Jacobean stage at just this time, someone whose identification as Mercury is quite indisputable: Frank Quicksilver in *Eastward Hoe*. 'Mad Quicksilver', 'my nimble-spirited Quicksilver', 'my runagate Quicksilver'; a 'bragging boy', a 'prodigal coxcomb', a 'crackling bavin' with 'that quick braine of yours': once again, all these might describe the Fool *as he appears*, the dancing irreverent urchin jester. Ben Jonson's Quicksilver is Mercury envisaged on the London stage in 1605. Shakespeare undoubtedly saw, and probably admired, his cantankerous friend's *pièce de théâtre chymique*, with its Quicksilver and Golding, its volatility and fixation, its presiding Touchstone. How typical that the play he was himself writing in 1605 should rebound those alchemical meanings multiplied and refined a hundredfold. I do not want to embark on a comparison of *Eastward Hoe* and *King Lear*, though it is interesting to note in passing how the opposition of good and evil is in both cases manifested in a duplex daughter-figure – Touchstone describes Gertrude and Mildred just as Lear *ought* to have described Goneril and Cordelia, 'the eldest of a proud ambition and nice wantonnesse, the other of a modest humilitie and comely sobernesse'.[105] Some five years after these plays, Jonson's *Alchemist* and Shakespeare's *Tempest* seem almost to constitute an exchange: antithesis and thesis in a dialogue on magic – Dr Subtle the phoney magician whose magic deludes, Prospero the genuine magician whose magic transforms. Might not this dialogue have a prelude, an opening exchange, in the different uses of alchemy in *Eastward Hoe* and *King Lear*? Both plays 'harness' alchemy as a particular kind of momentum, incorporating alchemical terminology as a dramatic imagery of change and transformation. But each writer finds what he needs in a different part of the alchemical lexicon. The alchemical language of *Eastward Hoe* derives from the laboratory. Quicksilver 'works on' Golding, 'evaporates' into lawlessness, is fixed as sublimate in the 'crucible' of prison: this is an imagery of chemical reaction which nowhere insists on its potential esoteric signification. 'Light' gold, 'blanched' copper, 'current' metal, 'Touchstone, touch me still': an imagery of coining and assaying entirely apt for the play's setting in Goldsmith's Row. Jonson's is an exoteric, practical alchemy: Quicksilver is no *anima*, only common quicksilver. But Shakespeare's *Lear* – with its sinking King and uprising Dragon; its double daughter as the balm and serpent of double Mercury; its precious made vile and body made no body; and, finally, its Fool as the secret presence of Mercurius himself – all of this reaches down into the occult and esoteric layers of alchemical meaning, where chemical reactions figure forth magical and psychic propositions. Jonson embraces 'chymistry', Shakespeare 'true, religious Alchymie'.

Sendivogius' Mercury (1604) and Jonson's Quicksilver (1605) shed, then, an unexpected light on their precise contemporary, the Fool in *Lear*. He is Shakespeare's wonderful envisaging of Mercury. He is chemically mercurial, like Jonson's Quicksilver – nimble, volatile, crazy, piquant. Yet he is also, like Sendivogius' Mercury, 'the true Mercury':

> *Alchemist* Do but tell me if thou art the true Mercury, or if there be another.
> *Mercury* I am Mercury, but there is another.[105]

He is Cordelia disguised as her opposite, a winged messenger from her, *anima* as familiar spirit. And thus, by both of these routes, the Fool becomes the figure to whom they both lead: the *god* Mercurius. The 'wicked' Mercury was god of vagabonds and rogues (Autolycus was 'lytter'd under Mercurie'[106]): who better to accompany Lear into the storm which turns him into a vagabond and 'rogue forlorn'?[107] The more elevated Mercury was a god of revelation – described by Virgil as '*deorum hominumque interpres*' (spokesman of gods to men)[108] and by Jonson as 'president of language' and 'god of eloquence'.[109] This role of the god Mercury was not omitted from the alchemical formulation of Mercury: Michael Maier promised that Our Mercury 'will make you a witness of the mysteries of God and the secrets of Nature'.[110] Nor was it omitted from Shakespeare's mercurial Fool. He is the truth-teller, quite relentless in revealing to Lear the truth of his situation. As the 'great wheele' runs downhill, the Fool alone can speak truth in the now-compulsory language of nonsense. As the King becomes a Fool, the Fool becomes a god of revelation. They are painful revelations at first, but finally become those promised above by Maier, when Lear comes to 'take upon's the mystery of things, As if we were Gods spies'.[111]

As well as having particular properties and presiding over particular activities, Mercury had distinctive visual characteristics constantly reproduced from Classical tradition. The Roman Mercurius was in itself a version of the Greek god, Hermes. 'To portray him, Roman artists drew upon representations of Hermes', giving him 'a lithe and graceful body ... a beardless face, and, for attributes, the *caduceus* and the winged *petasus*, with a purse in his hand.'[112] Again one might discern in the lithe and boyish figure of the Fool a kind of Mercurius in motley, wearing not a *petasus* but a coxcomb, carrying not a snake-entwined *caduceus* but a ribbon-festooned bauble. And the prototype of Hermes reminds us of something else: that Mercury was, like him, a patron of journeys. 'Hermes was above all thought of as the god of travellers, whom he guided on their perilous ways.' He was indeed the guide for one especially perilous journey: 'he was charged with conducting the souls of the dead to the underworld.'[113] Is this perhaps the central purpose, the gravest burden, of the Fool's mercurial role: to conduct Lear on a journey to the 'underworld'? Alchemically, Hades represents the blackness of putrefaction: the spirit rises up to the 'heaven' of the vessel, the mortified body is consigned to the 'hell' beneath. Our bodies 'undergo oppression and are enclosed in Hades', where they are 'fettered and afflicted in darkness'.[114] This visit to the underworld is also a journey inwards, that visit to 'the interior of the earth' where, in the words of the vitriol acrostic, the 'hidden stone' is to be found. So Lear, accompanied by the Fool as Hermes Psychopompos (the Conductor of Souls), journeys out into the storm and madness. His descent into blackness, his going down on the Wheel, is finally a journey down into the unplumbed darkness of the self, the 'element below' the conscious mind, the dragon world beneath the kingly surface. There Lear will meet, in the Fool's words, 'Lear's shadow' – his opposite, or his 'shade' in the Hades of the self. The journey is both a death and discovery: something hidden and precious is to be found down there. Here is Jung's description of the descent into Hades in terms of psyche:

> Theseus and Peirithous descended into Hades and grew fast to the rocks of the underworld, which is to say that the conscious mind, advancing into the unknown regions of the psyche, is overpowered by the archaic forces of the unconscious: a repetition of the cosmic embrace of Nous and Physis. The purpose of the descent as universally exemplified in the myth of the hero is to show that only in the region of danger (watery abyss, cavern, forest, island, castle, etc.) can one find the 'treasure hard to attain' (jewel, virgin, life-potion, victory over death). The dread and resistance which every natural human being experiences when it comes to delving too deeply into himself is, at bottom, the fear of the journey to Hades.[115]

Shakespeare's hero journeys 'deeply into himself' – a journey outwardly expressed as passage through a 'region of danger', the storm-tossed heath. He discovers there the 'treasure hard to attain': a new self, a healed vision, a returning daughter. He becomes an adept in the secret art which 'can make vile things precious'. Shakespeare knew all that Jung is saying here, but his access to that knowledge was not through the language of psychiatry. It was through the language of alchemy. The Fool as 'runagate quicksilver', the goblin Mercurius, is also Hermes Psychopompos. He conducts Lear to the depths, leading him, like the Hermes who leads Orpheus in Rilke's poem,[116] through 'the strange unfathomed mine of souls'. He is, in Rilke's words,

> the god of faring and of message,
> the travelling hood over his shining eyes,
> the slender wand held out before his body.

Lear in the storm, like the traveller through the underworld, would be lost without 'the slim god's endlessly gentle contact'.

The fire

We had been preparing all along for the Storm to break. The stripping-down of Lear must lead to nakedness. The caustic mortification – Goneril's 'sharpe-tooth'd unkindnesse', the 'pestilent gall' of the Fool's revelations – must eventually break him. When Regan says,

> O, Sir! you are old;
> Nature in you stands on the very verge
> Of her confine.[117]

we hear a threatening hint of something about to explode and disintegrate. Lear too cries, 'O sides! you are too tough; will you yet hold?'[118] An intolerable pressure without and within: finally the rigid King cracks –

> No, you unnaturall hags,
> I will have such revenges on you both
> That all the world shall – I will do such things,
> What they are, yet I know not, but they shalbe
> The terrors of the earth. You thinke Ile weepe;
> No, I'le not weepe;
> I have full cause of weeping, but this heart
> Shal break into a hundred thousand flawes
> Or ere Ile weepe. O Foole, I shall go mad.[119]

The syntax collapses ('the world shall – I will'); the threats fail ('what they are yet I know not'); the last attempt at control ('No, Ile not weepe') breaks into a hundred thousand pieces; and Lear at last surrenders. 'I shall go mad.' These are the last words of the old King to his court, his last edict. Moments later, he has stalked out, abjuring the castle for the heath, the shelter of flattery for the storm of truth. The castle doors close behind him and the Fool.

A stage direction is very precisely placed in that crescendo of Lear's. Three lines from the end, between the words 'weeping' and 'but', is the direction 'Storm and Tempest'. The first rumble of thunder is heard in the play, and the accompanying image is unmistakable: the shattering of something hard and brittle into a 'hundred thousand' fragments. Lear enters the Storm, with its 'all-shaking thunder',[120] to be broken into pieces. His famous apostrophe to the Storm[121] invokes it as an agent of total destruction. It must destroy the kingdom which has abandoned him – 'spout Till you have drench'd our steeples, drown the cockes'. It

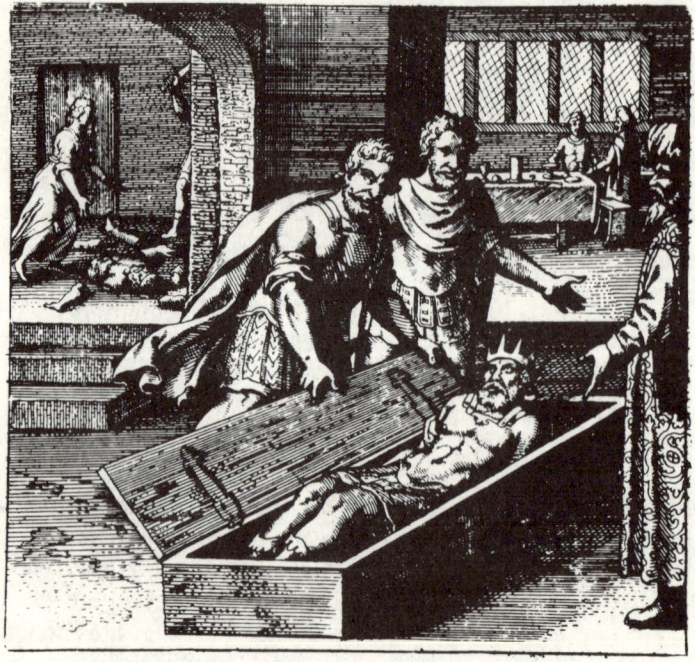

17 From Michael Maier, *Atalanta Fugiens* (1618)

must destroy the mind with its 'thought-executing fires' and nature with its 'oake-cleaving thunderbolts'. Above all it must negate all fertility and fruition:

> Strike flat the thicke rotundity o' th' world!
> Cracke Natures moulds, all germaines spill at once
> That makes ingratefull man!

And, because the kingdom, the mind, the natural world and the furtherance of the species are all finally entailed in the King, Lear's injunctions are all one prayer: for the Storm to destroy the King. He surrenders to the Storm as he could never have done to anyone human:

> then let fall
> Your horrible pleasure. Heere I stand, your slave,
> A poore, infirme, weake, and dispis'd old man.

He calls for the Storm to shake 'covert and convenient seeming' to pieces, to break out 'pent-up guilts' from their 'concealing continents': again it is Lear whose illusions must be shattered, whose guilts must be exorcised. The Storm is what makes the King nothing. It

> tears his white hair
> Which the impetuous blasts, with eyeless rage,
> Catch in their fury and *make nothing of*.[122]

It is insupportable – 'man's Nature cannot carry Th'affliction nor the feare' – and unendurable – 'the tiranny of the open night's too rough for Nature to endure'.[123] The King-made-nothing is a naked beggar, a fugitive outcast, the nadir of that Wheel which has the King as its zenith:

> Poore naked wretches, where so ere you are,
> That bide the pelting of this pittilesse storme,
> How shall your houselesse heads and unfed sides,
> Your loop'd and window'd raggednesse, defend you
> From seasons such as these? O I have tane
> Too little care of this. Take physicke, Pompe;
> Expose thy selfe to feele what wretches feele,
> That thou maist shake the superflux to them,
> And shew the heavens more iust.[124]

This is compassion of the truest kind – a 'suffering with'. Lear has become a 'poore naked wretch': he has renounced shelter – 'rather I abiure all roofes, and chuse To wage against the enmity o' th' ayre'[125] – and so exposed himself 'to feele what wretches feele'. We see that the successive denuding of the King has been also an unveiling of hidden truths: he has 'tane too little care' of the true nature of the kingdom. This is one sense in which he makes the prophesied encounter with his 'shadow': the wretch is the shadow of the king; beggary is the hidden underworld, the Hades of the kingdom. This, the outer form of his descent, continues further when he meets Poor Tom:

> Is man no more then this? Consider him well. Thou ow'st the worme no silke, the beast no hide, the sheepe no wooll, the cat no perfume. Ha? Here's three on's are sophisticated; thou art the thing it selfe; unaccomodated man is no more but such a poore, bare, forked animall as thou art. Off, off, you lendings! Come, unbutton heere.[126]

The Storm's 'all-shaking thunder' is the ultimate solvent, for this immense compassion which floods through Lear is a dissolving of categories, a recognition of the ultimate parity of all creatures. Lear rejoins the universal community from which he had stood aloof. The King has become a naked man, and man is no more than a two-legged animal, comrade to the worm, the sheep and civet. Lear has journeyed to the underworld and found it peopled with creatures; he has become nothing and found 'the thing it selfe'.

There is also the inner storm of his madness: 'Thou think'st 'tis much that this contentious storme invades us to the skin', but 'the tempest in my mind Doth from my senses take all feeling else'.[127] This inner tempest is likewise 'all-shaking'. Madness dismantles his mind into kaleidoscopic fragments, undifferentiated ideas, perceptions, reminiscences and hallucinations, jostled into random utterance –

> Nature's above Art in that respect. Ther's your presse-money. That fellow handles his bow like a crow-keeper: draw mee a cloathier's yard. Looke, looke, a mouse. Peace, peace! this peece of toasted cheese will doo't. There's my gauntlet; Ile prove it on a gyant. Bring up the browne billes. O well flowne bird; i' th' clout, i' th' clout: hewgh! Give the word.[128]

This speech might nowadays be labelled 'stream of consciousness', and this would catch its key quality, which is a flowing. One idea runs into another, boundaries of meaning are ignored. Lear's mad speeches announce a liquefaction of distinctions. Jung writes:

> The essence of the conscious mind is discrimination; it must, if it is to be aware of things, separate the opposites, and it does this *contra naturam*. In nature the opposites seek one another – *les extrèmes se touchent* – and so it is in the unconscious.[129]

Lear's fall to madness is an immersion in the indiscriminate world of the unconscious. It precisely complements his 'outer' fall – the King becoming a beggar becoming an animal, which is similarly a negating of categories, a liquefaction of hierarchy. These two storms – the storm of the elements and the tempest of the mind – combine to perform a total dissolution. Lear disappears: the King is lost in the embrace of the kingdom, the mind immersed in the ocean of the unconscious. The Storm makes the King nothing.

*   *   *

'This being done, let it be put in a furnace.' 'Set it over the fire of putrefaction and generation.' 'Burn then the brazen body with an exceeding great fire.'[130] Lear is fed into the Storm as matter into a furnace. The Storm with its 'sulph'rous and thought-executing fires', its 'wrathfull skies' emitting 'such sheets of fire, such bursts of horrid thunder', is an Athanor wherein the King as Raw Stuff is 'burned upon the fire of the art'.[131] Its supreme violence makes even the rigid King capitulate – 'heere I stand, your slave' – just as the alchemist's furnace provides the 'Heate of mighty Coaction' which will break even 'mineralls that be hard of liquefaction'.[132]

Two storms work their disruptions in Lear, the macrocosmic storm of elements and the inner tempest of the mind. So there are two fires in alchemy combining to destroy the Raw Stuff. The outer heat of the furnace – 'fire elementall' – is complemented by the inner secret fire, the *ignis innaturalis*. 'Fire against Nature', says Ripley, 'must doe thy bodies woe' –

> Therefore make fire thy glasse within,
> Whych burneth the bodie much more than fire Elementall.[133]

The fire of the furnace envelopes and intensifies the fire within. So Norton in the *Ordinall*:

> Then worketh inward heate naturall,
> Which in our substance is but Intellectuall:
> To sight unknowne, hand maie it not feele,
> His working is knowne to few Men and sield;
> And when this heate naturall moved be shall
> By our outward heate artificiall,
> Then Nature excited to labour will not cease.[134]

Sendivogius describes the same relationship between the two fires. The 'intrinsecal fire' receives 'nourishment' from the 'extrinsecal fire' of the furnace: thus 'Art purifies by a twofold heat'.[135] This seems to tie in with Lear in the Storm. The Storm envelopes him, it 'invades us to the skin', it provides the outer form of his dissolution – nakedness, beggary, bestiality. It is (to borrow that apt alchemical phrase of Queen Elizabeth's) 'the furnace of affliction'.[136] But the Storm encloses, and enacts in a cosmic arena, that central and secret dissolution within Lear. His madness is the fire within the glass, the tempest in the mind, a disintegration which 'in our substance is but intellectuall'.

The effect of both storms in *Lear* is, we have seen, to explode distinction and hierarchy, to level everything to a chaotic parity. The Storm makes Lear 'nothing' in that it merges him with all things: it removes the outer *form* which distinguishes the King from all other animals and the relevant utterance from the swirl of possible utterances beneath consciousness. This removal of form is precisely what the alchemist's fire accomplished. Philalethes describes the Raw Stuff being 'set to our fire to digest' until

> all together become a Broth, which is a mean substance of dissevered qualities, between the Water and the Body, till at length the Body burst asunder and be reduced into a powder, like to the atoms of the Sun, black of the blackest and of a viscous matter.[137]

Once again the language of alchemy provides an exotic but totally apposite terminology to apply to *Lear*. The mind of the maddened Lear becomes a *broth*, a fulmination of 'dissevered qualities': it bursts asunder into 'atoms'. Few pieces of writing could stand so well in the company of Shakespeare's 'storm scene' as these words of the anonymous Philalethes. Scorched and disintegrated by the 'fire of the art', he continues, the Stuff enters the blackness of putrefaction:

> This reduction of the Body ... ingenders so venemous a Nature, that truly in the whole World there is not a ranker Poyson, or stink, according as the Philosophers witness: And therefore he is said to cast his fell venem from his poysoned bulk; inasmuch as the exhalations are compared to the Invenomed Fume of Dragons.... And indeed it is a wonder to consider (which some Sons of Art are eye-witnesses of) that the fixed and most digested Body of Gold should so rot and putrefie, as if it were a Carcass.... This allegorically is called Death, for as a man will resist violence, which intrencheth upon his life, as long as he can, but if his Enemies are many and mighty, at length they grow too mighty for him, and he begins to fail both in strength and courage ... so our Body or Man, the Sun, like a strong champion doth resist long, till he be wounded, and bleed as it were all over, and then dies, at whose death blackness doth begin to appear.[138]

So Lear's resistance is finally overcome in the Storm. He sinks into eclipse, blackness, the alchemical death. Later Cordelia will look back on the Storm, and her father at its mercy: ' 'Tis wonder that thy life and wits at once had not concluded all.'[139] In a sense, though, they did conclude. The old, brittle, corrupt Lear died in the Storm. He arrived at that state which Cordelia herself had planted in his destiny: 'Nothing.' Seeing this immense scene on stage, our chymicall spectator may have glimpsed the vision of the 'old Philosophers':

> They saw a Fog rise, and pass over the whole face of the Earth, they also saw the impetuosity of the Sea, and the streams over the face of the earth, and how these same became foul and stinking in the darkness. They further saw the King of the Earth sink, and heard him cry with eager voice, 'Whoever saves me shall live and reign with me forever in my brightness on my royal throne.'[140]

We see that, even as he drowns in the Storm, the alchemical King speaks of his restoration. For this breakdown is the gate to new life: 'no generation without corruption'. This operation, says Philalethes, is called 'Extraction of Natures and Separation ... also Reduction to the First Matter, which is Sperm or Seed'. Lear's reduction to nothing reveals the 'sperm or seed' of a new Lear. His dark plea to the Storm – 'Cracke Natures moulds, all germaines spill at once' – goes unheeded. The germens are not spilled: they are activated.

The blackness

In the 'hell-blacke night' of the Storm, strange qualities begin to surface in Lear. As he and the Fool turn towards the hovel which offers them 'some friendship ... gainst the Tempest', Lear says:

> My wits begin to turne.
> Come on my boy. How dost my boy? Art cold?
> I am cold my selfe. Where is this straw, my Fellow?
> The Art of necessities is strange,
> And can make vilde things precious. Come, your hovel.
> Poore Foole and knave, I have one part in my heart
> That's sorry yet for thee.[141]

This is a vital speech. In the first line, Lear resigns himself for the first time to his madness: he has forseen it before – 'I shall go mad' – but now it is happening. In the next line, Lear shows for the first time a concern for someone other than himself. It is the moment of axis: madness and enlightenment, division and generation are simultaneous, inextricable, symbiotic. As the Wheel of Lear's fortunes reaches its nadir, it does not stop turning. The only way it can go now is upwards. As Edgar says, 'The lamentable change is from the best. The worst returnes to laughter.'[142] That frail human phrase – 'How dost my boy?' – is the turning point. We glimpse another Lear, gentle and concerned. As his

18 From Daniel Stolcius, *Viridarium Chymicum* (1624)

heart breaks 'into a hundred thousand flawes' he finds that 'one part in my heart that's sorry yet for thee'. It is then that Lear speaks of the mysterious transforming quality of his suffering, the alchemy of hardship which 'can make vile things precious'. He means, perhaps, that the vile hovel offers a shelter more precious than any palace. Necessity perceives a gift, where luxury breeds indifference – Lear as pampered King was prey to the very sin he castigates his daughters for: ingratitude. Gloucester finds a similar paradox in the midst of his ruin:

> I have no way, and therefore want no eyes;
> I stumbled when I saw. Full oft 'tis seene
> Our meanes secure us, and our meere defects
> Prove our commodities.[143]

But the vile thing made precious is not only the hovel, it is Lear himself. His words to the Fool, his mercurial boy, are the first hints of preciousness within the corrupt and deluded King.

Lear's great oration to the 'naked wretches' also takes place outside the hovel. In this speech too, we see how his loss of self, his disappearance into formlessness, is also a beginning enrichment: the compassion of his prayer promises the feeling and seeing Lear that will emerge. 'Take physicke, Pompe; Expose thy selfe to feele what wretches feele'. His exposure is called a medicine, physic for the sick King. He ends his speech with a plea for wealth to be redistributed to the wretched of his kingdom – 'that thou maist shake the superflux to them'. The word 'superflux' catches up the image of physic and makes it Paracelsist. Chemically, the superflux (literally, 'that which flows from above') signifies the drops of condensing vapour which flow down from the still-head to be collected, as distillate, in the receiver below. It is, in Paracelsist terms, the quintessential *arcanum* extracted from crude material. Lear takes his medicine but he also gives it, for this speech, uttered at the nadir of his kingship, becomes paradoxically his first assumption of the full responsibility of kingship. 'O I have tane too little care of this,' he cries. The King becoming a beggar is seen as an act of redemption. He brings *himself* down to the kingdom like a descending quintessential rain, a superflux of healing virtues. The compassion of the King becomes an act of alchemical benediction. Once again, the reduction and ruin accomplished by the Storm is seen to contain the promise of restoration and enrichment.

While Lear delivers his prayer outside the hovel, a strange encounter takes place in the unseen interior of the hovel. The Fool has entered into its darkness and now rushes out in terror, crying, 'Come not in heere, Nuncle; here's a spirit, helpe me!'[144] The 'spirit' is Edgar, disguised as Poor Tom, the 'Bedlam beggar' possessed by the 'foule Fiend'. Lear's first words to Poor Tom are: 'Didst thou give all to thy daughters? And art thou come to this?' He assumes an immediate kinship with Poor Tom, indeed he sees in Tom his own mirror-image – a naked madman – and so insists that their situations are identical, that 'nothing could have subdued Nature To such a lownesse but his unkind daughters'. In becoming Poor Tom, Edgar has assumed 'the basest and most poorest shape That ever penury in contempt of man Brought neere to beast'.[145] This is what Lear sees before him, the shape he recognizes as his own. Soon he is calling Tom his 'noble philosopher', a 'learned Theban', 'good Athenian',[146] instating him as a source of 'unaccommodated' truth.

As Bedlam beggar, Poor Tom releases into the play a stream of devils and demons. Flibbertigibbet, Smulkin, Modo and Mahu, Fraterretto, Hoppedance – the very names seem to bubble up from some underworld of the English language. The Fool's first terrified reaction announces Tom as 'a spirit' – if so, he is a spirit of darkness, an ambassador from the 'Prince of Darkenesse' himself. He receives daemonic messages – 'Fraterretto cals me, and tells me Nero is an angler in the Lake of

Darknesse.'[147] The sun-god Apollo, by whom Lear swore at the beginning of the play, is eclipsed. Lear himself, a crude despot like Nero, is immersed in a lake of darkness. If the Fool offers truth in the guise of folly, Poor Tom offers truth in a garb of blackness. 'The King when he enters into his Bath pulls off his robe, and gives it Saturn, from whom he receives a Black Shirt.'[148] So Philalethes describes the onset of putrefaction: the King (gold) receiving a garment of black from his opposite, Saturn (lead). Lear's identification with Poor Tom is not only a perception of his own state as naked beggar, and thence as 'a poore, bare, forked animall': it also identifies the King as a thing of darkness, indeed the very 'Prince of Darkness' to whom Poor Tom is bound. Lear, immersed in the dissolving *balneum* by Goneril and the Fool, made nothing by the furnace of the Storm and the secret fire of his madness, now receives the 'black shirt' of putrefaction.

But the Storm is, as we have seen, a turning point, a nadir pointing upwards, a place of exchange. Madness has brought the first hint of enlightenment, the tragic fall has assumed a redemptive medicinal meaning, 'physicke' for king and kingdom. And so the encounter with Poor Tom, entailing Lear's final loss of form and entry into blackness, has also its opposite meaning. For Poor Tom is Edgar, bound upon his own Wheel, the first-born made nothing – 'poore Tom! That's something yet: Edgar I nothing am.'[149] And Edgar it is who rises up at the end of the play to sustain 'the gor'd state'. Edgar becomes King, and so when Lear encounters Poor Tom, he meets not only nakedness, bestiality and blackness: he meets, clothed in these qualities, the King's Son. In the extremity of the Storm, at the intersecting nadirs of their separate Wheels, the dying King at last finds the King's Son. The dynastic seed is sown. So there is a double irony in Gloucester's shocked words at seeing Lear and the Bedlam beggar converse – 'What! hath your grace no better company?' The conventional irony is that Gloucester is speaking of his own son, Edgar. The more profound irony is that the lunatic conversation between the mad Lear and the naked Tom is in truth a discourse between kings. The old King has become nothing in order to find the new King.

*   *   *

The alchemist's reduction of the Raw Stuff to nothing – its mortification, eclipse, dissolution into a broth of dissevered qualities – marks, as in *Lear*, the turning-point of the circular *opus*. Ripley writes that

> When the bodie is from his first forme alterate,
> A new forme is induced immediatly,
> For nothing being without all forme is utterly.[150]

Nature, in other words, abhors a vacuum. As metallic form (shape, substance, identity) is surrendered, so the seeds of a new form are sown. The same idea is expressed as the symbiosis of *solve* and *coagula* – 'between the solution of the bodie and congelation of the spirit, there is no distance of time or diverse work.'[151] So it is that putrefaction, which is the epitome and climax of the Stuff's surrender to formlessness, is universally described by the alchemists as the condition out of which new forms, new life, begin to grow. Ripley begins the fifth Gate of the *Compound* with the words:

> Now we begin the chapter of Putrifaction,
> Without which Pole no seed may multiply.[152]

And he defines putrefaction as

> of bodies the fleying ... a division of things. ...
> The killed bodies into corruption foorth leading,
> And after unto regeneration them abling.[153]

A flaying, a division, a death: and afterwards the regeneration. The alchemists perceived this paradox – putrefaction as a pre-requisite for growth – as the secret of all forms of natural generation. Basil Valentine writes in the *Twelve Keys*:

> All flesh be it Mans or Beasts yeildeth no increase or propagation, unless it be first putrified; also the seed when it is sown, and all that is under or belonging to vegetables cannot increase but by putrifaction; many insects and worms receive life, so that by meer putrifaction they attain a vivifying power and motion, which ought to be deservedly esteemed as a wonder above all wonders.[154]

This was the principle the alchemist sought to harness within the microcosmic vessel: to generate a Blessed Stone out of the decomposition of a Confused Mass. The alchemical formula, 'No generation without corruption', expresses this aspiration. Shakespeare too, I think, embraced this paradox in his depiction of Lear in the Storm – in the way that madness immediately hints at enlightenment, the fall becomes a medicine; above all, in the way that Poor Tom brings blackness and final dissolution to the old King, while being himself the seed of the new King. What the 'all-shaking' Storm breaks apart will eventually be pieced together again: the King will be healed, the gored state sustained. The Storm is, in the words of Sendivogius, 'the fire of putrefaction and generation'.

This whole theme of putrefaction and generation introduces a strange cluster of images into alchemical writing. We find associated with putrefaction – and also with the agents mercurial and fiery which induce it – an imagery of fertilizing: seed, sperm, womb, mother. Sendivogius speaks of the fire ejaculating the seeds of new matter: 'Fire separates,

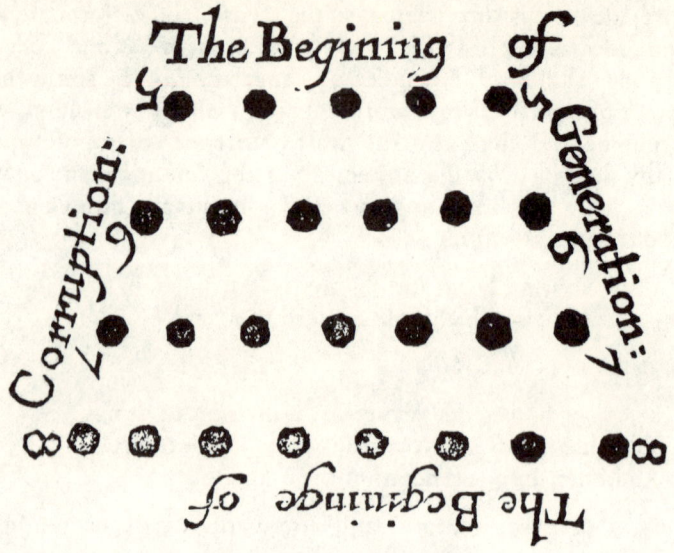

19 From Robert Fludd, *Mosaicall Philosophy* (1659)

cleanseth, digesteth, coloureth, and maketh all seed to ripen, and being ripe expels it by the sperm into places and matrixes.'[155] Thomas Tymme, in a similar vein, calls fire a 'parent' fertilizing the earth:

> Fire is the author of all formes, powers, and actions, in all the inferior things of nature, as the first cause, and carrying it selfe like the parent towards his offspring: which fier by his winde carryeth and conveyeth his seedes into the belly of the earth, whereby the generation or fruite is nourished, fostered, groweth, and is at last thrust foorth, out of the lappe or bosome of the elements.[156]

The fire which dissolves also fertilizes, pours seed into 'the belly of the earth'. Putrefaction thus becomes associated with matter in a receptive, feminine, womb-like state. So Sendivogius: 'the place or earth in which they are putrified is the Female'.[157] The lines from Donne's elegy, 'The Comparison', refer to this implied copulation of fire and earth:

> like the Chymicks masculine equall fire,
> Which in the Lymbecks warme wombe doth inspire
> Into th'earths worthlesse durt a soule of gold.[158]

So we come to the strange alchemical image of the Mother. Philalethes calls putrefaction the 'reduction to the first matter': a return of the Raw Stuff to primordial formlessness. This becomes a 'return to the mother' because, in the words of Thomas Vaughan, 'the First Matter was existent before man and all other creatures whatsoever, for she is the mother of them all'.[159] Paracelsus calls this *prima materia* the *Mysterium Magnum* and also characterizes it as a mother:

> This Great Mystery is mother of all the elements, and at the same time the spleen of all the stars, trees and carnal creatures. As children come forth from the mother, so from the Great Mystery are generated all created things. . . .[160]

The dissolution of the Raw Stuff thus becomes a return to the 'womb' of First Matter, there to be born again as the Stone. All this might help us to understand a curious turn of events in George Ripley's *Cantilena*. We met his protagonist at the beginning of the chapter – the ageing barren King, bewailing his hollow authority and infertile nature – and related him to the ailing King Lear. To redeem himself from his parlous condition, Ripley's King must perform a strange rite:

> By other meanes I cannot enter Heaven:
> And therefore (that I may be borne agen)
> I'le humbled be into my Mother's breast,
> Dissolve to what I was. And therein rest.

This, then, is his route to redemption. The King must be reduced to formless First Matter ('dissolve to what I was'): he must return to his 'mother's breast'.

> Hereat the Mother animates the King,
> Hasts his Conception, and doth forthwith bring
> And hide him closely underneath her Traine
> Till (of herselfe) sh'had made him Fleshe againe.
>
> 'Twas wonderfull to see with what a grace
> This naturall Union (made in one Imbrace)
> Did looke; and by a League both sexes knitt.

This vignette of alchemical incest is graphically insistent on the subject of rebirth. The King crawling beneath his mother's skirts is literally returning to the womb. He is smothered and enveloped: the eclipsing of the King, the onset of putrefaction and blackness, is depicted as an Oedipal darkness beneath the skirts, the darkness of the womb itself. The following verses balance putrefaction and generation:

> The Mother unto her chast Chamber goes
> Where in a Bed of Honour she bestowes
> Her weary'd selfe, 'twixt Sheets as white as snow
> And there makes Signes of her approaching Woe.
>
> Ranke Poison issuing from the Dying Man
> Made her pure Orient face look foule and wan:

> Hence she commands all Strangers to be gone,
> Seals upp her Chamber doore, and lyes Alone.[161]

A pregnancy and a death, confinement and interment, no generation without corruption.

The swallowing up of the male in an incestuous coitus is reminiscent of that other alchemical narrative, the *Visio Arislei*. There Beya embraced her brother Thabritius 'with so much love that she absorbed him completely'. Beya represents Mercury (*al baida* – the white one) and indeed the female in any alchemical *coniunctio* can be interpreted as Mercury. So too the King's mother in the *Cantilena* is Our Mercury. She must devour and swallow up the King before she can give birth to him: she is Mercury in both its dissolving and renewing aspects. Certainly 'mother' is one of the terms applied to Mercury. Sendivogius: 'in this Philosophical Work, the Mother of this thing is that Water of thine so often repeated [i.e. Our Mercury] and whatever is produced of that is produced by putrefaction.'[162] In his *Metallorum Metamorphosis*, Philalethes lists 'mother' among a whole range of synonyms for Mercury: others he gives have already been found in the language and patterns of *Lear* – 'our balm', 'true fire', 'venomous Dragon', 'a most common thing, and yet the most precious treasure'.[163]

We might appear to have strayed some way from the world of *Lear* in dwelling on this bizarre alchemical courtship between the King and his mother. Not so. I have tried to trace an analogy between the paradox of Lear in the Storm and the paradox of Raw Stuff in the extremity of dissolution and putrefaction. In both cases, a 'becoming nothing' is also a gestation of new form. In both cases a circular process reaches its nadir and begins an arduous return, a climbing up out of blackness. This has led us to the alchemical image of the mother, eclipsing and suffocating the King but also receiving him into the dark womb of rebirth: the mother as duplex Mercury. I want now to take that alchemical image and carry it back into *King Lear*, to find in the play that motif of 'the Mother'.

When Goneril first instructed Oswald to treat the King with 'negligence' – to begin, in other words, the relentless mortification of the King – she remarked: 'Old fools are babes again, and must be us'd With checks as flatteries.'[164] The 'Wheel of Generation' has turned – 'the yonger rises when the old doth fall' – and Lear has become a 'babe' again. Edmund has a similar perception of the 'correct' relationship between child and father – 'the father should bee as ward to the son, and the sonne manage his revennew'.[165] Goneril and Regan treat Lear with a harsh, bullying authority: they become a cruel 'mother' to him. This is exactly what the Fool tells Lear: 'since thou mad'st thy Daughters thy Mothers . . . thou gav'st them the rod, and putt'st downe thine owne breeches.'[166] Their cruelty to Lear is punitive: he has given them the 'rod' (an image that neatly includes his surrender of the sceptre of kingship to them) and they will now whip him with it. They will make him (again in the words of the Fool) 'an obedient father'. Among the catalogue of Lear's fatal reversals, making 'thy daughters thy mothers' is a vital one. His fall from kingly authority is mirrored in his fall from parental authority. He becomes helpless like a child.

By the alchemical notation we have assembled for the play, Goneril is the dissolving serpent-aspect of Mercury. Now she is also the 'cruel mother' who benights the alchemical King: the dark half of 'mother Mercury'. There is also, we have seen, an inner Mercury, a dissolving energy inside Lear: his madness. It first rises up as the Dragon and gains ascendancy as the secret fire, 'the tempest in my mind'. And here too we find this crucial image. In one of his first intimations of coming madness, Lear calls it *a mother*:

> Oh how this Mother swels up toward my heart!
> *Hysterica passio*! downe, thou climing sorrow!
> Thy element's below.[167]

In calling madness 'a fit of the mother' Shakespeare was using conventional medical terminology. The source of '*hysterica passio*' was believed to be the womb (the word 'hysteria' coming from Greek ὑστερα, womb). A medical pamphlet of 1605 explains that 'This disease is called . . . the Mother or the Suffocation of the Mother, because most commonly it takes them with choking in the throat; and it is an affect of the mother or wombe.'[168] In Forester's *Pearle of Practise* we find oil of white amber recommended as chemical remedy for 'the rising of the mother'.[169] But how beautifully Shakespeare works this current term into the poetic pattern of the drama. As Goneril is a cruel and punitive mother to Lear, his madness is a suffocating and eclipsing mother inside him. They combine to bring Lear to dissolution, which is done in the fire of the twofold Storm. Ripley's King says: 'I'le humbled be into my Mother's Breast, Dissolve to what I was.' Shakespeare's King is humbled and dissolved by his 'mother', at once the overbearing Goneril and the swelling fit of madness. All this is expressible by the alchemical name Mercury: serpent daughter, dragon madness, eclipsing mother: that half of Mercury which devours and dissolves.

Lear's journey into the 'hell-blacke night' of the Storm is, we now see, a journey into the primal darkness of the womb. Indeed, in the fit of his madness, Lear calls the vagina a 'hell', a 'sulphurous pit': it is a corridor of death and birth. As he stands outside the hovel, delivering that oration to the wretches that marks his immersion into formlessness, what lurks inside the hovel is Poor Tom, the new form of the King. Perhaps the hovel – at least to those eyes in the audience that saw the *Lear* stage as an alchemical landscape — actually depicts the womb itself, cradling in its dark interior the seed of the King reborn. It is the vessel of transformation, 'the Limbecks warme wombe'. The 'art of our necessities' does indeed make the vile hovel precious: it makes it the womb of new life.

The audience never sees inside the hovel. In fact, Lear himself does not enter there. It is the Fool who goes in and draws Poor Tom out onto the stage: the Fool bridges that synaptic gap between the King disappearing and the King rising. As the guiding Mercurius-Hermes, the conductor of souls to the underworld, he alone ventures into that last and darkest recess and draws forth the 'treasure hard to attain'. Led by alchemy, we have come upon many meanings for Lear in the Storm: he is journeying to the underworld, he is dissolved in the fire, he is eclipsed in *nigredo*, he is returning to the mother's womb. These might all be contained in one meaning: Lear's terrifying descent into the depths of his self. For just as Jung interprets the mythic descent into Hades as an advance into 'the unknown regions of the psyche', so too he characterizes the mother as a symbol for the unconscious: 'The mother stands for the collective unconscious, the source of the water of life.'[170] And:

> A man's unconscious is likewise feminine and is personified by the anima. The anima also stands for the 'inferior' function and for that reason frequently has a shady character; in fact she sometimes stands for evil itself.... She is the dark and dreaded maternal womb, which is of an essentially ambivalent nature.[171]

These psychological meanings invested by Jung in the symbol of the mother have their counterpart in the mother motif in *Lear*. Goneril as daughter-mother does indeed 'stand for evil itself'. And madness as a 'fit of the mother' is indeed the 'inferior function' breaking out of its confines — its 'element below' — to invade the self. Much of Jung's fascination with alchemy stems from his perception of its symbolism as a receptacle for psychic archetypes. For him, the alchemical *nigredo* was — like the hero's venture into Hades, like the archetypal image of the womb — a symbolic descent into the unconscious. He cites Michael Maier's stirring depiction of *nigredo* as a sinking King:

> Although that King of the Philosophers seems dead, yet he lives, and cries out from the deep: 'He who shall deliver me from the waters, and bring me back to dry land, him will I bless with riches everlasting.'

Jung glosses: 'the depths of the sea symbolize the unconscious state of an invisible content that is projected.'[172] The King is, as it were, sinking into the hidden psychic meanings of the symbol, beckoning the alchemist to follow. Perhaps the presence of alchemical meanings in *King Lear* might finally be seen as Shakespeare's reaching for a language to express mysterious ideas about the self, the unconscious, madness and wholeness. If alchemy provided, as Jung asserts, a vocabulary and even a method for penetrating the unconscious – if it is 'psychiatric' centuries before psychiatry existed – that is perhaps *why* it suffuses *King Lear*. *Lear* is a play that depicts giant revolutions, cosmic and dynastic upheavals, but at its heart is the story of a man – and a mind – hopelessly lost and miraculously restored. Perhaps what happens to King Lear is so like what happens to the alchemical King because both dramatist and alchemist were penetrating into an inner unknown. The symbols and patterns which the play seems to share with contemporary alchemy gave access to those mysteries.

The dew

Lear's madness must run its course: the hallucinated trial of Goneril, the topsy-turvy world of 'supper i'th' morning', the psychotic pastorale of Lear 'fantastically dressed with wild flowers', the sexual rage of 'Let copulation thrive!' and the vagina as 'sulphurous pit' (the hell that leads to the womb).[173] Lear's mind is a 'broth ... of dissevered qualities': he is 'cut to th' braines'. In Gloucester's richly alchemical phrase, he is a 'ruin'd peece of Nature'.[174] And when a devastating clarity breaks forth from this fragmented mind – 'O! matter and impertinency mix'd; Reason in madness' – that clarity is itself a fragmenting vision:

> See how yond Iustice railes upon yond simple theefe. Hearke in thine eare: change places, and, handy-dandy, which is the Iustice, which is the theefe? Thou hast seene a farmer's dogge barke at a beggar? ... And the creature run from the cur? There thou might'st behold the great image of Authoritie: a dogg's obey'd in office.
>
> Thorough tatter'd cloathes small vices do appeare;
> Robes and furr'd gownes hide all. Plate sinnes with gold,
> And the strong lance of iustice hurtlesse breakes;
> Arme it in ragges, a pigmy's straw does pierce it.
> None does offend, none, I say, none.[175]

These coruscating insights into the partiality of social organization (partial both as superficial and unequal) continue that dissolving of hierarchy which is so much a feature of the King's madness: they disclose, like the prayer to the wretches, a compassion which is also a nullification of the King. But all this 'sleep of reason' leads forward to the reawakening. It is Lear's purgatory, the blackness in which new forms gestate:

20 From Leonhardt Thurneisser, *Quinta Essentia* (1574)

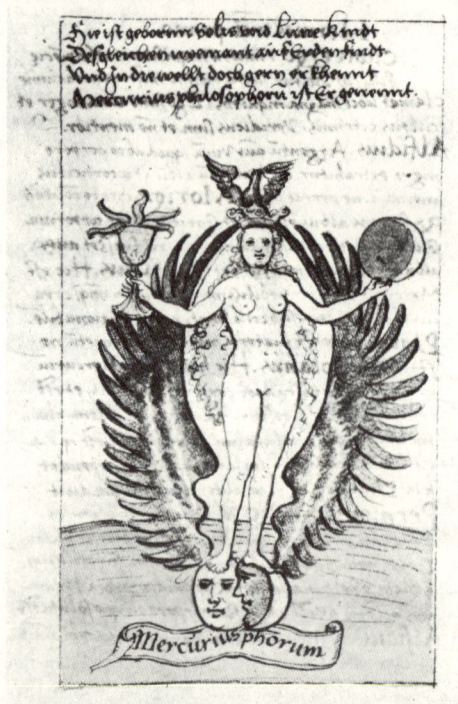

21 From anon., *Turba Philosophorum* (MS, sixteenth century)

> For like as soules after paines transitorie
> Be brought to Paradice where ever is ioyfull life,
> So shall our Stone (after his darknes in Purgatorie)
> Be purged and ioyned in Elements withouten strife,
> Reioyce the whitenes and beautie of his wife . . .
>> And thus by the gate of blacknes thou must come in
>> To light of Paradice in whitenes if thou wilt win.[176]

The purgation ends with a union, a rejoicing in the arms of the White Queen. The scene is set for the reunion of Lear and Cordelia.

Lear's estrangement from Cordelia – truth 'whipt out', the 'most choise, forsaken', the precious 'balm' accounted vile – marked the beginning of his fall. The promise of her return has hovered fitfully over the action – we saw how the Fool seems to sustain her presence and speak her truth, Cordelia in motley. There is the letter from Cordelia that Kent reads in the stocks, with its distant but definite promise that she will 'give losses their remedies'.[177] As Kent goes out to search for Lear in the Storm, he promises the Gentleman that 'from France there comes a power into this scatter'd kingdom', and leaves him with the instruction:

> If you shall see Cordelia,
> (As feare not but you shall) shew her this ring.[178]

And then at last, neither too soon nor too late, the promise is fulfilled: Act IV, scene iv, the French camp near Dover, 'Enter, with drum and colours, Cordelia, Doctor, Gentlemen and Soldiers.' Her triumphal re-entry, with a retinue incarnating her own healing efficacy.

The poetry spoken by the returning Cordelia expresses again and again her arrival as healer and redeemer:

> What can man's wisedome
> In the restoring his bereaved sense?

> All blest secrets,
> All you unpublish'd vertues of the earth,
> Spring with my teares! Be aydant and remediate
> In the goodman's distress.

> O you kind Gods,
> Cure this great breach in his abused Nature!
> Th'untun'd and iarring sense, O winde up
> Of this childe-changed father.

> O my deere Father! Restauration hang
> Thy medicine on my lippes, and let this kisse
> Repaire those violent harmes that my two sisters
> Have in thy reverence made![179]

Cordelia restores what is bereaved, heals what is broken, remedies distress: her tears descend on the blackened earth like dew, drawing forth its secret *arcana*, 'unpublish'd vertues'. She is the King's physic, indeed a Universal Medicine, the

> one daughter,
> Who redeemes Nature from the generall curse
> Which twaine have brought her to.[180]

She assumes at last her true role, the role defined in the first scene of the play: a balm for the sick King, that 'balsam' which Thomas Tymme calls 'the only mean to conserve our life . . . the only immediate putter away of sicknesses'.[181]

As restoring remedy, Cordelia confirms her opposition to the tearing caustic, Goneril and Regan, the 'twaine' whose curse she redeems. As they stripped Lear down to nothing and fed him into the fire, so Cordelia reclothes him – 'In the heaviness of sleep We put fresh garments on him' – and regenerates him:

> He wakes . . .
> How does my Royall Lord? How fares your Maiesty?

The address – 'royal Lord', 'your Majesty' – is itself a reinstatement. Lear believes he has woken from death: 'You do me wrong to take me out o' th' grave.' He is right: the old Lear is dead. He believes Cordelia is a spirit: 'Thou art a soule in blisse. . . . You are a spirit, I know.' Again he is right: Cordelia is the precious *anima* he cast out, now descending to revive him. Lear's frail awakening words confirm his regenerate nature:

> I am mightily abus'd. I should ev'n dye with pitty
> To see another thus. I know not what to say. . . .
> Pray, do not mocke me:
> I am a very foolish fond old man.

The tone of compassion and humility is matched in the touching gesture of his kneeling before Cordelia. The crystalline moment of his return is signalled with great delicacy:

> For as I am a man, I think this lady
> To be my childe Cordelia

The words grope until they touch. After all the dissolution of categories comes their perilous rebuilding: an act of dawning consciousness – man, lady, child – finally focusing on the face and speaking the name of Cordelia. Lear is awake. He made his daughters his mothers: Goneril to destroy him, Cordelia to give second birth to him.

* * *

Cordelia calls Lear a 'childe-changed father'. It is her return which completes the transformation of father by daughter, Raw Stuff by Mercury. She had been cast forth – the flight of the *anima mercurii* expelled as vapour from the corrupt chaotic Stuff. She had hovered in present-absent promise while the other Mercury, the caustic serpent and secret fire, had performed its ruinations. Now she descends to reanimate – to return the *anima* to – the eclipsed King and thus completes the double nature of Mercury, destroyer and restorer:

Mercury cures the holes which it has provoked.

Mercury is our field, in which the Sun rises and sets.

Our living western quicksilver, which has placed itself above the gold and vanquished it, is that which kills and quickens.

O blessed watery pontic form, that dissolvest the elements! ... For when, by the power of the water, the composition is dissolved, it is the key of the restoration; then darkness and death fly away.[182]

The echoes are unmistakable. As Mercury can 'cure the holes it has provoked', so Cordelia can 'cure this great breach' that her sisters have provoked. As Mercury is 'the key of the restoration', so Cordelia bids 'restauration hang thy medicine on my lippes': her mouth restores what Goneril's, with its 'sharp-tooth'd' property, had dissolved. Her kiss breathes life into Lear – the Mercurial spirit as 'divine breath'. 'All things have need of this spirit,' writes Thomas Tymme (quoting the Hermetic *Asclepius*), 'for it carryeth all things, and it quickneth and nourisheth all things, according to the dignitie of each thing it selfe.'[183] Cordelia quickens and nourishes the King with a queen's kiss: the 'divine breath' is perhaps her love itself. Her nature as loving rescuer could not be better expressed than by the words of Sendivogius' Mercury: 'Whatsoever is with me, I love as a Friend; and whatsoever is brought forth with me, to that I give nourishment; and whatsoever is naked I cover with my wings.'[184]

As Cordelia's kiss is air, her tears are water – 'vertues of the earth, Spring with my teares'. The image is of her tears as a life-giving water fertilizing the earth, generating in it secret and invisible 'virtues'. The alchemists too spoke of the return of the Mercurial spirit as a 'celestial rain', or 'dew' (Norton's *stilla roris madidi*, Dee's *ros caeli*), a 'divine water' falling from the 'heaven' of the vessel to reanimate the blackened earth of the Stuff. The *Turba Philosophorum* speaks of 'The Dew' which 'is joined to him who is wounded and given over to death'. The *Hydrolithus Sophicus* of 'our Pontic and Catholic water, which in its refluent course irrigates and fertilizes the whole earth',[185] the *Tractatus Aureus* of condensing vapours which 'watered the earth as fertilizing dew, and washed our bodies, which became more beautiful and white'.[186] Philalethes writes of these healing waters as something patiently awaited 'for indeed a long time' –

> so they will seem to the Artist who attends the fire day by day, and yet must wait for the fruit with Patience, till the Heaven have showred down upon the Earth the former and latter Rain: yet be not out of heart, but attend until the compleatment, for then a large Harvest will abundantly recompence all thy toyl.[187]

Such encouragement in the midst of darkness is offered by the Fool and Kent, those who keep alive the promise of Cordelia. The fulfilment of the promise is an occasion of great beauty, on stage or within the chemical theatre of the vessel:

> There rises upward a mist of dazzling whiteness, whence there is distilled upon the earth a pure, sweet, and fragrant dew, that softens it and stirs up great winds at its centre; these winds bear our Stone upward, where it is endowed with heavenly virtue.[188]

As descending Mercurial dew, Cordelia brings life to the sleeping, corpse-like body of her ruined father. Exactly this scenario is found in Lambspringk's *De Lapide Philosophico*. The Father (another of our alchemical Kings) has devoured his son (another version of the incestuous *coniunctio*) and now prays God to restore his son (i.e. for himself to be reborn):

> God hearkens to his prayers,
> And bids the Father lie down and sleep.
> Then God sends down rain from heaven
> To the earth from the shining stars.
> It was a fertilizing silver rain,
> Which bedewed and softened the Father's Body.
> Succour us, Lord, at the end,
> That we may obtain Thy gracious gift!
> The sleeping Father is here changed
> Entirely into limpid water,
> And by virtue of this water alone
> The good work is accomplished.
> There is now a glorified and beautiful Father,
> And he brings forth a new Son.[189]

Again the imaginations behind *King Lear* and the alchemical writings of the day touch hands: Cordelia's tears as 'fertilizing silver rain'; the sleeping Father transformed and reborn.

It is a *spurned* Cordelia that returns and revives. This is a measure of her special love, that it is somehow redoubled by the injuries Lear did her. The characterization of Cordelia as *anima mercurii* plays on this very paradox, making Lear's rejection of her an exact requisite for his transformation. It is *only* by being expelled that the Mercurial spirit is released into its transforming role: this 'expulsion out of its habitation by Vulcan'[190] is a violent and fiery affair, but its promised end is redemptive. The casting-out of Cordelia is the very operation which makes her a spirit, a quintessence, an *arcanum* for the sick King: 'Medicine must be reduced to air, that it may readily be ruled by the stars. Can a stone be lifted by the stars? No, unless it be volatilized.'[191] Truth, we see, must be 'whipt out' before it can prevail, the precious spirit accounted vile before it can transform: the paradoxes of *Lear* are alchemical paradoxes, the shape of its tragic destiny an unfolding *opus*. Lear's route to wholeness was assured by that first act of 'hideous rashness'. It was a breakdown, a breaking-apart of the four elements to reveal the spiritual fifth. After her flight, that quintessence now returns to heal the broken King. As Norton writes, in language redolent of Cordelia:

> That cheefe liquor
> Was *Aqua Vitae* Elixir to succour;
> For she was spirituall, and would revive
> Dead things fro death to live;
> Shee was Quintessence, the fift thing.[192]

Cordelia might, finally, be visibly recognizable as the *anima mercurii* represented in Thurnheisser's *Quinta Essentia* (1574) or as the *Mercurius philosophorum* from the sixteenth-century German manuscript *Turba* (figs. 20 and 21). Perhaps the most 'Cordelian' image of Our Mercury is to be found in *De Alchimia* (fig. 22) – the crowned princess cradling the dragon, mounted on a pedestal that confirms her identity both as volatile spirit (wings) and life-giving water (the gold and silver fountains). This emblematic portrait is mobilized in the other illustration from this text: here the maiden pours the water of life over the sleeping King, from whose mouth issues a son (or spirit, or himself reborn). This emblem might illustrate the lines of Lambspringk about the falling 'silver rain' and the sleeping father who 'brings forth a new son'. These are perhaps some of the faces and attributes of Cordelia as she figured on the retina of Shakespeare's imagination. Ripley's *Cantilena* also provides us with a visual image of the *anima mercurii*. As the Mother finally delivers the King into second life – 'from the shade of Night . . . doth a Ruddy Nature spring T'enjoy the Merry Scepter of a King'[193] – his condition as *corpus subtile* is expressed in an heraldic alchemical vision:

> Foure Elements, Brave Arms and polish'd well
> God gave him: In the midst whereof did dwell

22 From anon., *De Alchimia* (MS, sixteenth century)

> A Crowned Maid, ordained for to be
> In the fifth Circle of the Mystery.
>
> With all delicious Unguent flowed she
> When purged from Bloody Menstruosity:
> On every side her Count'nance brightly shone,
> She being adorn'd with every precious Stone.

This 'crowned maid' (*virgo redimita* in the Latin original) is Mercury as quintessence ('in the fifth circle of the Mystery'). Like the Mercurial princess of the *De Alchimia* illustration, she is represented nursing the Dragon (here 'a Lyon greene') which is the chaotic matter she redeems (as well as being an image of her own 'other nature' as dissolving Mercury):

> A Lyon Greene did in her Lapp reside
> (the which an Eagle fed) and from whose side
> Blood gushed out: the virgin drunck it upp,
> Whilst Mercury's hand did office of a Cupp.
>
> The Milk (admir'd) she hastened from her Breast,
> Bestowed it frankly on the Hungry Beast,
> And with a Sponge its face she likewise dry'd
> Which her own Milke had often madefy'd.

This has its esoteric details (the *lac virginis*, milk of the virgin, is yet another synonym for Mercury as fertilizing dew), but let us consider it for a moment as a bizarre scenario. The dying monster in the lap of the crowned maid, she drinking up the blood from his wounds, suckling him at her breast, washing his brow. It is an image of the dying being nursed into new life. So Cordelia cradles the exhausted Lear – the wrathful Dragon who cast her out at the beginning of the work – and restores him with a kiss. Ripley concludes:

> Thus she triumphantly of Kings is Chiefe,
> Of Body's sick the only Grand Reliefe;
> Such a Reformist of Defects, that shee
> Is worshipped by Men of each degree.
>
> To Priests and Kings she yields an Ornament
> The sick and needy sort she doth content:
> What Man is he will slight so Rich a Store,
> As drowns the very thought of being Poore?[194]

As Cordelia 'redeems the general curse' and 'gives losses their remedies', so this Mercurial princess is 'the only grand relief' and 'reformist of defects'. Lear has slighted so rich a store once but will never do so again.

The Stone

Ænigma Regis.
**Hie ist geboren der Keyser aller ehren/
Kein höher mag vber jn geboren werden.**

23 From anon., *Rosarium Philosophorum* (1550)

'The great rage, You see, is kill'd in him.' The reawoken Lear is tender and serene:

> Be your teares wet? Yes, faith. I pray, weepe not:
> If you have poyson for me, I will drinke it.

> Do not abuse me... You must beare with me.
> Pray you now, forget and forgive:
> I am old and foolish.[195]

Compassion and humility: the virtues seeded in the Storm now brought to fruition by the 'silver rain' of Cordelia's tears. Lear never again lets Cordelia out of his sight: the precious is now revealed to him as precious. When we meet them next they are prisoners in the hands of Edmund's army. His misfortune as King has not yet ended, but the transcendent 'inner royalty' of the man quite overshadows his outer helplessness. As the all-powerful King of the play's beginning was decrepit and deluded, so the powerless King now committed to prison is unassailable and all-seeing.

> Come, let's away to prison;
> We two alone will sing like Birds i' th' Cage:
> When thou dost aske me blessing, Ile kneele downe
> And aske of thee forgivenesse: so wee'l live,
> And pray, and sing, and tell old tales, and laugh
> At gilded butterflies, and heere poore rogues
> Talke of court newes; and wee'l talke with them too,
> Who looses and who wins; who's in, who's out;
> And take upon's the mystery of things,
> As if were Gods spies: and wee'l weare out,
> In a wall'd prison, packs and sects of great ones
> That ebbe and flow by th' Moone.[196]

This is a transcendent speech: blessing and atonement, prayers and songs and stories. His words transform the prison that awaits them. It becomes a palace filled with the buzz of court gossip, an orchard full of golden butterflies, a temple in which to penetrate divine mysteries. The 'wall'd prison' is made into a whole self-contained world. Lear's love for Cordelia performs the same alchemy that Donne writes of in 'The Good-morrow':

> For love, all love of other sights controules,
> And makes one little roome, an every where.

Lear and Cordelia stand before their captors indissoluble and untouchable, like Donne's lovers, who say: 'Let us possesse one world, each hath one, and is one.'[197] The Wheel has taken Lear from palace to hovel and now returns him to a prison richer than any palace, an 'every where' filled with Cordelia. We will, he says, 'take upon's the mystery of things': the occult will become manifest, the spirit will be revealed in matter. Thus exalted as *corpus subtile*, they will be incorruptible and 'weare out ... packs and sects of great ones That ebbe and flow by th' Moone'. They will outlive all temporal kings, whose materiality makes them prey to shifting tides and transience, 'dull sublunary' kings.[198] As they are about to be led away Lear renews his promises:

> Upon such sacrifices my Cordelia,
> The Gods themselves throw incense. Have I caught thee?

> He that parts us shall bring a brand from heaven,
> And fire us hence, like foxes. Wipe thine eyes;
> The good yeares shall devoure them, flesh and fell,
> Ere they shall make us weepe.[199]

Three promises: the influx of godly virtues; indissoluble unity; impregnable strength — precisely the qualities demanded of a true King; precisely the qualities invested in the Philosophers' Stone.

This is the true restoration of Lear. He reigns over that kingdom of the mind which has undergone such division, usurpation and anarchy — a whole revolution of madness. He is not actually reinstated as King until a few moments before the end of the play, when Albany delivers up the kingdom:

> What comfort to this great decay may come
> Shall be appli'd. For us, we will resigne
> During the life of this old Maiesty,
> To him our absolute power.[200]

Lear ascends once again the throne he had squandered in the opening scene of the play. He is 'every inch a king'. At this point, in the words of the dying Edmund, 'the Wheele is come full circle'.

*   *   *

Lear and Cordelia side by side at the centre of the stage, suffused with a magical power that confounds their misfortune, is Shakespeare's masterly dramatization of that favourite alchemical emblem of the Stone — the Red King and the White Queen inseparably linked in the Chemical Wedding. The version from the *Rosarium* (fig. 23) bears the legend '*Aenigma Regis*': the enigma of kingship. The Stolcius emblem (fig. 24) is labelled '*Rebis*': two things joined into one transcendent nature. Both represent the royal pair standing over the venomous dragon, showing their conquest of chaotic materiality: 'the Dragon and his wrath' overcome. The *Rosarium* shows in the background a Pelican, symbol of the circular and self-devouring route of the *opus*: ' 'twas this flesh begot those pelicane daughters.'[201] The Stolcius has a globe full of numbers and shapes and the pair hold geometric instruments — instruments, like the chemical apparatus of alchemy, with which the Renaissance magus might probe 'the mystery of things', and 'spy' on God as manifest in mathematical harmony.[202]

Among the many descriptions of this *mysterium coniunctionis* in alchemical literature, a passage from the *Tractatus Aureus* of 1600 is especially appropriate to *Lear*, because it represents the royal pair as King and *daughter*:

24 From Daniel Stolcius, *Viridarium Chymicum* (1624)

When we marry the crowned King to our red daughter, and in a gentle fire, not hurtful, she doth conceive a son, conjoined and superior, in it, and he lives by our fire. But when thou shalt send forth fire upon the foliated sulphur, the boundary of hearts doth enter in above it, let it be washed from the same, and the refined matter thereof extracted. Then he is transformed, and his tincture by help of the fire remains red as flesh. But our son, king-born, takes his tincture from the fire, and death even and darkness, and the waters flee away. The Dragon, who watches the crevices, shuns the sunbeams, and our dead son will live; the King comes forth from the fire and rejoices in the espousal; the occult treasures will be laid open and the virgin's milk whitened. The son, already vivified, is become a warrior in the fire and over the tincture super-eminent. For this son is himself the treasury, even himself bearing the Philosophic Matter. Approach, ye sons of Wisdom, and rejoice; let us now rejoice together; for the reign of death is finished and the son doth rule, and now he is invested with the red garment, and the purple is put on.[203]

The King's Son is, of course, the King himself reborn and triumphantly reinstated, clothed in 'the red garment', the Stone attained. In this incestuous *coniunctio*, the King again makes 'his daughter his mother': 'she doth conceive a son, conjoined and superior', the daughter giving second birth to the father, as in *Lear*. We find in this remarkable piece of alchemical prosody many of the symbols and processes which, in the course of this chapter, have been associated with *Lear*. The fire which transforms, but first brings 'death even and darkness'; the life-giving waters that 'flee away' and then return (as 'virgin's milk') to nourish and vivify the reborn King; the Dragon lurking in the 'crevices' of matter, finally overcome by the 'sunbeams' of Sol rising. Finally, 'the King comes forth from the fire and rejoices in the espousal' – Lear released from his 'wheel of fire' to stand beside, 'espoused' to, his daughter Cordelia. 'The occult treasures will be laid open': King and daughter 'take upon's the mystery of things, As if we were God's spies'.

In the other *Tractatus Aureus*, Grasshoff's published in 1625, the depiction of the final *coniunctio* offers further tantalizing echoes of Lear and Cordelia. There is again a reviving of the King with special waters:

> As he had passed the dog days in great heat, he was extremely thirsty, spent and weary; wherefore he humbly requested me to bring him some water from the river. . . . After slaking his thirst with a deep draught, he returned to the chamber, requesting me to shut the door carefully so that no one might disturb him or rouse him from his sleep. So he slept a few days, and then called me back to open the door again. He looked much more beautiful, ruddier and more royal, and said that this water was very precious and full of virtue.[204]

This seems to suggest the awakening of Lear: the 'spent and weary' King, the water 'precious and full of virtue' that revives him, the healing sleep ('Our foster-nurse of Nature is repose, The which he lackes,' says the Doctor to Cordelia; sleep 'will close the eye of anguish'[205]) Now the renewed King of the *Tractatus* steps forth, 'more beautiful, ruddier and more royal' than before. The element of personification is particularly strong in this parable, and so we are treated not only to an emblematic vision of the King donning 'that most precious scarlet garment', but also to his 'behaviour' as a human King. When he speaks, the alchemist-narrator is 'filled with awe by the majesty and persuasive speech of the King'. He marvels at the bearing of this transcendent King: 'he bore himself so kindly, so gently, and so humbly, that I could not help reflecting that these virtues are the most glorious ornaments of the great.'

These are surely our own reactions to Lear's 'prison speech': awed by its majesty persuasiveness, touched by its gentleness and humility. Again we see how these rudimentary vignettes of alchemical allegory seem to offer skeletal versions of certain moments in *King Lear*, suggesting that *Lear* might be seen as a refined and powerful exploitation of the intrinsic dramatic potential of alchemy. It is in this *parabola* from the *Tractatus Aureus* that the White Queen sums up the *opus* in terms that echo the process of tragedy in general and *Lear* in particular: she says that God (in *Lear*, 'the gods')

> has power to set up and pull down kings. He makes rich and poor as He wills. He has killed and raised again. I was great and was brought low ... after death life has been restored to me.[206]

So the gods in *Lear* are at once destructive ('As flies to wanton boyes, are the we to th'gods; They kill us for their sport'); presiding ('The gods are iust, and of our pleasant vices Make instruments to plague us'); and healing ('O you kind Gods, Cure this great breach in his abused nature').[207]

With the final reinstatement of Lear as King, we look back over the whole sweep of the play. The pattern of his fortunes might be summed up by any number of epigrammatic synopses of the *magnum opus*:

> It has its birth in the earth, its strength it does acquire in the fire, and there becomes the true Stone of the ancient Sages.
>
>> His life and body are both devoured
>> Until at last his soul to him ristored
>> And his volatile Mother is made one,
>> And alike with him in his own Kingdome.
>> Himself also vertue and power hath gained
>> And far greater strength than before attained.
>> In old age also doth the Son excell
>> His own Mother, who is made volatile
>> By Vulcan's Art.
>
> These Imperfect Bodies are not reducible to Sanity and Perfection unless the contrary be operated in them; that is, the Manifest be made Occult, and the Occult Manifest: which operation, or contrariation, is made by Preparation. Therefore they must be prepared, Superfluities in them removed, and what is wanting supplied; and so the known Perfection inserted in them.
>
> I am old, debilitated and ill, my surname is Dragon; because of this, I have been shut up in a tomb so that I may attain the Royal Crown and enrich my family.

Out of the gross and impure one there cometh an exceeding pure and subtle one.[208]

We might not readily see *King Lear* in these terms, because they are a language we no longer use. It is a language which belongs to the early seventeenth century (with one exception, these extracts appeared in works published between 1597 and 1613). The language of *Lear* belongs likewise to that time, however portable to other eras it has since proved to be. Language as a way of naming is both effect and cause of a way of seeing. If the meaning of a play is what any given audience sees in its enactment, then *King Lear* is, among many other things, the most extraordinary alchemical myth ever written.

* * *

We have not, however, quite finished the story of *Lear*. The Wheel has indeed come full circle, Lear is restored to the throne, and 'after the Eclipses in rednes with glorie', is 'King to raigne upon all mettals and Mercurie'.[209] But all this takes place amid no rejoicing. Instead, 'all's cheerlesse, darke and deadly.'[210] Lear does not even hear the words which announce his final restitution, because that cruel thing has happened which Dr Johnson confessed 'so shocked' him that he scarcely 'endured to read again the last scenes of the play':[211] the strangling of Cordelia by Edmund's Captain. These last moments are a return which grimly echoes the outset: Lear is King again at the centre of the stage, his three daughters are ranged around him, only this time they are all dead.

> Cordelia, Cordelia, stay a little. Ha:
> What is't thou saist? Her voice was ever soft,
> Gentle and low.

The soft voice Lear strains to hear says just what it said at the beginning of the play: nothing. He hunches over her body, knowing she is dead but willing her to live:

> O you are men of stones;
> Had I your tongues and eyes, Il'd use them so
> That heaven's vault should crack. She's gone for ever.
> I know when one is dead, and when one lives;
> She's dead as earth. Lend me a looking-glasse,
> If that her breath will mist or staine the stone,
> Why then she lives.[212]

This might be a crescendo of redemptive alchemical images. 'You are men of stones!' might echo Dorn's 'Transmute yourselves into living philosophical stones'. Breath staining the Stone might suggest the influx into matter of that 'divine breath ... which vivifies all things that exist'.²¹³ But the crescendo is quite deadened, possible meanings breaking against deadly reality and dispersing. The spirit is 'dead as earth'.

What are we to make of this marring of Lear's triumph, this severing of the conjoined pair into a forlorn duality of hanged woman and broken-hearted old man? It seems to amputate the play's redemptive pattern at the very moment when the pattern is completed. It seems the Stone turns out be the kind that always falls crushingly back on Tantalus (a story often used to allegorize the futility of alchemy). Or that Lear has returned from the underworld only to lose at the last moment, like Orpheus, the Euridyce he brought back. To the characters around Lear and Cordelia – Kent, Albany, Edgar: those still remaining in 'this tough world' – this is how it seems. Yet the whole impetus of this final scene has been to place around Lear and Cordelia a kind of magic circle, a 'wall'd prison' that excludes this tough world and replaces it with another, a *novus orbis* like that Donne's lovers achieved – 'She is all States, all Princes I, Nothing else is.'²¹⁴ What then is happening inside this circle? We can say that Lear stakes everything on the possibility that Cordelia is still alive –

> If that her breath will mist or staine the stone,
> Why then she lives.

> This feather stirs; she lives! if it be so,
> It is a chance which does redeeme all sorrowes
> That ever I have felt.²¹⁵

Mist on a looking-glass, the stirring of a feather: Cordelia's kiss has been the breath of life to Lear and now he looks for nothing else but her breath, joying in its imagined presence, grieving over its actual absence. Lear's very last words insist that Cordelia *is* breathing – 'Do you see this? Looke on her, looke, her lips, Looke there, looke there!' He dies then in joy, all sorrows redeemed: he dies seeing Cordelia's breath. Breath and spirit are one and the same, so his last words are a vision of Cordelia's spirit issuing from her. He calls for everyone else to look: this is his last command as King. Those who remain on 'this great stage of Fooles' will say that he was wrong, that no breath came to stain the stone. This is only as much as to say that spirit is invisible. Nature, says Paracelsus, 'is not visible, though it operates visibly; for it is simply a volatile spirit, fulfilling its office in bodies'.²¹⁶ Its office completed, itself revealed, it has no further need of the body. Cordelia's death is indeed 'the promis'd end':²¹⁷ it makes the occult spirit manifest, which revelation is also the moment of Lear's death. The moment is felt as 'tragedy' by all except Lear himself.

The re-enthroned Lear has made just one speech before his death. His 'restoration address' to the kingdom consists of these words:

> And my poore Foole is hang'd: no, no, no life?
> Why should a dog, a horse, a rat have life,
> And thou no breath at all? Thou'lt come no more,
> Never, never, never, never, never!
> Pray you, undo this button. Thanke you Sir.
> Do you see this? Looke on her, looke, her lips,
> Looke there, looke there![218]

It is an extraordinary speech for a new King to make. It is full of nothingness: no edicts are made, no laws are passed, no subjects of the kingdom mentioned other than an absent Fool, a dog, a horse and a rat. It is somehow ethereally negative – 'no life', 'no breath', 'no more' – and contains what must be the most minimal iambic pentameter ever written. Lear does not so much die as disappear. His last speech is an evaporation, a mist which momentarily sustains an illusion that it is Lear on whom the mantle of new kingship has fallen. When the mist clears and Lear is gone, we see that it is really Edgar, the discovered King's Son, who assumes the vacant crown. He will 'rule in this realme, and the gor'd state sustaine'. The rule of the new King begins: the old King's task is completed by his death. The gene of royalty has been safely delivered. 'Let us now rejoice together,' says the alchemist, 'for the reign of death is finished, and the son doth rule.' His mystical optimism would see this accomplished at the end of *King Lear*, and we who are not mystics – not even optimists – cannot deny that we carry away from the play a strange sense of rejoicing.

Coda: Thomas Tymme's chymicall pathology

Alchemy opens, to borrow its own metaphor, a 'secret entrance' into the heart of *King Lear*. I have called this reading of *Lear* an alchemical 'reconstruction', believing that I am restoring (rather than importing) meanings to the play. In other words, I think alchemy played a central role in Shakespeare's own thinking about *Lear* the play and Lear the man. It is partly to approach this awe-inspiring business – the actual creation of *King Lear* – that I conclude the chapter with a brief 'coda', focusing on a particular Paracelsist feature of the play.

Among many possible approaches to *King Lear*, there is the pathological. Lear goes mad. In Jacobean medical parlance, he suffers a 'fit of the mother', *hysterica passio*. A more modern study of Lear as 'case-history' diagnoses 'senile dementia with attacks of acute mania'.[219] The word 'insanity' simply means sickness, a loss of *sanitas* or health, and like all other sicknesses Paracelsus held strong views on its causes and

25 Creede's device. From the title-page of Thomas Tymme, *The Practise of Chymicall Physicke* (1605) [enlarged]

treatment. According to him, lunacy spelt the victory of '*viehischer Verstand*', literally the 'bestial intellect'. It marked 'the subjugation of man and his divine spirit by his low animal instincts'.[220] Certainly this general conception of madness as something ascending from the lower instinctual reaches of the self is echoed in *Lear*:

> Oh how this Mother swels up toward my heart!
> *Hysterica passio*! downe, thou climing sorrow!
> Thy element's below.[221]

Throughout the play, this engulfing ascent is echoed by an imagery of man bestialized – 'a dogg's obey'd in office', 'think a man a worme', 'humanity must perforce prey upon itself like monsters from the deep', etc. In both king and kingdom, *'viehischer Verstand'* wins a terrifying though temporary victory.

Paracelsus also drew analogies between this uprising of madness and the volatile ascent of certain chemical vapours. There is a 'sulphur of vitriol' which lies dormant in the body, but, if ignited, it rises in fumes to the brain. Overpowering the 'cells which control reason', it 'stupifies, intoxicates, corrodes'.[222] It is tempting to suggest that when Lear speaks of the 'sulphurous and thought-executing fires' of the storm, there lurks a covert Paracelsist analogy with the sulphurous and stupefying fumes of madness. This might seem a tenuous connection, until one turns to that English Paracelsist text, the *Chymicall Physicke* of Thomas Tymme. He too writes of these vapours which 'doe hurt the braine, dazeling, dulling or troubling the spirits ... by fumes which are sulphurus and stupefactive'. What is more, he describes their effects in terms of *a storm*:

> Even as the vapours and exhalations Sulphurous ... carried up out of the earth into the ayre and cloudes doe cause fiery Meteors, Corruscations, Lightnings, Thundrings, Comets and such like: even so also in our bodyes, from the ... divers tartarous, sulphurous and nitreous fumes with the which our bowels doe abound, the like meteors are produced. For such fuming matter ... being stirred up by an immoderate and feaverous heate, being at the last lifted by and carried into the braine and therein set on fire, stirre(s) up meteors, long madnesses, burning Phrenzies, setled melancholies, dotings, paines of the head, falling sicknesses and many such like.[223]

The connections Tymme makes are exactly the same as Shakespeare's in *Lear*, where the macrocosmic storm of the elements figures forth the 'tempest in my mind' of madness. The similarity is still more suggestive when one remembers that Tymme's *Chymicall Physicke* was published, in London, the year Shakespeare was writing *King Lear*. Quite possibly, this passage influenced Shakespeare's whole conception of the Storm as the 'arena' for Lear's madness. And perhaps, impressed by Tymme's description of symptoms in 'the little world, man' which 'represent the winds, raynes and earthquakes of the earth', Shakespeare half-consciously echoed his words in describing how Lear 'strives in his little world of man to outstorm The to-and-fro conflicting wind and rain'.[224]

Tymme's analogy between storm and madness traces back (through Quercetanus) to Paracelsus himself. In his *Liber de Caducis* ('On Falling Sickness'), Paracelsus compares epilepsy to a thunderstorm, because both develop in a 'shell' and burst when they are 'ripe': 'when the earthquake matter contained in it has matured, an earthy thunder spells imminent destruction.'[225] This conception of madness as a mounting pressure within, finally shattering its 'shell', again chimes in with the *passio* of Lear:

> O sides! you are too tough;
> Will you yet hold?
>
> This heart
> Shal break into a hundred thousand flawes.[226]

As Lear speaks those last words, the first crack of thunder is heard. The shell explodes into flaws: the fit and the storm are upon him. In the words of Paracelsus:

> Then thunder breaks out, moving heaven and earth, when the eyes are flashing and nothing but fire is sensed by them. As thunder splashes rain, so the patient brings out froth. The flash as well as the wind develops pressure which may break walls and disrupt everything. It is the same power which beats, breaks and curves the limbs.[227]

The storm in Lear — an 'all-shaking thunder' to 'strike flat' the self and the world — sounds much like this harrowing epileptic tempest of Paracelsus.

'*Wie die Kranckheit ist, also ist auch die Artzney*': as the disease, so also the physic. If the presentation of Lear's mental 'disease' has its strong Paracelsist echoes, so too has the physic which cures it. Cordelia is herself called Lear's 'balm' — i.e. that Paracelsist balsam which Tymme calls a 'quickening Nectar', a 'permanent and certain spiritual water of life' which can 'repair and increase' the afflicted spirit of the patient 'by reason of the simpathy and common likeness therewith'. Is there, Tymme asks,

> any pain or griefe that would be asswaged? This medicine shall be thy mittigating anodine and most healthsome Nepenthes. . . . What more speedy altering medicine can be found, which is able to correct a distemperature, then this most temperat remedy?[228]

All this seems to describe Cordelia's healing potency — her 'restoring his bereaved sense' and repairing 'violent harms' — describes it, too, in terms which catch something of the poetic atmosphere of Shakespeare's reunion scene. Cordelia as 'medicinall balsam' is, of course, totally entwined with her alchemical identity as redeeming refluent Mercury — indeed Paracelsus goes as far as to say that 'all physic and remedy is Mercury' and that 'the doctor who purges, consolidates and cures is Mercury'.[229] This last reminds us that, standing beside Cordelia in this scene, is the Doctor himself. He too has his Paracelsist traits. His first words are a prescription:

> Our foster-nurse of Nature is repose,

> The which he lackes; that to provoke in him,
> Are many Simples operative.[230]

Thus the Doctor invokes Nature as the healing agent, entrusts the patient to her 'foster-nurse' sleep, and recommends medicinal herbs ('simples operative') for sedation. In this he follows a fundamental tenet of Paracelsus:

> Nature is the physician, not you; from her you take your orders, not from yourself; she composes, not you. See that you learn where her pharmacies are, where her virtues are written, and in what boxes they are kept; not in *Mesue*, not in *Lumine*, not in *Praeposito*.[231]

True physic is not found in the traditional pharmacopoeias, but is disclosed by Nature herself. As Tymme puts it:

> Nature is the beginning and, as it were, the first moving of all curing; because without the strength and vigor of Nature, all medicine is unprofitable and vaine. . . . The Lord hath created medicine out of the earth, and he that is wise will not abhorre it.[232]

It is after the Doctor's 'prescription' that Cordelia prays for secret medicinal virtues to 'spring from the earth'. She knows where Nature's pharmacies are.

*   *   *

It is hardly surprising to find Paracelsist tones in a play so suffused with alchemy as *Lear*. Paracelsism was integral to the alchemical 'package' of Shakespeare's day, the tangibly therapeutic aspect of alchemy's overall promise. While the *filius artis* revealed spirit and redeemed base matter, the chymicall physician extracted *arcana* and healed sick bodies. But if I am right in suggesting Tymme's *Chymicall Physicke* as the specific source and inspiration for the 'storm scene' in *Lear*, this brings us tantalizingly close in time and place to Shakespeare at work. One might even suggest that it was when leafing through that text, fresh on the London bookstalls of 1605, that Shakespeare was first struck by the possibility of alchemical metaphor to express psychic events and processes in the 'little world, man'. For all its arcane abstractions, Tymme's Paracelsist alchemy is so much about the human world – about mind and body, sickness and wholeness, wounds and remedies. 'Let no man thinke,' he says,

> that I meane the transmutation of metalls, as if such transmutation were the cheefe medicine of a man's body: but know rather that in man (which is a little world) there lye hidde the mynes of

imperfect metals, from whence so many diseases do growe, which by a good faithful and skilful Phisitian must be brought to Gold and Silver, that is to say, unto perfect purification.[233]

It would be somehow typical of Shakespeare's genius for seizing and using, if the vast alchemical drama of *King Lear* had grown from this single seed: a perception of alchemy as a way into the vexed mind of his king, a language to chart its breakdown and healing. From there, of course, he ranged far and wide into more esoteric territories. He found – probably in Ripley's *Compound* – the alchemical motif of the Wheel, and took it for the pattern, the very shape, of his play. More important still, he found in the alchemical resonances of 'the King' a way in which Lear's whole story could be presented, through a constant undertone of metaphor and symbol, as a *magnum opus* of transmutation. Through it all, Shakespeare never lost touch with the human centre of his drama. Tymme's is not the most subtle exposition of the 'human centre' of alchemy, but it may just be the one that caught Shakespeare's eye in 1605. It is partly for this reason that I quoted Tymme's definition of transmutation at the head of this chapter.

All this is pure conjecture, and must in no way displace the assembled evidence of *Lear*'s general debt to alchemy. But we have seen how useful Tymme's *Chymicall Physicke* may have been to Ben Jonson in this same year: how the 'chemical morality' of *Eastward Hoe* is neatly encapsulated in Tymme's description of how the 'liquor and running nature' of Quicksilver can be 'made solid and firme' and so 'perfectly clensed'. I have also suggested that in his poem, 'The Dissolution', John Donne drew on Tymme's doctrine of 'first' (or 'simple') elements; and, more particularly, used Tymme's exposition of the modified status of fire as the lynchpin of the poem's conceit. Now, finally, these pages of crabbed Gothic type, this work so long abandoned up one of the culs-de-sac of learning, might be claimed as a formative participant in the creation of *King Lear*. Looking at its title-page, one is struck by the curious printer's device which adorns it (fig. 25). This shows a naked princess, with crown and flowing hair. She is strongly reminiscent of those emblems of Our Mercury as feminine apparition, also naked and crowned, which I associated with Cordelia (figs. 20–2). Since the printer of the *Chymicall Physicke* – Thomas Creede[234] – was also responsible for the *Mirror of Alchimy* (which features the same device on its title-page), this alchemical overtone is not inappropriate. More strictly, however, this *virgo redimita* represents Truth, as the surrounding legend makes clear. '*Viressit vulnere veritas*' means, literally, 'truth grows green through injury'.[235] Truth, in other words, is refreshed and fortified by the trials it must undergo (as a plant 'grows green' by being cut back). The device expresses this visually by the divine hand, issuing from a cloud to whip the princess:

Truth with a scourge. This is, one suddenly sees, an image at the heart of *King Lear*. Cordelia *is* this Mercurial princess: she is actually described in the play as *'truth whipped out'*.[236] She returns, her truth and love redoubled by this scourging, to bring new 'green' life to her mortified father. As one stares at this rough woodcut on a yellowing page, one is perhaps seeing the very image of Cordelia as Shakespeare first saw her.

· 8 ·

Shakespeare's Chemical Theatre

The Secret Cave

'Shee'le Natures Way and Secret Cave
And Tree of Lyfe detect.'

William Backhouse – *The Magistery*

Lear leads us, chronologically and otherwise, into the strange world of Shakespeare's last plays. The four works generally grouped as the 'last plays' or 'romances' – *Pericles, Cymbeline, The Winter's Tale* and *The Tempest* – were apparently composed between 1608 and 1611. They belong together not only in time, but by virtue of a guiding pattern they seem to share. 'At the heart of each of these plays,' writes Traversi, 'lies an organic relationship between breakdown and reconstruction, the divisions created in the most intimate human bonds ... and the final healing of these divisions.'[1] This rhythm of division and healing is one we discerned in *King Lear*, the alchemical pulse of *solve et coagula*. In each of these last plays, furthermore, it is a relationship between a King and his child (Pericles and Marina, Cymbeline and Guiderius/Arviragus, Leontes and Perdita, Alonso and Ferdinand) which most fully incarnates this rhythm of breakdown and reconstruction. The child lost, the king stricken, the child found again: this is the recurrent pattern. Again one sees *King Lear*, and particularly Lear and Cordelia, as prototypical. It is not surprising, then, to find alchemical language in these plays, adhering to these figures of broken king and 'mercurial' child ('This peece of tender Ayre, thy vertuous daughter'[2]), and infusing its peculiar tones into the moments of rebirth, reunion and transformation scattered through the plays:

> The diamonds of a most praysed water
> Doth appear to make the world twice rich.[3]

> The benediction of these covering heavens
> Fall on their heads like dew.[4]

2. From anon., *Geheime Figuren der Rosenkreuzer* (1785)

> This is an Art
> Which do's mend Nature: change it rather, but
> The Art it selfe is Nature.[5]

> Who was most marble, there changed colour.[6]

> Does not the Stone rebuke me
> For being more Stone then it? O royall peece,
> There's magick in thy maiesty.[7]

> Those are pearles that were his eies:
> Nothing of him that doth fade,
> But doth suffer a Sea-change
> Into something rich and strange.[8]

I am not about to embark on any alchemical analysis of the last plays. Alchemy is present as part of an overall magical or mystical tone, rather than as a dominant motif as in *Lear*. I want instead to focus on one very curious scene, and to suggest that its alchemical elements contribute not only to what is presented, but also to the very special *type* of presentation which makes the last plays so unique theatrically. This scene – *Cymbeline*, IV, ii, probably composed in 1609 – features six characters: Imogen, daughter of King Cymbeline; Posthumus, her banished husband; Cloten, her lustful step-brother; Guiderius and Arviragus, her long-lost brothers; and Belarius, the brothers' foster-father.[9] The setting is a cave in the wilds of Pembrokeshire, where for twenty years Belarius and the lost princes have lived. By a series of coincidences and confusions, Imogen arrives unwittingly at this cave. She has, not unlike the Lear who abjures all roofs for the 'enmity of the ayre', fled 'back to nature'. She stumbles to the cave lost, hungry and exposed; she has 'made the ground my bed' and learned how 'foundations flye the wretched'.[10] There she finds hospitality and refreshment. The Welsh wilderness is unexpectedly populated: the lost way leads to a hidden place, a court in the rocks or a hovel in the storm. Nobility is discovered flourishing amid Nature, and with it blooms a philosophy, that ascetic vision of enrichment through hardship which Belarius expresses:

> Stoop, boyes! This gate
> Instructs you how t'adore the Heavens, and bowes you
> To a morning's holy office. The gates of monarchs
> Are arch'd so high that giants may jet through
> And keep their impious turbonds on, without
> Good morrow to the sun. Haile, thou faire Heaven,
> We house i'th'rocke, yet use thee not so hardly
> As prouder livers do.[11]

Already one senses an alchemical formulation. The princes house *in* the rock. We see how deeply into Nature Imogen has travelled: she has penetrated into matter itself. *Visita interiora terrae rectificando invenies occultum lapidem*; the alchemist promised that the bodily ardours and psychic perils of this journey into 'the interior of the earth' would be rewarded by finding there the 'hidden stone'. So Imogen discovers in the rock not only shelter and nourishment, not only noble 'sparkes of Nature' in humble guise,[12] but also her sequestered royal brothers. She goes into the rock and fetches out the lost King's Son. Recognition and restoration only happen in the fullness of time — the end of *Cymbeline* is a positive riot of *dénouements*[13] — but it is here at the cave that the momentous discovery is made. The cave is, like the hovel in *Lear*, the place of exchange. Lear found his 'son', 'brought neere to beast', in the hovel: this discovery entailed his own rescue, the redemption of the King. So too in the last plays: 'the King shall live without an Heire, if that which is lost be not found.'[14] The King's Son, the King reborn, is by familiar alchemical notation the Stone itself: 'The Son, already vivified, is become a warrior in the fire and over the tincture super-eminent. For this Son is himself the treasury, even himself bearing the Philosophic Matter'.[15] Housed in the rock is the 'treasury' of Cymbeline's new life, the 'philosophic matter' of rebirth.

Since the cave contains an undoubtedly alchemical secret, we might pause to look more closely at its nature and presence as a setting. Here we are lucky, for thanks to the inveterate scribblings of Simon Forman we have a tantalizing glimpse of Shakespeare's own production of *Cymbeline*, probably at the Globe in 1611.[16] Forman recalls how Imogen 'turned her self into mans apparrell' and how she 'chanchsed to fall on the Cave in the wods wher her 2 brothers were'. This offers just one visual clue: a cave *in the woods*. Forman stresses this, for he describes a later action near the cave as happening 'in the wods'. It seems very likely that trees — stylized or actual — were placed on the stage, as they doubtless were to represent locales like the Forest of Arden. The cave itself would have been placed (like the hovel in *Lear*) in the recessed inner stage, centre back, and represented by a fairly elaborate 'prop'. Stage caves, hermit cells, etc. of this sort are mentioned in the papers of the Elizabethan impresario Philip Henslowe. We have, at least, a skeletal image of this *mise-en-scène*. To flesh it out, we need only turn to some popular alchemical emblems of the day.

One of these is the Rosicrucian 'Mountain of the Adepts' (fig. 26), bearing the date 1604.[17] Its central figure, a naked and emaciated hermit, does quite literally 'house i'th'rocke'. One can almost hear him telling us, as Belarius does, that 'this twenty yeeres, this rocke and these demesnes have been my world'.[18] The demesnes of this Hermetic hermit are, we see, populated with alchemical creatures — dragon, lion, crow, eagle, incubating hen. Also featured is various apparatus suggesting the

27 From Stefan Michelspacher, *Cabala* (1616)

breakdown of substance: Sol and Luna *in balneo* on the left, the Athanor belching fumes on the right. Indeed the surrounding brickwork and tower shape suggest the whole rock as a furnace. It is the ascetic's 'furnace of affliction', and the orb and crown atop the rock suggest the paradoxical enrichments of his hardship. As Belarius says of the princes:

> I'th'cave wherein they bow, their thoughts do hit
> The roofes of Palaces, and Nature prompts them
> In simple and lowe things to prince it.[19]

In the foreground of this emblem, we see three people arriving at this house in the rock. One is blindfolded, the others apparently chasing a hare. They are, like Imogen, embarked on a journey: lost in one sense (blindfolded) but secretly guided (by the animal) to this dwelling in the heart of Nature. In another illustration of the alchemical 'rock' (fig. 27) – found in Stefan Michelspacher's *Cabala* (1615) – similar figures appear, one blindfolded and another pursuing a fox down its hole. 'The sons of knowledge', says the *Musaeum Hermeticum*, 'go forward along the royal road which Nature prescribes in all her operations.'[20] This is the road along which Imogen stumbles, and her goal is 'dramatized' alike by playwright and alchemist: inside Michelspacher's mountain is a courtly room, with inlaid floors and tall windows, and in it a naked king and queen, the conjoined opposites of the Stone. The same redeeming royal secret is housed in the cave of *Cymbeline*.

A rather different version of this cave-emblem is Heinrich Khunrath's (fig. 28). This is, in a highly stylized way, a 'cave in the wods', surrounded by trees, birds and deer. It is a place where people arrive, enter and are illuminated. Above its rocky entrance is the warning, *'Procul hinc abeste profani'* ('get thee hence all profaners'). The cave is no place, in the words of Shakespeare's Arviragus, for 'those who worship durty Gods'.[21] Khunrath is quoting the cry of the Sibyl in Virgil's *Aeneid* – '*O procul este profani*' – uttered before a 'deep rugged cave' that is the entrance to Hades itself. This reminds us again of Lear, his journey through the 'hell-blacke night' of the Storm to find his son in a hovel. To enter this alchemical cave, this palace in the rock, is once more a quest into the underworld of Nature and self to discover, in Virgil's words, 'truth sunk in depths of earth and gloom'.[22] 'Truth' to Khunrath would mean hidden alchemical-mystical wisdom, the truth encoded as a stone made of spirit and dramatized, as in *Cymbeline*, as a king revivified.

Khunrath's cave appeared in his *Amphitheatrum Sapientiae Aeternae*: its first public edition was in 1609, the year that Shakespeare was writing *Cymbeline*. In it, or at least in the type of emblem reproduced here, Shakespeare found an apt alchemical setting to stress visually the alchemical meanings of the scene Perhaps, more than that, he was annex-

28 Entrance to the amphitheatre. From Heinrich Khunrath, *Amphitheatrum Sapientiae Aeternae* (1609)

ing a particular *type* of theatre. The very title of Khunrath's text reminds us of the intrinsic theatricality of alchemy: it suggests the *opus* as an 'amphitheatre', where 'eternal wisdom' is revealed in the guise of chemical events. Inscribed beside the mouth of this cave is a rubric about a 'divine mystery' which 'siezes all spectators [*spectatores omnes*] . . . with admiration and love'.[23] Khunrath's cave is already halfway towards a stage set: Shakespeare completed the transition. The alchemist was a 'speculative' – literally, watching – chemist: the changes he witnessed in the glass theatre of his vessel became a mirror for changes inside him. Is this the kind of theatre to which Shakespeare aspired in these last plays: a 'chemical theatre' of transformations enacted and received?

For certainly there are transformations to be witnessed within this symbolic 'stage-scape'. The cave and its environs are the setting for one of the most bizarre of all Shakespearean scenarios, one that draws deeply on alchemical esoterica. The events which conspire to create it are briefly as follows. First Imogen, 'heart-sick' for her estranged husband Posthumus, swallows a potion. She believes it to be a 'cordial' to 'drive away distemper', but we know it to be a sleeping draught: there is, as the physician who distilled it has told us,

> No danger in what shew of death it makes,
> More then the locking up the Spirits a time,
> To be more fresh, reviving.[24]

She falls into a death-like sleep on the cave floor: the first half of the equation is complete. Now the base Cloten appears on the scene in lustful pursuit of Imogen. He is dressed in the clothes of her husband Posthumus (for reasons apparently pathological: 'with that suite upon my backe I will ravish her'[25]). Before he can commit this dastardly deed, however, he encounters Guiderius, quarrels and (discreetly off-stage) has his head lopped off. Meanwhile Arviragus has discovered Imogen, apparently dead: 'the Bird is dead that we have made so much on.'[26] Shakespeare required two corpses for the strange vision he is about to disclose and now (by means rather more effective than my truncated version suggests) he has got them. Then follows the funeral: as Forman saw it in 1611, the brothers 'laid her in the wods & the body of cloten by her in her love's apparell'. Doleful and 'ingenious' music sounds, a poignant obsequy (the famous 'Feare no more the heate o'th'sun') is sung, flowers are strewn over the bodies, and the brothers depart. Imogen awakes. She sees the decapitated corpse, sees the garments of Posthumus on it, and assumes it is indeed he who lies beside her. After a speech of anguish and appeal, the distraught princess delves her hands in the gaping neck of Cloten-Posthumus, smears her face with his blood 'that we the horrider may seeme to those who chance to finde us', and falls fainting on top of him (the stage directions instruct that she 'falls *on* the body').[27] There the action freezes, this extraordinary spectacle before our eyes: in the woods before a cave, a bloodstained princess and a headless prince lying together as in a sexual embrace.

What is Shakespeare intending here? Ingeniously, without either of them dying, Imogen and Posthumus are presented 'dead' onstage. What we see is, then, a symbolic death. The cave setting is once again a polar point of exchange: it is where the love between Imogen and Posthumus reaches that nadir which is also (in Ripley's phrase) the 'perfect meane of profound alteration'. What began as estrangement and bitterness will end with reconciliation and love. As always in *Lear* and the last plays, the route to wholeness goes by way of dismemberment. This weird scenario presents a 'valley of death' through which their love must pass. Ripley's *Compound* provides the alchemical itinerary:

> Our medicine if we exalt right so
> It shalbe thereby nobilitate.
> That must be done in manners two:
> From time the parties be disponsate,
> Which must be crucified and exanimate:

> And then contumulate both man and wife,
> And after revived by the Spirit of Life.[28]

His cumbersome Latinisms clearly disclose that strange deathly embrace called the *coniunctio* or 'Chymicall Wedding'. For the matter to become 'nobilitate' (ennobled, transformed), its opposing halves (sulphur and mercury, body and spirit, man and wife) must first be 'disponsate' (betrothed, Latin *desponsus*), then 'exanimate' (killed, robbed of *anima*), then 'contumulate' (buried together), then finally revived. This surely provides a reading for this frozen moment on stage, with its blend of elegiac melody and gruesome sexuality. It is Shakespeare's virtuoso presentation of the 'chemical wedding' of Imogen and Posthumus: a funereal coitus as the fulcrum of putrefaction and generation, division and healing. 'Burie each one in other within their grave, Then equally betwixt them a marriage make.'[29] One sees now a quality other than Jacobean melodrama in Imogen coating herself with the corpse's blood. As the scene's concluding icon hovers on the retina, its visual information superceding the narrative that composed it, the woman's bloody face and hands inescapably suggest that she has, mantis-like, devoured the head of her mate. This is Shakespeare's version – as near as he could get on the public stage – to those 'dramatizations' of *coniunctio* found in alchemical allegories like the *Visio Arislei*: 'Then Beya mounted upon Thabritius and enclosed him in her womb, so that nothing at all could be seen of him any more. And she embraced him completely into her own nature and divided him into indivisible parts.'[30] This swallowing up of the male into the body of his sister (Imogen and Cloten are also step-brother and sister) will prove to be his redemption: Prince Thabritius is reborn after putrefaction in a glass 'prison'. So Posthumus, here devoured in the bodily and 'sulphurous' form of Cloten, will return from exile and hatred to stand beside his Mercurial princess. This promise is made in appropriately riddling alchemical tones when a vision of Jupiter appears to Posthumus in prison, and announces that his 'miseries will end' when 'a Lyons whelpe shall, to himself unknown, without seeking finde, and bee embrac'd by a peece of tender Ayre.'[31] A soothsayer later glosses this oracle: the lion's whelp is Posthumus Leonatus (*leo-natus* = lion-born) and the tender air is *mollis aer* = *mulier* = his wife Imogen. The more esoteric interpretation – matter dissolved and regenerated by the 'embrace' of *anima mercurii*[32] – remains unspoken, but one can find tantalizing echoes of it in a work by Michael Maier published in 1618. In his *Themis Aurea*, Maier interweaves the alchemical *coniunctio* with the Greek legend of Deucalion and Pyrrha, whose solving of Themis' riddle redeemed mankind from the deluge sent down by Jupiter (also the presiding deity of the *Cymbeline* oracle). Maier writes:

> Mankind was generated by two stones, the Male and the Female,

29 From Michael Maier, *Atalanta Fugiens* (1618)

 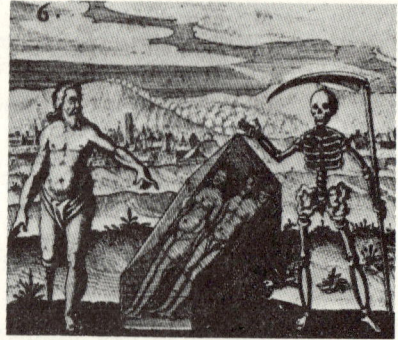

30 From Johann Daniel Mylius, *Philosophia Reformata* (1622)

F. Solvtio Perfecta III.	F. Pvtrefactio IV.

31 From Georgius Anrach, *Pretiosissimum Donum Dei* (MS, seventeenth century)

whence proceeds the wonderfull multiplication of that Golden Medicine. For the man Deucalion and his wife Pyrrha are the Gabritius and Beia, the Sun and the Moon, which two by projection of their Specifick Stones can multiply even to a thousand. Pyrrha within is ruddy, not unlike the colour of Flesh, although her outward garments are white ... Deucalion is a Lyon, not in body, but spiritually; not in shape, but operation; because he is so cruel to his wife that he kils her, and then bewraps her with his bloody mantle. But very few have attained the true knowledge of the Oracle, since most men apprehend it to be only a History, and thence draw some wholesome Morals, which here have no place, neither were they ever intended.[33]

Strange flashes of the chemical wedding as performed in *Cymbeline*: the lion and his mate; the deathly embrace and the bloody mantle (as the corpse of Cloten is called Imogen's 'bloody pillow'); the oracle whose hidden promise is redemption, that mankind 'be restored and multiplied' (Maier), that Britain 'flourish in peace and plenty' (Shakespeare).[34]

It has been said of *Pericles* – probably the immediate predecessor of *Cymbeline* – that certain of its scenes affect us as 'pictures more than drama'.[35] This moment from *Cymbeline* epitomizes this iconic quality. And as a 'picture', it is a supremely alchemical composition. If its setting owes much to the image and signification of the cave in contemporary alchemical illustration, then the foreground figures draw equally potent echoes from the many current emblems of the chemical wedding (figs. 29–31). And with this importing of alchemy's visual terminology may come an idea of effect: that this scene, like those emblems, not only imparts an occult meaning, but also wields an occult power; that the story of Posthumus and Imogen is, as Maier says of Deucalion and Pyrrha, no mere 'History' full of 'wholesome Morals'. To the 'true, religious Alchymist', his texts were devotional and his emblems were talismanic icons possessed of the same magical 'inward spirit' as matter itself. Does this three-dimensional emblem of Shakespeare's aspire to a similar potency? It is generally thought that it was for the exclusive Blackfriars, rather than the rumbustuous Globe, that these last plays were primarily intended.[36] One must place this bizarre alchemical scenario in a hushed, candlelit auditorium, surround it with sylvan scenery, the scent of strewn flowers, the dying echoes of the music. One does not know how long the pause was held, or what echoes and archetypes stirred in the watcher's mind. But one might say he was witnessing a moment of pure chemical theatre.

Magister ludi

> The purpose of playing ... is to holde, as twere, the Mirrour up to nature.
>
> William Shakespeare – *Hamlet*

> When a Man looketh in a Glass, there is the reflection of his image, which if you go to touch with your hands, you find nothing tangible but the Glass wherein the person looked. So also from this matter must be drawn a visible spirit which nevertheless is impalpable. That very same spirit, say, is the Radix of the Life of our Bodies, and the Mercury of the Philosophers.
>
> Basil Valentine – *The Twelve Keys*

All this looks forward to Shakespeare's riddling swan-song, *The Tempest*. Many have seen its presiding magus, Prospero, to be Shakespeare's incarnation of his own magical powers as dramatist. If Prospero's 'so potent Art' can conjure storms and scupper kings, then Shakespeare's art exercises the same magic presidency over events, actors and spectators. If the island is Prospero's magic circle, the playhouse is Shakespeare's. It was almost in awe of his own powers that Shakespeare wondered, in the Prologue to *Henry V*:

⟨ UT MAGUS ⟩

32 'Truly a magician.' Engraving of Shakespeare by Charles Turner (1824)

> Can this cock-pit hold
> The vastie fields of France? Or may we cramme
> Within this woodden O the very caskes
> That did affright the ayre at Agincourt?[37]

There is one point in *The Tempest* where Prospero does quite literally turn his magical prowess to playmaking. This is when he mounts the betrothal masque for Ferdinand and Miranda. The players that enact this delicate interlude are spirits at Prospero's command:

> *Ferdinand* This is a most maiesticke Vision, and
> Harmonious charmingly. May I be bold
> To thinke these spirits?
>
> *Prospero* Spirits, which by mine Art
> I have from their confines call'd to enact
> My present fancies.[38]

Here again the punning notion of 'art' as at once a magical and theatrical skill. Within the magic island of his theatre, the dramatist gives shape and substance to invisible spirits. The vision performed, it is summarily dissolved:

> Our Revels now are ended. These our actors,
> As I foretold you, were all spirits, and
> Are melted into ayre, into thin ayre.[39]

Throughout his career, Shakespeare had delighted in the ambiguous presence, the insubstantial substance, of theatre. A play was brilliantly there, yet a mirage. The players take their leave at the end of *A Midsummer Night's Dream*, saying:

> If we shadowes have offended,
> Thinke but this and all is mended:
> That you have but slumbred heere
> While these visions did appeare.[40]

Perhaps it *was* just a midsummer night's dream, after all. Hamlet too is haunted, even galled, by this paradox of potency and weightlessness:

> Is it not monstrous that this Player heere,
> But in a fixion, a dreame of passion,
> Could force his soule so to his own conceit
> That from her working all his visage warm'd;
> Teares in his eyes, distraction in's aspect,
> A broken voyce, and his whole function suiting
> With formes to his conceit? And all for nothing![41]

The powers of the player are positively supernatural: he conjures gestures, tears and poetry out of thin air. He manifests the invisible, the nothing of 'a fixion, a dreame', in the visible sphere of theatre, just as Prospero calls up invisible spirits 'from their confines' to 'bestow them on the eyes' of the spectator. The play, as dream made visible, is perhaps not so far from that 'spirituall fixt thing' that the alchemist sought to witness in the theatre of his vessel. 'In one glasse must be done all this thing,' says Ripley, 'like to an Egge in shape'[42] – a glass O, an amphitheatre of strange and tonic revelations, like the 'wooden O' of the playhouse. And what we see revealed there, the alchemist and dramatist both insist, is something hidden inside ourselves. At the close of his revel-ending valediction, Prospero says: 'We are such stuffe as dreames are made on.'[43] He includes us all. *We* are the Raw Stuff out of which these dream-plays are drawn, like the Mercurial spirit out of matter. If we wonder why we keep on coming back to Shakespeare, this last little alchemical flourish is perhaps his own explanation.

Notes

Where full bibliographical details are not given in these Notes the reader should refer to the Bibliography

1 The Hunting of the Alchemist

THE MODERN IMAGE

1. Bacon, F., Book II, fol. 32(v).
2. Taylor, *The Alchemists*, p. 13.
3. Bacon, F., Book I, fol. 22(v).
4. JD I 39. The title 'Loves Alchymie' was not apparently Donne's. It first appears in the first collected edition – *Poems by J.D.* – published posthumously in 1633. All MSS of the poem bear the title 'Mummye' – viz. mummy or mumia. 'Whatsoever when killed has the power of healing diseases. Hence mumia of the elements is the balsam of external elements' (HAWP II 375).
5. A huge number of alchemical texts, of which the *Testamentum* is a celebrated example, was attributed to the Spanish mystic Ramon Lull (1232?–1315). It is thought that none of these is genuinely by Lull, and that they are works of the fourteenth and fifteenth centuries fathered on to him. See Ferguson, *Bibliotheca*, sv Lullius; Stillman, pp. 291f. The authors of these works are correctly styled 'pseudo-Lull(ius)': I use the Anglicization, Raymond Lully, almost invariably used in Shakespeare's day.
6. Hoefer, Ferdinand, *Histoire de la Chimie* (Paris, 1842–3). Cited Stillman, p. 292.
7. Stillman, pp. 422, 393.
8. Bacon, F., Book II, fols 32(r)–32(v). To cast Bacon in the role of '*contra-chymicus*' is not to deny the presence of alchemical suppositions in his own philosophy. See Gregory, J. C., 'Chemistry and Alchemy in the Natural Philosophy of Sir Francis Bacon' (*Ambix*, Vol. II (1946), p. 93f); and Rees, Graham, 'Francis Bacon's semi-Paracelsian Cosmology' (*Ambix*, vol. XXII (1975), p. 81f).
9. Descartes, IV, principium CCIII.
10. Pascal, *Pensées*, II, 77. '*Je ne puis pardonner a Descartes.*'
11. HAWP I 289. These words derive rather tortuously from Paracelsus. They are Waite's translation from Baron Tschoudy's 'Hermetic Catechism' in *L'Etoile Flamboyant* (Hamburg, 1785). This in turn is a version of the so-called Vatican Manuscript of Paracelsus. Sendivogius, MH II 86, writes: 'Nature is not visible, though she acts visibly; she is a volatile spirit who manifests herself in material shapes, and her existence is in the will of God.'
12. Bloomfield, TCB 313. On the life and work of Bloomfield, see Schuler. He

is not to be confused (as DNB does) with his younger contemporary Miles Bloomfield (1525–1603).
13 As for instance Newton in Coleridge's summary: 'Newton was a mere materialist. Mind, in his system, is always passive – a lazy Looker-on on an external world.' (Letter to Thomas Poole, 1801: Penguin, *Selected Poems and Prose*, p. 26).
14 This celebrated text is by Eirenaeus Philalethes (or Philaletha), whose true identity has been the subject of much debate. Certain informed contemporaries – e.g., William Cooper, who published some of his works – identify him as George Starkey, others as John Winthrop. He is not to be confused with Eugenius Philalethes, *nom-de-plume* of Thomas Vaughan. See Ferguson, *Bibliotheca*, sv Philaletha; Waite, *Lives*, pp. 187–200. The *Introitus Apertus* was first published in Latin, Amsterdam, 1667, but the English edition, published by Cooper in 1669, claimed to be taken from the original MS.
15 Anthonie, p. 19.
16 The 'vitriol acrostic' appears on the title page of Salamon Trimosin, *La Toyson d'or* (Paris, 1613); in Johann Bringer (ed.), *Azoth, sive Aureliae Occultae Philosophorum* (Frankfurt, 1613); and in Daniel Stolcius, *Viridarium Chymicum* (Frankfurt, 1624). See Read, *Prelude*, pp. 104–5, 270.
17 'Scito quod hanc scientiam habere non poteris, quousque ... in corde omnem corruptionem deleas.' *Aurora Consurgens* (MS, fifteenth century), I, x, parable 5. Cited JCW XII 270.
18 Valentine, *Antimony*, p. 26.
19 Dorn, *Philosophia Meditativa*, TC I 472. Cited JCW XII 255. This tract first appeared in Dorn's *De Naturae Luce Physica* (Frankfurt, 1583).
20 Dorn, *Speculativae Philosophiae Gradus Septem*, TC I 267. Cited JCW XII 269. This tract comprises Part Two of Dorn's *Clavis Totius Philosophiae Chymisticae* (Lyons, 1567).
21 JCW XII 476.
22 Jung defined archetypes as 'a priori, inborn forms of intuition ... perception and apprehension' (*Instinct and the Unconscious*, JCW VIII) and (following Burckhardt) as 'primordial images'. See Fordham, pp. 24–8.
23 JCW XII 23, 34.
24 Tymme, sig. A3(r). See also Robert Fludd, *Mosaicall Philosophy* (London, 1659), p. 175: 'It was the Spagericall or high Chymicall virtue of the Word, and working of the Spirit, that the separation of one region from another was effected.'
25 Donne, 'An Anatomie of the World' (1611), JD I 237.
26 JCW XII 228.
27 TCB (Prologomena) sig. B4(v).

THE SATIRICAL IMAGE

28 Lyly, sig. C3(v): II, iii, 27–9.
29 Plat, p. 86.
30 Nashe, *The Terrors of the Night* (1594), sig. E3(r), TN I 367.
31 Plattes, Gabriel, 'A Caveat for Alchymists', in *Chymical, Medicinal and Chyrurgical Addresses made to Samuel Hartlib, Esquire* (London, 1655). Cited Geoghegan, p. 98.
32 Jonson, *The Alchemist*, IV v 58.

33 Chaucer, Fragment VIII, 635–6. See below, n. 46.
34 Norton, Ch. I, TCB 17.
35 Chaucer, VIII, 728, 730.
36 Lodge, sig. I2(v).
37 Scot, Book XIV, p. 355.
38 Ibid., p. 353.
39 This figure is provided by the astrologer William Lilly (1602–81) in the *History of his Life and Times* (1715). Cited Rowse, *Forman*, p. 265.
40 DNB, sv Lambe.
41 Dee spent much energy rebutting various slanders against his magic, notably his petition to King James (4 June 1604) offering himself 'willingly to the punishment of death' if the 'name of Conjurer, or Caller, or Invocator of Divels' could be proved against him. He delivered a similar petition to Parliament (this time in doggerel verse) four days later. The mathematician and astronomer Thomas Hariot (1560–1621) was called the 'master' of Raleigh's alleged 'Schoole of Atheisme' in a pamphlet of 1592 (see below, n. 75). Baines's testimony against Marlowe in 1593 (see below, n. 60) attributed to Marlowe the belief that Moses 'was but a Iugler and that one Heriots being Sir W Raleighs man can do more than he'. Hariot was (with William Warner and Robert Hues) one of 'the Earle of Northumberland's three magi' (Aubrey, sv Hariot). 'In those darke times,' said Aubrey *a propos* another mathematician, Thomas Allen, 'Mathematician and Conjurer were accounted the same things.'
42 Sendivogius, sig. A3(r).
43 Jonson, *Alchemist*, II, iii, 29–30, 42–6.
44 Scot, Book XIV, p. 353.
45 Plat, pp. 85–6.
46 There is no reason to doubt the Yeoman's assertion that the rogue Canon of the *Secunda Pars* of the Tale is distinct from the Canon, his master, described in the Tale's prologue and *Prima Pars*. Scot says (p. 353) that by alchemy 'some cousen others and some are cousened themselves': this is perhaps Chaucer's distinction between the two Canons.
47 Chaucer, VIII, 1161–4.
48 Jonson, *Eastward Hoe*, IV, i, 214–16, 221–8.
49 The word 'test' in its current meaning of 'examination' derives from Latin *testum*, an earthen vessel. In the assayer's laboratory, the test or cupel was a small circular vessel in which the gold or silver to be assayed was placed together with lead. Subject to the heat of the furnace (often called the 'test furnace'), the lead formed an oxide, litharge, which dissolved the oxides of any impurities in the gold or silver. The test being porous (often made of bone ash), the dissolved impurities were absorbed into it, leaving the purified gold or silver at the bottom of the vessel. See OED, sv 'test', and also the reconstruction of a sixteenth-century assayer's laboratory at the Science Museum, London.
50 Dorn, p. 319.
51 HAWP I 158. The nine books *De Natura Rerum* were written in 1537, first published in Latin in 1573, and translated into English by J. F(rench?) in 1650: 'Of the Nature of Things', sigs M ff of *A New Light of Alchymy*.
52 Jonson, *Eastward Hoe*, IV, i, 230–7.
53 Jonson, *Alchemist*, I, i, 114.
54 Jonson, *Eastward Hoe*, I, i, 38–41.
55 Robertson, H. M., *Aspects of the Rise of Economic Individualism*,

pp. 178–9; Keynes, J. M., *A Treatise on Money*, II, 156–7. Cited Knights, p. 40.

56 Bedo and Buckley were examined at the Tower on 28 July 1570 by Justice John Southcote and Thomas Stanley, treasurer of the Royal Mint. Their statements are recorded in State Papers (Dom.) Eliz. 1570, vol. LXXI, no. 63. See Hart, pp. 391f. Another case from the State Papers (vol. CCLXXI, no. 103) concerns an alchemist named Scory, 'sometyme dwelling in Petticoate Lane'. In July 1599, evidence was brought against him and John Beish, a metal-worker, 'concerning the counterfeiting and making of dollers in Turkie'. The forged coins were to be circulated in Turkey by a merchant named Mallarie, who had trading contacts there. Cited BJ X 94–5.

57 Della Porta, pp. 160ff. The eighth chapter of this fifth book (pp. 174ff) may have been Buckley's source: it tells how 'without loss in weight, nor yet the stamp being hurt, gold and silver may be diminished'. The powder whose recipe Buckley sold to Bedo may well have been the 'powder of brimstone' recommended by Della Porta. I quote from the English trans (1658): this was taken from the expanded edition of 1584, not the original one of 1558 which Buckley would have possessed.

58 These were seized by the authorities from Buckley's chambers at New Inn, Oxford. Astromancy is divination by means of the stars. Gematria is a Cabbalistic numerology 'in which numerical values assigned to Hebrew letters were subtly and intricately calculated' (French, p. 112). Dr Dee was 'especially addicted' to gematria.

59 Poole's seditious sentiments were spoken to one Humphrey Gunstone, in the pitch-dark dungeon called 'the Limboes', between 10 and 11 p.m. on 25 July 1587. Gunstone's testimony of same to the authorities is catalogued in State Papers (Dom.) for 1599. See Eccles, Mark, 'Marlowe in Newgate', Times Literary Supplement 9 September 1934, p. 604. Also Eccles, *Christopher Marlowe*, pp. 36–7. Eccles identifies Poole as a Cheshireman, brother-in-law of Sir William Stanley.

60 From the 'Note containing the opinion of one Christopher Marly concerning his damnable Judgment of Religion and scorn of God's word' sent by Richard Baines to the Privy Council in 1593. BM Harleian MS 6848, f. 185. On Baines, see Eccles, *Christopher Marlowe*, p. 100.

61 Marlowe, *Tamburlaine*, Part II, IV, ii, 59–64.

62 Hester, *Key of Philosophie*, sig. E7(v).

63 Lodge, sig. I2(v).

64 The statute issued by Henry IV in 1404 proclaimed that 'none shall henceforth use to multiply Gold or Silver, nor use the Craft of Multiplication: and if any the same do, and be thereof attaint, that he incur the pain of felony' (*The Statutes of the Realm*, ed. Luders, II, p. 144. Cited Taylor, *The Alchemists*, p. 102). Similar decrees were issued throughout Europe: by Pope John XXII in 1317, Charles V of France in 1380, the Council of Venice in 1418, *et al.* It was possible to obtain a royal licence to practise alchemy: see Taylor, loc. cit., for the text of a licence in 1445.

65 Plat, p. 82.

'A MULTIS AMATUR ALCHYMIA'

66 Plat, p. 82: '*Amultis amatur Alchimia & tamen virgo est.*' Cf, Philalethes, *Fons*, MH II 262: 'Our Water is a most pure virgin, and is loved of

many, but meets all her wooers in foul garments, in order that she may be able to distinguish the worthy from the unworthy.'
67 'Englishmen' is not strictly accurate: Dee was Welsh and Seton (supposedly the true author of Sendivogius' *Novum Lumen*) Scottish.
68 Moffett, Thomas, *Nobilis, or A View of the Life and Death of a Sidney*, trans. and ed. V. B. Heltzel and H. H. Hudson (1940), p. 75. Cited French, p. 127.
69 French, Ch. 6, 'John Dee and the Sidney Circle.'
70 Leicester's father – John Dudley, Earl of Northumberland – employed Dee as tutor in the early 1550s.
71 Aubrey, sv Herbert, Mary.
72 Ibid, sv Dyer.
73 Bodleian MS Ashmole 1420, p. 328. Cited Sargent, p. 112.
74 Aubrey, sv Raleigh.
75 'Philopater, Andreas' (identified as Robert Parsons), *An Advertisement written to a Secretarie of my L Treasurers of Ingland by an Inglish Intelligencer* (1592), p. 18. This was an abbreviated translation of Philopater's *Responsio ad Elizabethae Edictum* (1592). The commission convened at Cerne Abbas on 21 March 1594, headed by Thomas Howard, Viscount Bindon.
76 Raleigh, I, i, Ch. 7, p. 71.
77 Aubrey, sv. Raleigh.
78 Evelyn, John, *Diary and Correspondence*, ed. Bray (1850), I, 368–9 (20 September 1662). Cited Rattansi, p. 122. Lefevre (Lefebvre, Lefebuvre) published in 1665, at the King's command, a *Discours sur le Grand Cordial de Sr Walter Rawleigh*.
79 Lacey, p. 352.
80 BM, Add. MSS 6178, fol. 14. Cited Shirley, p. 53. According to a Restoration diarist, the Rev. John Ward, Raleigh's chemical 'operator' in the Tower was a man called Sampson. See Rattansi, p. 122.
81 Alnwick Castle MSS, U, I, 3. Accounts of Henry Taylor, 'Clarke of Kitchens', February 1606–February 1607. Cited Shirley, pp. 60–1.
82 These include Severinus' *Idea Medicinae Philosophicae* (1571); the *Triga Chemica*, ed. Barnaud (1599); the 3rd edition (1614) of Sendivogius' *Novum Lumen Chemicum*; Ruland's *Lexicon Alchemiae* (1612). All these will be of interest in the course of this book. Among the more general magical works are Della Porta's *Magia Naturalis* (edition of 1585) and Giordano Bruno's Lullian work, *De Lampade Combinatoria* (1588). See Shirley, pp. 64–6.
83 Davies, p. 85: *Hymnes of Astraea*, I. There are 26 'hymns', each with three stanzas, acrostic on ELISA-BETHA-REGINA.
84 MS Ashmole 1445, VIII, fol. 38(r). Cited Taylor, *Thomas Charnock*, p. 172. Taylor's article is a rich insight into the world of a sixteenth-century alchemist.
85 Sargent, p. 99.
86 Cambridge University Library MS Dd 3, 83, art 6. This unpublished work survives in MS copy by the other Bloomfield, Miles, who says of his namesake: 'in Alchimistri & Distillation he hath not left his lyke in this nation.' Schuler, pp. 80–2, quotes the dedication to Elizabeth *in extenso*.
87 State Papers (Dom.) Eliz., vol. LXXV, no. 66. Cited Hart, p. 390. The contents of the three 'glasses' were described as follows: 'The firste is of *Sol* prepared and dispersed. The seconde is of *Luna* devided and dispersed. The thirde is of *Mercury* made homogeniall.'

88 MS Ashmole 1421, Norton's 'Preamble' to the *Key of Alchimie*. Cited Reidy, p. 68. The context suggests that the great grandfather spoken of is the Thomas Norton who wrote the *Ordinall of Alchimy* in 1477, though no such passage appears in the 6th chapter of the *Ordinall*.
89 Shakespeare, *Richard II*, II, i, 41, 46.
90 TCB 483. This was in the early 1590s, after Dee's return from Bohemia. At this time Dee wrote his 'prefatory verses' to Ralph Rabbards' 1591 edition of Ripley's *Compound of Alchymy* (see Chapter Two): this too was dedicated to Queen Elizabeth.
91 Dee, John, *The Compendious Rehearsal*, p. 19. Cited French, pp. 38–9.
92 Hume, V, xliv, p. 426.
93 By Gabriel Plattes, Edith Sitwell and E. J. Holmyard, respectively.
94 State Papers (Germany), vol. VI, no. 21, ff. 45–7. Cited Sargent, p. 109.
95 Harvey, p. 293. Nashe, *Have with you to Saffron-Walden*, sig. H4(r), TN III 52. See also Harvey, pp. 68–9: 'I wondred to heare that Kelly had gotten the Golden Fliece, and by vertue thereof was sodenly advaunced into so honorable reputation with the Emperours maiestye.'
96 Bald, p. 88.
97 Jonson, *Alchemist*, IV, i, 82, 89–90.
98 Fletcher, John, *The Fair Maid of the Inn*, IV, ii.
99 BM Sloane MS 3188, f. 9. Reproduced French, plate 10. Dee kept detailed 'minutes' of his seances: these, known as his Spiritual Diaries, were published by Meric Casaubon in *A True & Faithful Relation of what passed for many Years Between Dr: John Dee . . . and Some Spirits* (London, 1659). This extraordinary text details a variety of exotic 'dialogues' between May 1583 and September 1607. Earlier actions, between December 1581 and May 1583, are recounted in a series of six '*Libri Mysteriorum*' catalogued among Dee's MSS.
100 TCB 478.
101 Dee's diary entry for 10 May 1588: Halliwell, p. 27.
102 Evans, p. 199
103 Waite, *Lives*, p. 155
104 Evans, pp. 223, 227 n. 5.
105 Ibid., p. 226.
106 Letter from Burghley to Dyer, May 1591. Cited Sargent, p. 115.
107 TCB 484.
108 Evans, pp. 227–8.
109 Waite, *Lives*, p. 156. It is Evans, p. 204, who styles Hajek Rudolf's 'senior alchemical adviser'.
110 Dee's diary entry for 19 December 1586. Halliwell, p. 22. See also TCB 481.
111 Cited Sargent, p. 103.
112 Evans, p. 226.
113 TCB 481.
114 Shakespeare, *A Midsummer Night's Dream*, V, i, 210–11.

2 Three English Alchemical Texts

THE MIRROR OF ALCHIMY

1 TCB (Prolegomena) sig B4(v).
2 This anthology, *De Alchemia*, also contains the Latin version of the

Mirror of Alchimy – viz. *Speculum Alchemiae Rogerii Bachonis* – but the evidence remains strong for the French version as an immediate source for our text. The entry in the Stationers' Register (26 May 1593) reads: 'Thomas Scarlet. Entred for his copie . . . to be printed in Englishe / The Myrrour of Alkamye of Roger Bacon a most excellent Philosopher, to be *translated out of French into Englishe*' (Arber, II, 631). The conscientious translator, anonymous throughout, appears to have had both versions to hand. Bookseller Scarlet surrendered his copyright to Richard Olive, probably in 1596, and Thomas Creede printed it for the latter in 1597. For other transactions between Scarlet, Olive and Creede in 1596, see Arber, III, 68, 72.

3 Stillman, p. 271, finds 'nothing in it that is characteristic of Roger Bacon's style or ideas', and assigns it the fourteenth century. He cites Little's and von Lippman's consideration of it as among Bacon's '*pseudepigrapha*'. Read, *Prelude*, p. 24, ascribes it to pseudo-Roger Bacon, but dates it to the thirteenth century. Among Bacon's genuine observations on alchemy is a famous passage in the *Opus Tertium* (*c.* 1268) describing 'this duplex science of alchemy (that is, theoretical and practical)'. Stillman, pp. 262–5, quotes this *in extenso*.

4 Holmyard, pp. 78–9, 95–8. An abridged version of the *Emerald Table* was discovered by him in the Second Book of the *Element of the Foundation* ascribed to Jabir-ibn-Hayyan. Latin versions of the *Emerald Table* appear to have been current in Europe by 1200. The translation contained in the *Mirror* was the first English version of this seminal text to be published.

5 TCB (Prologomena), sig. Br(r): 'He was the first Christian Philosopher after Morienus, who . . . because he was the first that transplanted the Chemicall Muses from remotest Parts into his own Country, is called Garland, *ad Coronam Hermeticam & Poeticam*.' The name Hortulanus – deriving from Latin *hortulus*, a (small) garden – is explained in the *Mirror*, p. 17, as 'so-called for the Gardens bordering upon the sea-coast'. This is probably an esoteric rather than geographic formulation. The alchemist's desire to 'grow' or 'generate' new forms of matter is often depicted in gardening terms: metallic seed, philosophical tree, chemical pleasure-garden (*viridarium chemicum*), etc. Cf. Maier, *Atalanta*, Emblem VI, sig. E(r), where the alchemist is portrayed as a husbandman who 'sows his gold in white earth'.

6 TCB 372. The 175-line description of the 'experience Which cleped is Alconomie' in Gower's *Confessio Amantis* (Book IV) is the first literary notice of alchemy in England (bar a glancing reference to 'experimentis of alconomye' in *Piers Plowman*, A, XI, 157).

7 See Holmyard, pp. 61–4.

8 Bacon, R., p. 1.

9 Ibid., loc. cit.

10 It is the Philosophers' Stone (a translation of *lapis philosophorum*) rather than the Philosopher's Stone, as it is usually misnamed. See OED, sv 'philosopher', for a discussion of the term's evolution.

11 MH I 97. The authorship of Siebmacher is not certain: see Ferguson, *Bibliotheca*, II, 383–5. The work appeared first in German, *Wasserstein der Weysen* (Frankfurt, 1619): in Latin in the 1625 edition of MH; and in English, *The Water Stone of the Wisemen*, trans. J. H.(ughes?), p. 77 ff of *Paracelsus His Aurora* (London, 1659).

12 Gratacolle, William, *The Names of the Philosophers Stone*, trans. H.P.

(London, 1652). Cited Read, *Prelude*, p. 128.
13 Read, loc. cit.
14 Bacon, R., p. 6.
15 'Our gold is not common gold.' Anon., *Rosarium Philosophorum* (Frankfurt, 1550). Cited JCW XII 78. This is not the same as the *Rosarium Philosophorum* attributed to Arnald of Villanova.
16 Bacon, R., p. 19.
17 Cf. Tillyard, *Elizabethan World Picture*, p. 73: 'Gold was king of metals, the sum of all metallic virtues; and alchemically it was a mixture of the elements in a perfect proportion. The same perfect proportion in the human body caused health.'
18 Bacon, R., p. 2.
19 'Ut compleatur per artificium quod a natura est relictum incompletum', Arnald of Villanova, *Rosarium Philosophorum*. Cited, from MS Cambridge Corpus Christi 99, p. 149, by Duncan, 'Alchemy in Chaucer', p. 648.
20 Jonson, *Alchemist*, II, iii, 135–9.
21 Shakespeare, *Winter's Tale*, IV, iv, 95–7.
22 Philalethes, *Introitus*, Ch. 1, MH II 165.
23 Norton, Ch. 5, TCB 67.
24 Sendivogius. I use Waite's trans of this passage (MH II 115) in preference to French's (p. 58), which is jumbled and obscure.
25 Backhouse, William, *The Golden Fleece or The Flowre of Treasures* (MS Ashmole 1395, pp. 1–223). Cited Josten, p. 6, n. 29. See also Ibid., pp. 32–3. Backhouse made his translation from the French version of the *Aureum Vellus*, viz. *La Toyson d'or, ou La Fleur des Thresors* (Paris, 1613).
26 Backhouse, TCB 342.
27 Bacon, R., pp. 32–3, 35.
28 Ibid., p. 33.
29 Ibid., p. 15.
30 Philalethes, *Manuductio*, MH II 249.
31 Norton, Ch. VII, TCB 103–5.
32 Bacon, R., p. 11.
33 Ibid., p. 38.
34 Ibid., p. 19.
35 The actual term 'quintessence' does not appear in the *Mirror*, but this 'Soule' which quickens and 'raiseth up' the matter is much the same. This whole aspect of alchemy derives (via Hermetism) from Greek neo-Platonism, but the Platonist's distinctions between soul (*anima*), spirit (*spiritus, pneuma*), essence (*essentia*), etc., are generally ignored by the alchemist.
36 Grasshoff, MH I 14, quoting Ricardus Anglicus.
37 Bacon, R., p. 2.
38 Ibid., pp. 2–3.
39 Sendivogius, p. 9.
40 *Tractatus de Generatione Metallorum*, MH I 251–2. According to its title-page in MH, this tract was first 'published by a German sage in the year 1423'.
41 Philalethes, *Introitus*, Ch, 1, MH II 165.

42 Ripley, *Compound*, sig. B(r).
43 See Duncan, 'Alchemy in Chaucer', pp. 636–7. Another reference at this time is found in the Patent Rolls for 1329, where Johannes de Rous and William de Dalby are said 'to know how to make silver metal', and to have been summoned before Edward III to demonstrate their skills (Taylor, *Alchemists*, p. 101). The often repeated legend that Raymond Lully visited England in 1330 and performed transmutations for Edward III (TCB 467) has no basis in fact, but perhaps expresses the transmission of 'Lullian' alchemy into England at this time.
44 Ripley, *Compound* (Epistle to Edward IV), sig. L3(r).
45 Ripley, *Cantilena*, verse 2: 'Through Roman Countreys as I once did passe.' Ripley, *Compound* (Epistle): 'learning in Italy', 'the Universitie of Louvayne.' TCB 458: Ripley, says Ashmole, 'gave yearely to those Knights of Rhodes £100,000 towards maintaining the war (then on foot) against the Turks.'
46 The medieval MS is styled *The Compende of Alkemie*, reminding us that 'compound' is in the sense of 'compendium'.
47 See Bibliography. Marginalia on this MS have been identified as the handwriting of Thomas Charnock, whose death in 1581 thus provides a *terminus ad quem* for this translation. Ripley's original Latin text – *Georgii Riplaei Omnia Opera* (Cassel, 1649), pp. 421–6 – is reproduced in JCW XIV 274ff.
48 The *Vision* was included in Rabbards edition of the *Compound* (sig. *4(r)). All the works mentioned here are to be found in TCB, pp. 374–92, except the *Medulla Alchemiae* ('Marrow of Alchemy') of which Ashmole prints only the nineteen verse Preface. A sixteenth century version of the Ripley Scrowle by James Standysh is in the BM: a detail from it is in Rola, plate 65.
49 Ripley, *Compound*, Gate IV, sig. E3(v).
50 Ibid., I, sig. D(r).
51 Ibid., II, sig. D3(r).
52 Ibid., VIII, sig. H4(v); XII, sig. K3(r).
53 Taylor, *Alchemists*, pp. 113–14.
54 Many sequences are depicted in the *Philosophia Reformata*, but this one stands out by virtue of each emblem being labelled to identify the operation it depicts. See the reproductions in Rola, pp. 98–100: the sequence begins at Emblem 5, labelled *Calcinatio*, and ends at Emblem 16, *Multiplicatio*. The final 'gate', projection, is not depicted, and a curious banquet scene styled '4 Gradus' is inserted between *Putrefactio* and *Congelatio*. Otherwise Ripley's itinerary is followed exactly.
55 I use the lay term 'Raw Stuff' rather than the more correct 'Prime Matter' (*prima materia*) because the latter term has a whole range of meanings which are apt to become self-cancelling. See Ch. 4, part ii, 'The Cloudy Voice', on the interchangeable meanings of alchemical terminology.
56 Ripley, *Compound* (Preface), sig. B3(v).
57 Ibid., loc. cit. Norton – Ch. V, TCB 62 – calls the Stone 'Microcosmus' (*sic.*).
58 Ripley, *Compound*, I, sig. D(r).
59 Ibid. (Epistle), sig. M(v).
60 *Genesis* I, ii.
61 Ripley, *Compound* (Preface), sigs B4–B4 (v).

62 Quercetanus, *Originall and Causes*, fol. 17(v).
63 Ripley, *Compound*, I, sig. C2(v).
64 Ibid., II, sig. D(v).
65 Ibid., II, sig. D2(r).
66 Quercetanus, *Originall and Causes*, fol. 17(v).
67 Ripley, *Compound*, II, sig. D2(r).
68 Ibid., loc. cit.
69 Ibid., III, sigs D3(v)–D4(r).
70 Ibid., III, sigs E(r)–E(v). I have conflated Ripley's distinctions to get at the basic point. He in fact mentions four types of fire – 'Naturall, Innaturall, against Nature, also Elementall' – but the vital distinction is between an 'outer' and an 'inner' fire. Cf, Norton – Ch. V, TCB 61 – where an 'inward heate naturall, which in our Substance is but Intellectual, to sight unknowne' works in tandem with the 'outwarde heate artificiall' of the furnace.
71 Ibid., IV, sig. E2(r).
72 Ibid., IV, sig. E2(v).
73 Ibid., IV, sig. E3(r).
74 Ibid., IV, sig. E3(v). The 'Chymicall Wedding', *coniunctio oppositorum*, is the centrepiece of the *opus*. The attained Stone is often represented as a conjoined King and Queen, and in this sense the Chemical Wedding is an emblem of triumphant harmony. The alchemists themselves insist, however, that this *coniunctio* must first yield to putrefaction and death. There are, then, two 'images' of the Chemical Wedding: a funereal coitus and an exultant rebirth.
75 Ibid., V, sigs E4(r)–E4(v); II, sig. D2(v).
76 Ibid., V, sig. F(r).
77 Ibid., V, sig. E4(r).
78 Quercetanus, *Originall and Causes*, fol. 18(r) (incorrectly paginated 14).
79 Ripley, *Compound*, V, sig. F(v).
80 Ibid. (Recapitulation), sig. K3(v).
81 Ibid., V, sig. F(r).
82 Ibid., VI, sigs G2(v)–G3(r).
83 Quercetanus, *Originall and Causes*, fol. 18(r).
84 Ripley, *Compound*, VI, sig. H(v).
85 Ibid., VII, sig. H2(v).
86 Ibid., VIII, sig. H3(v).
87 Quercetanus, *Originall and Causes*, fol. 18(r).
88 Ripley, *Compound*, XII, sig. K2(r).
89 Ibid., II, sig. D3(r).
90 Ibid. (Recapitulation), sig. K3(v).
91 Ibid., IV, sig. E3(v).
92 Ibid.(Recapitulation), sig. K3(v).

THE HIEROGLYPHICALL UNIT

93 A 'thousand years' is in fact an understatement. The earliest extant alchemical writings – the *Physika* and *Mystika* of pseudo-Democritus, the *Gold-making of Cleopatra*, the writings of Zosimos of Panoplis, etc. – date back to AD 100–300.
94 Aubrey, sv Dee.
95 DNB, sv Dee.

96 Dee's *Testamentum* is printed by Ashmole (TCB 334). It is the 'third and last' of a series of letters to John Gwynn: the first two have apparently not survived. Josten suggests (Dee (Introduction), p. 90) that this was the lawyer and fellow of St John's, Cambridge, John Gwyn (or Wynn), d. 1574. St John's was Dee's old college.
97 MS Ashmole 1486, V, 1–2. The same MS contains Dee's transcription of Ripley's 'vade mecum'.
98 Worsop, Edward, *A Discoverie of Sundrie Errours and Faults daily committed by Landemeaters* (London, 1582), sig. G3(v). Cited French, p. 6.
99 See Taylor, E. G. R., *Tudor Geography* (London, 1930), pp. 76–139, and Elton, G. R., *England under the Tudors* (London, 1955), pp. 334–7, on Dee as geographer and navigational adviser. See French, Ch. 7, 'John Dee and the Mechanicians', and Yates, *Theatre*, pp. 6–8 et passim.
100 Dee, 'Mathematicall Preface' to *The Elements of Geometrie of the most auncient Philosopher Euclide*, trans. Sir Henry Billingsley (London, 1570), sig. Ciii(v). Cited French, p. 107. On mathesis, see Yates, *Bruno*, pp. 296–7. George Peele's eulogy of the 'Wizard Earl', Northumberland, includes this useful definition: 'Mathesis.... That admirable mathematic skill, Familiar with the stars and zodiac, To whom the heaven lies open as her book.... Following in the ancient reverend steps of Trismegistus and Pythagoras.'
101 See French, pp. 97–103. See also Yates, *Bruno*, pp. 151–5, on Bruno's similar interpretation of 'the Copernican diagram as a hieroglyph of divine mysteries'. As they point out, Copernicus himself was aware of the Hermetic-Platonic overtones of his theory. See Copernicus, *De Revolutionibus Orbium Caelestium* (1543), ed. Thorn (1873), pp. 16–7.
102 See French, pp. 110–19. Dee's first recorded 'angelic conference' was on 22 December 1581. Before the advent of Kelley, his 'skryer' was one Barnabus Saul. Some of Dee's sigils are on show in the Renaissance Corridor of the BM.
103 French, p. 1.
104 The 1591 reprint was at Frankfurt. The defence (in the *Discourse Apologeticall* to the Archbishop of Canterbury) was prompted by criticisms made by Andreas Libavius in *Tractatus duo Physici* (1594).
105 Tymme translated Calvin's commentary on St Paul's Epistle to the Corinthians (London, 1577) and on Genesis (London, 1578) and Ramus' *Discourse of the Civill Warres of France* (London, 1574).
106 MS Ashmole 1459, pp. 469–71 (Epistle), 472–9 (Forespeech), 479–81 (Introduction). The patron Thomas Baker has not yet been identified. 'My purpose in this dedication,' wrote Tymme (p. 471), 'is not to procure you into the Laborinth of Alchimists practise, whereunto all that have entred with unwashed hands have hurt themselves ... but rather to allure you to like that wch I love.'
107 *De Priscorum Philosophorum Verae Medicinae Materia* (St Gervais, 1603) and *Ad Veritatem Hermeticae Medicinae* (Paris, 1604).
108 *Mathematicall Preface*, fol. 2. Cited Josten, Introduction to *Monas*, p. 91. I should point out that Tymme himself does not appear to be aware of this stipulated translation: his references to 'this Hieroglyphicall Monas' and 'his Monas Hieroglyphicall' suggest that he might have left the word untranslated in his title.
109 Dee, fol. 12(r) (Josten, p. 155).
110 Ibid., fol. 7(r) (Josten, p. 135).

111 Ibid., fol. 3(r) (Josten, p. 119).
112 See Josten's Introduction, pp. 102–3, for a fuller account of the monad's visual constituents.
113 Tymme, *Epistle Dedicatorie*, MS Ashmole 1459, p. 469.
114 On the significance of the combination of Mercury and fire, see Ch. 4, part ii, on the 'dragon Mercury'.
115 Tymme, *Epistle Dedicatorie*, MS Ashmole 1459, p. 469. In the monad, Tymme continues, '☿ hath the first in the top & the last in the foote, the cross going between, wch signifies the dejecting and humiliation of ♀ before his Exaltacon'.
116 See, e.g., Norton, Ch. V, TCB 77; Lambspringk, MH 1 302; Grasshoff, MH I 48; Philalethes, *Metamorphosis*, MH II 243; *et al.*
117 Bacon, R., p. 9.
118 Yates, *Bruno*, p. 68, referring to Ficino's *De Vita Coelitus Comparanda* (1489).
119 Paracelsus, *Astronomia Magna oder die Gantze Philosophia Sagax*, I, Ch. 6 (Huser, II 376).
120 Agrippa, I, p. 33. This chapter, treating 'Of the Spirit of the World' and how it 'unites occult vertues to their subjects', consciously echoes Ficino's conception of *spiritus mundi*. Agrippa's magical aspirations, like Dee's, went far beyond the scope of Ficinian *magia naturalis*, but retained its basic assumptions.
121 Ibid., loc. cit.
122 Dee, fol. 7(r) (Josten, p. 135).
123 Ibid., fol. 17(v) (Josten, p. 177).
124 Virgil, *Aeneid*, IV, 356.
125 'Id est, Matrimonii Terram; sive Influentalis coniugii, terrestre signum.' Dee, fol. 7(r) (Josten, p. 135).
126 Ibid., fol. 17(v) (Josten, p. 177).
127 Bacon, R., p. 16.
128 At the beginning of the 14th theorem: 'this whole magisterium depends upon the Sun and the Moon, of which a long time ago that thrice-great Hermes admonished us, when he asserted that the Sun is its father and the Moon its mother.' Dee, fols 14(v)–15(r) (Josten, pp. 165–7).
129 Bacon, R., p. 16.
130 See Yates, *Bruno*, pp. 2–6, 21–2, for a succinct summary of the origins of Hermetism.
131 See JCW XII 299–302; JCW XIII 73. Zosimos instructs: 'Hasten down to the shepherd and bathe yourself in the great cup.' The shepherd is the Hermetic god, Poimandres, and the great cup the χρατήρα (krater) which appears in the *Corpus Hermeticum*: 'Dip and wash thy self thou that art able in this cup or bowl. . . . As many therefore as understood the proclamation and were baptized or dowsed into the Minde, these were made partakers of Knowledge and became perfect men', Everard, *Pymander*, p. 161.
132 The earliest published version of the *Tractatus Aureus* appeared in the *Ars Chemica* collection (Strasburg, 1566). It was published separately, with commentary by 'Dominicus Gnosius of Belgium' at Leipzig, 1600(?) and 1610. William Salmon translated it in 1691: *The Golden Work*, pp. 179ff of *Medicina Practica*. On various other alchemical and magical treatises ascribed to Hermes Trismegistus, see Thorndike, II, Ch. XLV; and Ferguson, *Bibliotheca*, s.v. Hermes.
133 The latter phrase is from Ashmole's ode to Backhouse 'Upon his adopting

of me to be his Son' (1651). MS Ashmole 36, f. 241(v)–242(r). Cited Josten, p. 7.
134 Raleigh, Preface, sig. B6(r).
135 Ficino, *argumentum* before *Pimander*, I. Cited and trans. Yates, *Bruno*, p. 14. It was not until 1614 that the philologist, Isaac Casaubon, correctly dated the *Hermetica* – post-Plato, post-Christ – and thus exploded what Yates calls 'the great Egyptian illusion'.
136 Agrippa, III, p. 471.
137 Everard, *Asclepius*, p. 17.
138 Everard, *Pymander*, p. 21.
139 I have preferred Yates's translation of this passage (*Bruno*, p. 136) to J.F.'s of 1651 (Agrippa, II, p. 316).
140 Trithemius, letter '*ad amicum quendam*' in Gohorry, Jacques, *De Usu & Mysteriis Notarum* (Paris, 1550), sig. Iii(v). Cited Josten, Introduction to *Monas*, p. 109.
141 Agrippa, II, p. 175.
142 Yates, *Rosicrucian Enlightenment*, p. xi, describes the Rosicrucian manifestos broadcasting 'the combination of Magia, Cabala and Alchymia as the influence making for the new enlightenment'. On the contribution of 'Christian Cabala' to the evolution of Renaissance magic, see Yates, *The Occult Philosophy*, Chs I–VII.
143 Dee, fol. 14(r)–14(v) (Josten, pp. 163–5).
144 Ibid., fol. 18(r) (Josten, p. 179). This interpretation of Dee's words about Oedipus is Josten's.
145 Ibid., fols 7(r)–7(v) (Josten, pp. 135–7).
146 Dee, *General and Rare Memorials perteyning to the perfect Arte of Navigation* (London, 1577), sigs ei–ei(v). Cited Josten, Introduction to *Monas*, p. 92.
147 TCB 482–3.
148 Khunrath, *Von Hylealischen Chaos*, p. 204. Cited JCW XII 124.
149 Khunrath, *Amphitheatrum*, p. 6, quoting Dee, fol. 7(r). See Josten, Introduction to *Monas*, p. 135; Yates, *Rosicrucian Enlightenment*, p. 38.
150 Waite, *Lives*, p. 159.
151 Khunrath, *Chaos*, p. 185. Cited JCW XII 254.
152 Ibid., p. 170. Cited JCW XIV 56.
153 Ibid., pp. 274f. Cited JCW XII 275.
154 The word 'alcohol' derives from Arab '*al-kohl*', the fine black powder used as mascara. Alcohol is, in other words, the mist or 'powder' that is distilled from wine. All distillates were deemed to some extent curative ('quintessential'), and this particularly pleasant one was styled *aqua vitae*, the water of life. Liqueurs of all sorts are an alchemical heritage.
155 Dorn, *Speculativae Philosophiae Gradus Septem*, TC I 307. Cited JCW XIV 54.
156 Dorn, *Philosophia Meditativa*, TC I 460. Cited JCW XIV 54. This idea of the inner 'spark' of divinity is powerfully expressed by Dee, who addresses the *Monas* to those 'in whom there blazes fiery strength and a heavenly origin' (*Monas*, fol. 23(r); Josten, p. 199).
157 Dorn, *Physica Trismegisti*, TC I 423. Cited JCW XIV 94.
158 Dorn, *Speculativae Philosophiae*, TC I 267. Cited JCW XII 269.
159 Dee, Prefatory Verses; in Ripley, *Compound*, sig *2. Rabbards praises Dee's *Monas* in his dedication (to Queen Elizabeth) and includes Kelley's verse letter to 'G.S.', 'Concerning the Philosophers Stone'.
160 Ripley, *Compound*, sig. B4(r).

3 The Rise of 'Chymicall Physick'

PARACELSUS

Epigraph: (trans. Leishman) 'As a praiser and blesser he came like the ore from the taciturn mine.'

1. 'Paracelsus did that in physic, which Luther in divinity' (Codronchus, cited if not endorsed by Robert Burton, *Anatomy of Melancholy* (1625), p. 240). Levi, p. 8: 'the marvellous Paracelsus, always drunk and always lucid, like the heroes of Rabelais.' Rabelais himself practised physic (to du Bellay) and introduced himself on the title page of *Gargantua* (1534) as 'M. Alcofribas, Abstracteur de Quinte Essence.'
2. An alternative interpretation of 'Paracelsus' is as 'a fantastic version of von Hohenheim', relating German *Hohen*, heights, to Latin *celsus*, high, lofty. See Browning's notes to his dramatic poem, *Paracelsus* (1835) in *Poems* (Oxford, 1940), p. 64.
3. *Grosse Wundartznei*, III. Cited Pagel, p. 8.
4. Pagel, loc. cit.
5. *Sieben Defensiones*, IV *(Die Vierdte Defension: Von Wegen meines Landtfahrens)*. Huser I 258.
6. *Spitalbuch*. Cited Pagel, p. 14.
7. Johann Froben (1460–1527) was a pioneer of printing in Basle, producing the bilingual New Testament (ed. Erasmus) in 1516 and the Latin edition of Luther (1518–20). See Steinberg, S., *Five Hundred Years of Printing* (Harmondsworth, 1955, revised ed. 1974), pp. 48–52.
8. Cited Holmyard, p. 164.
9. Oporinus: letter to Solenander and Wierus, 26 November 1555. Cited Pagel, pp. 29–31.
10. Jonson, *Volpone*, II, ii, 125.
11. Butler, *Hudibras*, II, canto 3.
12. See, e.g., *Archidoxis*, lib. 4 (HAWP II 22): 'Quintessence is, so to say, a nature, a force, a virtue, and a medicine, once indeed shut up within things, but now free from any domicile.' *Opus Paramirum*, lib. IV (Huser I 109): 'These internal and invisible virtues are the fifth essences, of which one loth is equal to twenty pounds of the body.'
13. Joachim de Watt (Vadianus), Paracelsus' protector at St Gallen.
14. Of these, only the *Grosse Wundartznei* was published in his lifetime (two vols, at Augsburg, summer 1536).
15. Cited Holmyard, p. 165.
16. *Sieben Defensiones*, IV, Huser I 259.
17. *Das Buch Paragranum*, I (Philosophia), Huser I 210.
18. *Liber de Caducis*, Huser I 589f. Cited JCW XV 30.
19. Cited (without source) JCW XV 27.
20. *Paragranum*, III, HAWP II 148. (HAWP translates only the third of the four books of the *Paragranum*: the original – *Der Dritte Grundt der Medicin welcher ist ALCHIMIA* – appears in Huser I 219–25).
21. Ibid., I, Huser I 210.
22. Ibid., III, HAWP II 151–2.
23. Ibid., III, HAWP II 153.
24. Ibid., II (Astronomia), Huser I 213.
25. Ibid., III, HAWP II 149.
26. Ibid., III, HAWP II 150–1.

27 Cited JCW XV 10.
28 Pagel, p. 147.
29 *Opus Paramirum*, lib. IV (*De Matrice*), Huser I 78. 'Dieser Marcasith ist den Menschen Kranckheit darumb so hilfft er ihm.'
30 *Sieben Defensiones*, II (*Die Andere Defension: Betrieffendt die newen Kranckheiten*), Huser I 255.
31 *Von dem Naturlichen Dingen*. Ch. 9 (*Von dem Arsenico*), Huser I 1056: 'As long as it [sc. arsenic] lives, poison and remedy are close together. When its poison is subdued, it loses its power of physic.'
32 Ibid., Ch. 8 (*Vom Vitriol*), Huser I 1050–1. See also HAWP II 231–6 on 'the alchemical process and preparation of vitriol'.
33 *Paragranum*, I, Huser I 208.
34 *Liber Meteororum*, Ch. 4 (*Quid In Stellis De Viventibus Speciebus*), Huser II 80. See also *Labyrinthus Medicorum Errantium*, Ch. 5 (*Von dem Buch der Alchimen*), Huser I 271–2 (and HAWP II 165–8).
35 *De Natura Rerum*, VII (*De Transmutationibus Rerum Naturalium*), Huser I 900.
36 *Grosse Wundartznei*, II, 12.
37 *Labyrinthus Medicorum Errantium*, Ch. 5, HAWP II 165–6.
38 *Paragranum*, III, HAWP II 148.
39 *Liber de Caducis*, Huser I 589f. Cited JCW XV 30.
40 *Paragranum*, III, Huser I 220.

METAL-BREWING PARACELSIANS

41 Ibid., *Vorred* (Preface), Huser I 203. 'Spagyrus' is a 'practicioner in the Spagericall arte' (as John Hester styled himself, title-page of *Originall and Causes*). The term 'spagyric' derives 'from the Greek words σπαω and αγειρω, 'divide' and 'unite', corresponding to the alchemical *solve et coagula*' (Burckhardt, p. 20).
42 Severinus, Peter, *Idea Medicinae Philosophiae* (Basle, 1571), p. 39.
43 Penotus, sig. B3 (r).
44 Thurneisser, *Quinta Essentia* (Leipzig, 1574), subtitle: 'Die Höchste Subtilitet, Krafft und Wirking . . . der Medicina und Alchemia.'
45 On the Pharmacopoeia of 1618, see Urdang. He notes its innovations both general – 'being the first official drug formulary to be made obligatory . . . for all England, its recognition of the chemico-therapeutical movement was of the highest general importance' – and particular – the introduction of *tartarus vitriolatus* (potassium sulphate), *mercurius vitae* (a mixture of $SbOCl$ and Sb_2O_3), and *mercurius dulcis* (calomel).
46 *A Booke on the Nature and Properties as well of the bathes in England as of other bathes in Germanye and Italye* (Collen, 1562), fiii. In this same year William Bullein referred to 'Theophrastus Peraselpus' in his *Bulwarke of Defense against all Sicknes*. See Debus, *English Paracelsians*, pp. 55–7.
47 Baker, *The Newe Jewell of Health* (London, 1576). This was a translation (actually by fellow-surgeon Thomas Hill) of the second part of Conrad Gesner's *Treasure of Evonymus* (Zurich, 1569). Cited Debus, *English Paracelsians*, p. 57.
48 On the identity of Bostocke, see Debus, 'The Paracelsian Compromise', p. 77, n. 23.
49 B(ostocke), R(obert), *The Difference between the auncient Phisicke . . .*

and the latter Phisicke (London, 1585), Ch. 8 section 5. Cited Debus, 'Paracelsian Compromise', p. 79.
50 Ibid., Ch. 6. Cited Debus, loc. cit.
51 Urdang observes that the 'order and classification' of the proposed pharmacopoeia of 1585 includes the heading, '*Extracta, Sales Chemica, Metallica*'.
52 Dover Wilson and Chambers believed it was Herbert. See Schoenbaum, pp. 218–19, on the pros and (mostly) cons of this identification.
53 See Ch. 1, n. 68.
54 Hester was in practice by 1576, when George Baker wrote of him as a 'paynfull traveyler' (i.e. conscientious workman) in distillation (*Newe Jewell of Health*, sig IVf. Cited Debus, *English Paracelsians*, pp. 65–6). He was dead by 1593, when Gabriel Harvey (p. 80) honours the 'memorye of oulde Iohn Hester'.
55 BL Catalogue C 60 o 6. It has marginalia in the hand of Gabriel Harvey and is dated *c.* 1588.
56 Antimony was widely praised as a chemical remedy by Paracelsus, Quercetanus, Basil Valentine. *The Universall Medicine* by John Evans (London, 1642) is entirely devoted to 'the vertues of my magneticall or Antimoniall Cup.'
57 See Bibliography, sv Fioravanti. The first of these, *A Ioyfull Iewell*, was actually a translation by Thomas Hill, published after his death by Hester.
58 Pseudo-Paracelsus, p. 18.
59 Forester, p. 47.
60 This work was published posthumously in 1596. It claimed to be trans. from the German of 'the most learned Theophrastus Paracelsus'. For the purposes of bibliography, I have treated it as Hester's, since its origins are obscure.
61 Hester, sigs A6(r)–A6(v).
62 Quercetanus, *Originall and Causes* (Epistle to the Reader), sig. A3(r).
63 Forman, *Discourse of the Plague* (1593). Cited Rowse, *Forman*, p. 57. He was no great theorist, but his distilled 'strong waters', his disparagement of traditional urinalysis ('paltry piss'), and his general alchemical interests (e.g., his transcriptions of Norton's *Ordinall* and Bloomfield's *Blossoms*, both made in 1591), all give his physic a Paracelsist ring.
64 Anthonie ('4 ounces of this high medicine are sould for twenty shillings') and DNB, sv Anthonie.
65 Anthonie, sig. 3(v) et passim: 'Experiments and ocular testimonies' including certain 'Transmarine Testimonies'. Anthonie also published *Panacea Aurea* (1598) and an *Assertio* ('a short discourse for the Assertion of Chymicall Physick') in 1610. His success moved no less than three writers to publish counterblasts: Thomas Rawlin (1610), Matthew Gwinne (1611), and John Cotta (1623).
66 Fludd practised at Fenchurch Street and Coleman Street, and employed his own apothecary. In 1606, his licence was revoked by the Royal College of Physicians for his 'prating about himself and his chemical medicines and heaping contempt on the Galenic doctors', though he was later (1609) admitted as Fellow. See Godwin, pp. 7–8, and DNB, sv Fludd. Arthur Dee set up practice in London in 1603 and also fell foul of the RCP.
67 Bostocke, *The Difference*, Ch. 1. Cited Debus, 'Paracelsian Compromise', p. 80.
68 Penotus, sig. B(r). See also Hester, sigs E8(r)–E8(v).

69 Penotus, sig. A3(v).
70 Ibid., loc. cit.
71 Ibid., sig. A4(r).
72 Ibid., sig. B2(v).
73 Hester, sigs. E7(v)–E8(r).

SOME LITERARY REACTIONS

74 Puttenham, I, xxiv, sig. G3(r).
75 JD I 182. On its date, see JD II 140 and Bald, p. 100.
76 Paracelsus, *Paragranum*, I, Huser I 212.
77 Shakespeare, *Macbeth*, IV, iii, 214–5.
78 JD I 332.
79 Nashe, *Terrors of the Night*, sigs D4(v)–E2(v), TN I 363–6.
80 Jonson, *Alchemist*, II, iii, 229–33.
81 Jonson, *Volpone*, II, ii, 76, 94.
82 Fioravanti, *Ioyfull Iewell*, pp. 38–9.
83 Jonson, *Volpone*, II, ii, 59–62.
84 Cited Holmyard, p. 164.
85 Penotus, sigs A3(v), A3(r).
86 Ibid., sig. A4(r).
87 Jonson, *Volpone*, II, iii, 163–7, 151–6.
88 Shakespeare, *Romeo and Juliet*, IV, i, 94.
89 Ibid., II, iii, 8–18, 23–4, 27–8.
90 Paracelsus, *Von dem Naturlichen Dingen*, Ch. 9 (*Von dem Arsenico*), Huser I 1056.
91 Raleigh, I, i, Ch. 7, p. 71
92 Moffett, Thomas, *De Jure et Praestantia Chemicorum Medicamentorum*, (Frankfurt, 1584), p. 101. Cited Debus, 'Paracelsian Compromise', p. 90.
93 Shakespeare, *Romeo and Juliet*, II, iii, 53.
94 Shakespeare, Sonnet V, 9–11. For the text of the *Sonnets, Venus and Adonis* and *Lucrece* in original spelling, I have used the Nonesuch Press edition of *The Complete Works* (ed. Herbert Farjeon, 4 vols, London, 1953), vol IV.
95 Ibid., VI, 1–4.
96 Ibid., LIV, 14. I have followed conventional emendation of 'my' for 'by'.
97 Marlowe, *Tamburlaine* Part I, V, ii, 102–3.
98 Shakespeare, *Henry V*, IV, i, 4–5.
99 Khunrath, *Von Hylealischen Chaos*, p. 204. Cited JCW XII 124.
100 Shakespeare, *All's Well that Ends Well*, II, iv, 41–5.
101 Ibid., II, iii, 10–11
102 Ibid., II, i, 72–5.
103 Ibid., V, iii, 102.
104 Shakespeare, *Macbeth*, I, vii, 66–8.
105 See, e.g., *Titus Andronicus*, III, i, 17; *As You Like It*, III, ii, 153f; *Macbeth*, III, v, 25; et al.
106 Shakespeare, *Pericles*, III, ii, 26–38.
107 Harvey, p. 80.
108 It is curious that Hester's dedication is addressed (sig. A2(r)) to 'the right worshipfull Walter Raleigh esquier'. This may just be the chymist's gaffe, addressing Sir Walter as plain 'esquire'. It may, on the other hand, suggest that the work was completed before Raleigh was knighted in early 1584.
109 See DNB, sv Forester. The work he helped to transcribe was Barrow's

Briefe Discourse of the False Church (1590). A close associate of Barrow was John Penry, often identified as the chief author of the 'Martin Marprelate' tracts. This may add another dimension to Nashe's anti-Paracelsist sortie in *Terrors of the Night*: if his target there is Forester, he may be indulging in some residual Martin-bashing.

110 Forester, title-page. He praises Hester as one who 'spent much and indangered his body about such workes whereof many excellent men have enioyed the benefit'.
111 On the friendship of Field and Shakespeare, and of their fathers Henry and John, see Schoenbaum, pp. 27–9, 130–2; Rowse, *Shakespeare the Man*, pp. 56–8; Kirwood, A. E. M., 'Richard Field, Printer, 1589–1624' (*The Library*, XII (1931), 1–39).
112 Rowse, *Shakespeare the Man*, pp. 56–7. Forester's *Pearle* was entered in the Stationers' Register on 11 December, 1593 (Arber, II, 641).
113 Forester, Epistle to the Reader (unpaginated).
114 Shakespeare, *Venus and Adonis*, 443, 273–4.
115 Ibid., 27–8; Shakespeare, *Lucrece*, 1466.
116 Forester, sig *ii(r).
117 This company performed a brief season in June 1594 for Philip Henslowe at Newington Butts; Lord Hunsdon referred to them as his 'newe companie of players' in October; they played twice at the royal court at Greenwich in December. Schoenbaum, pp. 135–6.
118 Rowse, *Forman*, p. 289.
119 Nashe, *Terrors of the Night*, sig. A2(r), TN I 341.
120 Ibid., sig. C1(v), TN 1 351.
121 Chapman, *The Shadow of Night* (1594), dedication to Matthew Roydon – 'good Mat' – who had introduced Chapman to this noble company. For a highly original analysis of this poem in terms of Renaissance Cabbalism, Agrippan occult philosophy and Dürer's engraving *Melancholia* I, see Yates, *The Occult Philosophy*, Ch. XIII.
122 By Carey's sister (Margaret, Lady Hoby) to know whether Carey 'will live or die'. Rowse, *Forman*, p. 240.
123 Cited Ibid., loc. cit.
124 See, e.g. Hermann, p. 6: 'in curing the Pockes, there is nothing that dooth anie way profit . . . but only Quicksilver.'
125 Jonson, *Alchemist*, II, ii, 20–1.
126 Webster, *White Divel*, I, ii, 28–9.
127 Shakespeare, *Merry Wives of Windsor*, V, v, 60ff.
128 This particular location was first suggested by Leslie Hotson, *Shakespeare versus Shallow* (London, 1931).
129 Shakespeare, *Merry Wives of Windsor*, IV, iv, 87–8; II, iii, 97–8.
130 Ibid., III, i, 65–6.
131 Ibid., II, i, 67–9.
132 Ibid., III, v, 90–4, 114–25.
133 Quercetanus, *Originall and Causes*, fol. 17(v).
134 Hermann, p. 7.
135 Shakespeare, *Merry Wives of Windsor*, III, iii, 206: 'His dissolute disease will scarce obey this medicine.'
136 Forester, Ch. XII ('Of Man and the medicines that are made of him'), p. 78.

4 A New Chemical Light: Sources and Reflections

AN ALCHEMICAL RENAISSANCE

1 This suggestion is found written on the flyleaf of a copy of Basil's *Triumphwagen Antimonii* (edition of 1624) in St Andrews University library (Read, *Prelude*, p. 185).
2 The business about the 'table of marble' is recounted on the title-page of the *Last Will and Testament of Basil Valentine* (London, 1671).
3 Read notes on allusion in the *Triumphwagen* to '*die neue Franzosen Kranckheit*' (i.e., syphilis), a term not current till the sixteenth century. Stillman concludes, pp. 372–7, that 'all modern historians ... agree upon the post-Paracelsan character' of the Basilian *corpus*.
4 On Tholde, see Ferguson, *Bibliotheca*, II 246.
5 Eisleben, 1599 (in *Von dem Grossen Stein der Uhralten*); Zerbst, 1602; a piratical reprint which Tholde complained of in 1603; Frankfurt, 1618, trans. into Latin by Michael Maier (in *Tripus Aureus*).
6 *Von den Naturlichen und Ubernaturlichen Dingen* (1603); *De Occulta Philosophia* (1603); *Triumphwagen Antimonii* (1604).
7 Valentine, *Twelve Keys*, p. 232.
8 In Barnaud's *Quadriga Aurifera* (1599).
9 Valentine, *Twelve Keys*, pp. 267–8.
10 Ibid., pp. 254–5.
11 Ibid., pp. 242–3.
12 Read, *Prelude*, p. 186.
13 Valentine, *Twelve Keys*, p. 233.
14 Valentine, *Triumphant Chariot*, pp. 16–17.
15 Ibid., p. 27.
16 Ibid., p. 23.
17 A good example of Dorn's 'metaphysical Paracelsism' is his commentary on Paracelsus' *De Vita Longa*, part of which is extracted in JCW XIII 174–5.
18 See Shirley, p. 66. This, the third edition, was published at Cologne.
19 For a picturesque account of Seton and Sendivogius, see Read, *Humour and Humanism*, pp. 38–65.
20 Rowse, *Forman*, p. 197. Rowse wrongly identifies the 'little book' as a publication of 1594.
21 Read, *Humour and Humanism*, pp. 53–8.
22 Sendivogius, p. 9.
23 Sendivogius: I have preferred Waite's translation (MH II 83) to J.F.'s (sigs A5(r)–A5(v)).
24 Ibid., MH II 83. This is omitted by J.F.
25 Ibid., p. 41.
26 Ibid., pp. 40–1.
27 TC, title-page. The work is dedicated to Frederick, Duke of Württemburg.
28 Valentine, *Twelve Keys*, p. 266.
29 *Fama Fraternitatis*, title-page: 'Allegemeine und General Reformation der gantzen weiten Welt.' The English translation of the *Fama* and *Confessio* (by Thomas Vaughan, 1652) is reproduced, slightly modernized, in Waite, *Real History*, pp. 65f, and Yates, *Rosicrucian Enlightenment*, pp. 238f.
30 Valentine, *Twelve Keys*, p. 255.
31 *Fama* (Yates, *Rosicrucian Enlightenment*, p. 250).

32 'The most godly and highly illuminated father, our brother CR, a German, the chief and original of our Fraternity' (*Fama*, Yates, *Rosicrucian Enlightenment*, p. 239). According to the *Confessio*, he was born in 1378, died 1484, and his tomb was discovered in 1604.
33 *Fama* (Yates, *Rosicrucian Enlightenment*, p. 250).
34 The *ros caeli* epigraph appears on the verso of the title-page of the *Secretioris Philosophiae Consideratio Brevis* (the tract which opens the *Confessio*): the work itself is closely based on the *Monas*. The symbol of the monad appears on the title page of Andreae's *Chymische Hochzeit* (1616), and in the text, beside the poem '*Heut, Heut, Heut / Ist des Konigs Hochzeit*'. See Yates, *Rosicrucian Enlightenment*, pp. 37–9, 45–7, et passim.
35 *Fama* (Yates, *Rosicrucian Enlightenment*, p. 247); it is praised as a work which 'unfalsifieth'.
36 Ibid. (Yates, *Rosicrucian Enlightenment*, p. 243). On the wider influence of Paracelsus on the manifestos, see Waite, *Real History*, pp. 201–9.
37 Waite, *Real History*, p. 212, apparently citing Benedictus Figulus, author of *Pandora* (1608), as his source.
38 'The chimaera of the Rosie-Crosse, / Their seales, their characters, / Hermetique rings, / Their Jemme of riches, and bright stone, that brings / Invisibilitie' (Jonson, 'An Execration upon Vulcan' (1623), BJ VIII 206).
39 See, especially, Andreae, pp. 171f: The Sixth Day. The narrator recounts the hatching, feeding, blackening, bathing, beheading and cremation of a bird.
40 Andreae, *Vita ab Ipso Conscripta*, I, p. 10. Cited Waite, *Real History*, p. 226, n. 2. '*Nuptiae Chymicae, cum monstrorum foecundo foetu, ludibrium.*'
41 Yates, *Rosicrucian Enlightenment*, p. 31.
41 Andreae, *Vita*, loc. cit.: '*Esther et Hyacinthus, comoediae ad aemulationem Anglicorum histrionum.*'
43 Yates's description of this embassy (*Rosicrucian Enlightenment*, pp. 32–3) is based on a contemporary account by E. Cellius, who recalls the 'English musicians, comedians, tragedians and most skilful actors', and their performance of *The History of Susanna*.
44 There seems to be evidence that the *Fama* was circulating in manuscript by 1610. One Adam Haselmayer saw it, somewhere in the Tyrol, in that year, and wrote a reply to it that was supposedly printed in 1612. The earliest extant copy of Haselmayer's reply, however, is that included in the published edition of the *Fama*, 1614. See Yates, *Rosicrucian Enlightenment*, pp. 41–2, 235. Even if one pushes the date back to 1610, one is still coming in at the tail-end of Shakespeare's career.
45 The *Arcana Arcanissima* is generally supposed to be Maier's first publication. It bears no date or place of publication, but Craven asserts (*Count Michael Maier*, p. 32) that 'it is believed to have been issued anno 1614 and printed at Oppenheim'. The earliest works whose dates are certain are the *De Circulo Physico* and the *Lusius Serius*, both published at Oppenheim in 1616. The matter of Maier's visit or visits to England has recently been reopened. It had been assumed that Maier was in England in 1616 (see Debus, *Renaissance Chemistry and Robert Fludd*, p. 15; Holmyard, p. 186). This was presumably based on the dedication of Maier's *Lusus Serius*, which is signed '*ipso ex Anglia reditu, Pragam abituriens, anno 1616, mensi Septembri*' ('having returned from England, then departing from Prague, September 1616). In 1979, however, a

parchment was located in the Scottish Record Office, Edinburgh, by Adam McLean, editor of the *Hermetic Journal*. It is, in effect, a Christmas card from Maier to King James, sent in 'celebration of the birthday of the Lord (as) we enter the new auspicious year 1612'. The main text consists of an elaborate calligram, in which an eight-part benediction on James forms the eight petals of the rose, and a series of letters picked out in gold form a cross within the rose. The gold cross reads: *VIVE IACOBE DIV REX MAGNE BRITANNICE SALVE TEGMINE QUO VERE SIT ROSA LAETA TUO.* I make this (differing from Mr McLean): 'Hail and long live James, divine king of Great Britain; under your protection may the rose be truly fertile.' See *Hermetic Journal*, no. 5 (Autumn 1979), pp. 4–7. If this document is genuine, it offers strong evidence that Maier was in England by late 1611, and that he was seeking to interest James in the 'secrets of the Rosie Crosse'. I am indebted to Dame Frances Yates for news of this new development, which strengthens her case for a specifically Rosicrucian influence on Shakespeare's very latest plays.

46 His first publication was the *Apologia Compendiaria Fraternitatem de Rosea Cruce* (Leiden, 1616).

47 I am in no way attacking Dr Yates's position in *Shakespeare's Last Plays*. Her approach is primarily historical rather than literary, 'contextual' rather than textual, and the Rosicrucian element is only one thread of her argument. I hope my book might be complementary to hers. She wonders 'whether Rosicrucian symbolism, or something like it, might already have been current before the actual publication of the Rosicrucian manifestos' (*Last Plays*, p. 91). I simply answer: 'Yes, there was alchemy.'

48 Bacon, R., (Preface), sig. A2(r).

49 Valentine, *Twelve Keys*, p. 241.

50 Ibid., pp. 238–9, 233–4.

51 Ibid., pp. 250–1.

52 Sendivogius: I have preferred Waite's trans. (MH II 82) to J.F.'s (sig. A4(v)), which confusingly renders *potentia* (potency, hence spirit) as 'possibility'.

53 Valentine, *Twelve Keys*: again I prefer Waite (MH I 352) to the verse rendition in the 1670 translation (p. 288).

54 Tymme, Epistle Dedicatorie to the lost *Monas* translation, MS Ashmole 1459, p. 469.

55 Ripley, *Compound* (Preface), sig. B4(v).

56 Mylius, p. 308.

57 Siebmacher, MH I 78.

58 Valentine, 'The Manual Operations', Ch. 3 ('Of the Spirit of Mercury'), in *Last Will and Testament*, p. 486.

59 Many of these *Decknamen* (cover-names) for Our Mercury go back to Arab formulations – e.g., *tair-abyad* (white bird), *uqab* (eagle), *lu ab al-qamar* (spittle of the moon), etc. See Metlitski, p. 82 et passim. These in turn reflect Greek nomenclature (see, e.g., the synonyms for Mercury found in Zosimos, listed by Holmyard, p. 25).

60 Siebmacher, MH I 80.

61 Valentine, *Twelve Keys*, p. 247.

62 *Aureliae Occultae Philosophorum Duo Partes.* TC IV, 501f. Cited JCW XIII 218. Thomas Vaughan translates the same passage (p. 177) and attributes it to 'one of the Rosy Brothers'.

63 Tymme, sig. H3(r): 'This Vitriol . . . possesseth a green sharp spirit of so great an acting and penetrating force that . . . it will dissolve metalline

bodyes. . . . And this is that greene Lyon which Ripley commendeth.'
Andrewes, TCB 279: the green lion 'soone can overtake the sun [viz.
gold] and suddenly can him devoure'. Bloomfield, however, TCB 312,
says: 'the Lyon greene, Which some Fooles imagine to be Vitrioll
Romaine', and Ruland has *leo viridis, quorundam opinione aurum'* – the
green lion, said by some to be gold; so beware of too fixed an
interpretation!
64 Ripley, *Vision*, in *Compound*, *4(r).
65 The term 'ruddy' is often used to describe gold in English alchemical
writings and the redness of the achieved Stone (Red Stone, Tincture, King,
Man, etc.) is clearly related to its gold-making potential. Brewer notes
(*Dictionary of Phrase and Fable*, p. 899) that 'in the old ballads red was
frequently applied to gold ('the gude red gowd') and in thieves' cant a
gold watch is a red kettle, and the chain a red tackle.'
66 Philalethes, *Ripley Reviv'd*, in Rola, pp. 24–5.
67 Dorn, p. 305, sv *'acetum philosophorum'*.
68 Ripley, *Compound*, III, sig. E(r).
69 *Rosarium Philosophorum*, cited JCW XII 79, from *Artis Auriferae* II 223.
70 Lambspringk, MH I 287.
71 Ibid., MH I 286.
72 Sendivogius, p. 74.
73 Grasshoff, MH I 12, quoting the *Summa Perfectionis* of Geber.
74 Ibid., loc. cit., quoting 'Lilium' (Alain de Lisle?).
75 Bloomfield, TCB 312–13.
76 Bacon, F., Book II, fols 65(v)–66(r).

'OF WIT AND OF ALCUMIE'

77 Lodge, sig. I3(r).
78 Jonson, *Alchemist*, II, iii, 182–92. He is apparently echoing Arnald of
Villanova. Vaughan (p. 173) cites an almost identical passage, attributing
it to 'Arnoldus de Nova' and adding that he in turn 'borrowed this from
the Turba' (i.e. the Arab text, *Turba Philosophorum*). I have not traced
the original.
79 Ibid., II, v, 40–4.
80 Ibid., II, iii, 83.
81 This is quoted by Philalethes, *Ripley Reviv'd* (Rola, p. 29) and
Manuductio, MH II 259 (in the latter case acknowledging it as Trevisan's
'parable').
82 Jonson, *Alchemist*, II, iii, 63.
83 Ibid., IV, ii, 46–7.
84 See Duncan, 'Jonson's Use of Arnald', pp. 435f. The three parallels are
between: *Alchemist* II, i, 38–40 and *Rosarium* II, Ch. xxviii (*Bibliotheca
Chemica Curiosa* I 675); *Alchemist* II, iii, 102–14 and *Rosarium* II, Ch.
xxix (BCC I 675); *Alchemist* II, v, 36–40 and *Rosarium* II, Ch. 1 (BCC I
667).
85 Jonson, *Alchemist* II, i, 90, 94–5.
86 Flamel, Nicolas, *His Exposition of the Hieroglyphical Figures*, trans.
Eugenius Orandus (London, 1624), p. 67. Cited Duncan, 'Jonson's
"Alchemist" ', p. 704.
87 By Herford and Simpson, BJ X 47.
88 TC I 7f. Herford and Simpson note brief parallels with other tracts in the
Theatrum Chemicum – Johannes Chrysippus Fanianus' *De Arte*

Metallicae Metamorphoseos (TC I 33f), Theobald de Hoghelande's *De Alchemiae Difficultatibus* (TC I 109f), and Bernardus Penotus' *Quaestiones et Responsiones* (TC II 129f). In their copious notes to *The Alchemist*, Herford and Simpson mention a wide spectrum of alchemical literature, but they do so to elucidate alchemical references, rather than to suggest actual sources that Jonson may have used. Among the texts they refer to are the following: classics like Geber's *Summa Perfectionis*, Ricardus Anglicus' *Correctorium Alchemiae*, Bernard Trevisan's *Responsio*; contemporary texts such as Andreas Libavius' *Commentationes Metallicae* (1597), Martin Ruland's *Progymnasmata Alchemiae* (1607) and his *Lexicon Alchemiae* (1612), Heerman Condeesyanus' (i.e. Johann Grasshoff's) *Harmoniae Inperscrutabilis Chymico-Philosophiae Decas* (1625); Paracelsist works including Paracelsus' *Manuale de Lapide Philosophico*, George Baker's *The New Phisicke & the Old Phisicke* (1599), Francis Anthonie's *Medicinae Chymicae* (1610). They also show how frequently Jonson quotes verbatim from Martin del Rio's *Disquisitiones Magicae* (1599), drawing on del Rio's diatribes against alchemy.

89 Jonson, *Alchemist*, II, iii, 149, 153–4, 159–61.
90 Bacon, R., pp. 2, 9.
91 Jonson, *Alchemist*, II, ii, 24–8; Ripley, *Compound* (Recapitulation), sig. K4(r).
92 Ripley, *Compound* (Epistle), sig. M2(v).
93 Jonson, *Alchemist*, II, iii, 29–30.
94 Ripley, *Compound* (Recapitulation), sig. K3(v).
95 Jonson, *Alchemist*, II, iii, 167–8.
96 Sendivogius, pp. 20, 108.
97 It is probable that *Mercury Vindicated* was first performed on 1 January 1616. See BJ X 545–8.
98 Jonson, *Mercury Vindicated*, BJ VII 410–11.
99 Sendivogius, pp. 65, 67, 69.
100 Ibid., p. 63.
101 Jonson, *Mercury Vindicated*, BJ VII 412.
102 Sendivogius, p. 66; Jonson, *Mercury Vindicated*, BJ VII 413.
103 Sendivogius, op cit, p. 72; Jonson, *Mercury Vindicated*, BJ VII 414–15.
104 The friend was Howell: Jonson in turn retaliated with his 'Execration upon Vulcan'.
105 Lyly, sig. C4(v), II, ii, 86–8, 94–5; sig. E(r), III, iii, 20–1; sig. D(r), II, ii, 123–5.
106 Southwell, 'The Burning Babe', in *St Peter's Complaint* (1595). This was a posthumous publication (Southwell was hanged and quartered at Tyburn in 1595 for treason: he was a Jesuit). The poem was probably written before his imprisonment in 1592.
107 Harvey, p. 293.
108 Elegie VIII, JD I 91–2.
109 'Ecologue' (26 December 1613). JD I 133.
110 Shakespeare, Sonnet XXXIII, 3–4.
111 Shakespeare, *King John*, III, i, 77–80.
112 Shakespeare, *Sonnet* CXIV, 3–8; *Julius Caesar* I, iii, 158–60; *Antony and Cleopatra*, I, v, 35–7.
113 Davies, pp. 23–4: *Nosce Teipsum* (1599), Part II ('Of the Soule of Man and the Immortalitie Thereof'), section vii ('That it cannot be a Body').
114 Nashe, *Lenten Stuffe*, sigs K3(r)–K3(v), TN III 220–1.

NOTES TO PAGES 105–113

115 Marlowe, *Tamburlaine* Part I, V, ii, 102–3; Harvey, p. 68; Shakespeare, Sonnet LIV, 14; Nashe, Preface to Robert Greene's *Menaphon*, TN III 311; Middleton, 'The Golden Age' in *Father Hubburd's Tales* (1604); Donne, 'To E. of D. with six holy sonnets', JD I 317: on identity of 'E of D' (Richard Sackville, Earl of Dorset?) and date of dedication (c. 1609?), see JD II 226–7.
116 Sidney, sig B4(v)–C(r). This was written c. 1580.
117 Nashe, *The Terrors of the Night*, sig. A2(r), TN I 341.
118 Shakespeare, *Timon of Athens*, II, ii, 114–15.

5 Alchemical Patterns in Jonson and Donne

EASTWARD HOE: A CHEMICAL MORALITY

1 Jonson, *Eastward Hoe*, I, i, 149; IV, ii, 54; I, ii, 34–5.
2 Ibid., I, ii, 81–5.
3 Ibid., II, i, 120–1.
4 Ibid., I, i, 106–14.
5 Paracelsus, HAWP I 117–18.
6 Jonson, *Eastward Hoe*, SD after I, i, 131.
7 Ibid., II, ii, 35.
8 The word 'gas' was coined by the Belgian chemist J. B. Van Helmont (1577–1644). He derived it from Greek χαοσ, chaos (literally, abyss or chasm). Paracelsus calls his *arcanum* 'a chaos, clear, pellucid' (*Paragranum* III, HAWP II 153).
9 Jonson, *Eastward Hoe*, IV, i, 123–7.
10 Ibid., IV, ii, 328.
11 Tymme, sigs X(r)–X(v). See also ibid., sig. P3(v): 'Inconstant Mercurie ... always tendeth to his perfection, that is to say his coagulation and fixation.'
12 Jonson, *Eastward Hoe*, V, ii, 45–7.
13 Ibid., IV, ii, 68.
14 Grasshoff, MH I 49.
15 Ripley, *Compound*, II, sig. D3(r).
16 Jonson, *Eastward Hoe*, IV, ii, 251–3.
17 Ibid., V, v, 61–4, 82–3.
18 Jonson, *Alchemist*, I, i, 64–73.
19 Bloomfield, TCB 313. Paracelsus, *Archidoxis* IV (HAWP II 22): 'The quintessence ... once indeed shut up within things, but now free from any domicile.'
20 'Hermes', *Tractatus Aureus*, III (Regardie, p. 36).
21 Maier, *Symbola*, motto beneath the portrait of Morienus (Symbolum IV): '*Hoc accipe, quod in Sterquiliniis suis calcatur.*'
22 Jonson, *Alchemist*, I, iii, 91–2.

AURUM PALPABILE

23 Ibid., I, iii, 100–4.
24 Ibid., II, i, 78–9.
25 Ibid., I, i, 132–3.
26 Jonson, *Volpone*, I, i, 73.

27 Jonson, *Every Man in his Humour*, Prologue 23–4.
28 See Knights, Ch. 7, 'Jonson and the Anti-Acquisitive Attitude', which closely argues the basic assertion that 'from first to last one of Jonson's main preoccupations was acquisition'.
29 Jonson, *Alchemist*, I, i, 101–2.
30 Ibid., II, i, 38–40.
31 Jonson, *Volpone*, I, i, 26–8.
32 Keynes, J. M., *A Treatise on Money* II, pp. 158–9. Cited Knights, p. 41.
33 Jonson, *Alchemist*, II, i, 1–7.
34 Jonson, *Eastward Hoe*, III, iii, 25–32.
35 Scott, W. R., *Joint-Stock Companies* I, pp. 81–2. Cited Knights, p. 40, n. 4.
36 Jonson, *Alchemist*, II, i, 29–36.
37 Jonson, *Volpone*, I, i, 1–13.
38 Bloomfield, TCB 313.
39 Jonson, *Volpone*, I, i, 23, 25–6.
40 Bacon, R., p. 17.
41 Jonson, *Volpone*, III, vii, 270–2. Shakespeare plays on a similar anomaly between symbolic and financial gold in *Timon of Athens*. Gold, says Timon, 'will make Black, white; foul, fair; wrong, right; Base, noble; old, young. . . .' (IV, iii, 28–30). This transforming potency is ironically attributed to gold-as-money.
42 Ibid., I, v, 108–14.
43 Ibid., I, iii, 69–73.
44 Ibid., I, iv, 68–73.
45 Jonson, *The Divell is an Asse*, II, i, 1–10.
46 The phrase is Coleridge's.

LOVE'S ALCHEMY

47 On the dating – or rather undateableness – of the *Songs and Sonets*, see JD II 8–10. A twenty-year date-span of *c.* 1594–1614 would probably be safe enough. None of the 55 love-lyrics which comprise the *Songs and Sonets* was published in Donne's lifetime. They were first published in *Poems by J.D.* (London, 1633).
48 Montaigne, II, Ch. I ('Of the Inconstancie of our Actions'), p. 194. Florio's trans. (1603) was dedicated to Donne's friend and benefactor, Lucy, Countess of Bedford.
49 'Song' ('Sweetest love'), JD I 19.
50 'The Prohibition', JD I 67.
51 'Song' ('Goe, and catche a falling starre'), JD I 9.
52 'The Indifferent', JD I 12.
53 'The Canonization', JD I 15.
54 'Valediction: of my name, in the Window', JD I 27; 'The Legacie', JD I 20; 'Song' ('Sweetest love'), JD I 18; 'Valediction: of weeping', JD I 38.
55 'The Good-morrow', JD I 7.
56 'Womans Constancy', JD I 9.
57 Montaigne, II, Ch. I, p. 194.
58 Shakespeare, *Hamlet*, III, ii, 88–9.
59 Ibid., III, i, 62–3.
60 Ibid., I, v, 172.
61 Beckett, Samuel, *Proust* (London, 1931); paperback 1970, p. 8.

62 Shakespeare, *Hamlet*, III, i, 79.
63 Elegie XIX, 'Going to Bed', JD I 120.
64 'Valediction: of my name, in the Window', JD I 26.
65 'The Anniversarie', JD I 24; 'The Sunne Rising', JD I 11.
66 'The Good-morrow', JD I 7.
67 'Valediction: forbidding mourning', JD I 51.
68 'The Good-morrow', JD I 7.
69 First words of 'Breake of Day', JD I 23; 'The Expiration', JD I 68; 'The Flea', JD I 40.
70 'Valediction: of the booke', JD I 30.
71 'The Extasie', JD I 52; 'Valediction: forbidding mourning', JD I 50.
72 'Loves Growth', JD I 33.
73 Paracelsus, *Archidoxis* IV, HAWP II 22.
74 Bostocke, *The Difference*, Ch. 6. See Ch. 3, n. 49.
75 'The Good-morrow', JD I 7.
76 Norton, Ch. V, TCB 67.
77 Ibid., Ch. V, TCB 85.
78 'Corruption dwells where contrariety dwells.' Aquinas, *Summa Theologica* I, Quaest. lxxxv, Art. 6. Cited JD II 11.
79 Sendivogius: Waite's translation (MH II 115).
80 Tymme, sig. Q2(v).
81 Grierson suggests Plotinus (JD II 42), Helen Gardner Ebreo, (see *The Elegies and Songs and Sonnets*, ed. Gardner (Oxford, 1965), Appendix D.) 'The Extasie' is in JD I 51–3.
82 Letter to Sir Thomas Lucy, cited JD II 42.
83 'A Nocturnall upon S. Lucies day', JD I 44.
84 'To the Countesse of Bedford' ('Reason is our Soules. . . .'), JD I 190.
85 'An Anatomie of the World' ('The First Anniversary'), 56–8, JD I 233.
86 Tymme, sigs S3(r), S4(r) – Book II, Ch. 7: 'The vertue and preheminence of the Medicine Balsamicke.'
87 Ruland, p. 69, sv *balsamum*.
88 Shakespeare, *Venus and Adonis*, 25–8.
89 Ripley, *Compound*, III, sigs D3(v)–D4(r).
90 Ibid., IV, sig. E2(r).
91 Siebmacher, MH I 80.
92 Ripley, *Compound*, II, sig. D2(r).
93 Bonus, *Pretiosa Margarita Novella* (ed. Lacinius, Venice, 1546). Cited from Waite's translation (1894) slightly modified by Holmyard, pp. 143–4.
94 Sir Robert Drury, who became Donne's patron in early 1611. Elizabeth had died in December 1610: Donne had never met her.
95 'Anatomie of the World', 68–9. JD I 233.
96 Ibid., 11–13, JD I 231.
97 Ibid., 55–8, JD I 233.
98 Ibid., 74–8, JD I 233.
99 Valentine, *Twelve Keys*, p. 242.
100 'Anatomie of the World', 178–82, JD I 236.
101 Ibid., 401–5, JD I 243.
102 Ibid., 415–18, 425–6, JD I 243–4.
103 Lully, *De Tincturis Compendium, seu Vade Mecum*; in *Bibliotheca Chemica Curiosa*, ed. J. J. Manget (1702), I, p. 849.
104 'Elegie on the Lady Marckham', JD I 280.

105 JD I 64–5.
106 Tymme, sig. D(r).
107 Ibid., sigs G3(r)–G3(v).
108 'Anatomie of the World', 205–6, JD I 237.
109 The known dates of the latter two – 1611 and 1609 respectively – may help to locate the unknown date of 'The Extasie'.
110 'Resurrection, Imperfect', JD I 334.
111 *Holy Sonnets*, V, JD I 324; *The Litanie* I ('The Father'), JD I 338.
112 JD I 44–5.
113 *Frankenstein, The Fly*, and a host of Hammer films spring to mind.
114 Siebmacher, MH I 84.

6 Alchemical Bearings on *King Lear*

THE *LONGISSIMA VIA*

1 The translation is A. S. B. Glover's from the Greek text of the *Visions* in Berthelot, Marcellin, *Collection des Anciens Alchimistes Grecs* (1890). Cited JCW XIII 59–60. The complete text of the *Visions* (pp. 59–65) is followed by Jung's masterful commentary (pp. 66–108).
2 Jung asserts (JCW XIII 73) that the bowl-shaped altar is 'unquestionably related to the *krater* of Poimandres'. See above, Ch. 2, n. 131.
3 *Aenigmata ex visione Arislei*, in *Artis Auriferae* (Basle, 1593), I, 146–54. See JCW XII 327–37.
4 This version appears in vol. II of *Artis Auriferae* (pp. 246f) as part of the *Rosarium Philosophorum* (pp. 204–384). In this the prince is called Gabricus.
5 On the dating of this translation, see above, Ch. 2, n. 47.
6 Charnock, Ch. 2, TCB 291–2.
7 The parable is MH I 41–50. On the tortuous attribution of this text to Johann Grasshoff (author of *Der Kleine Bauer*, 1617), see Ferguson, *Bibliotheca*, sv Rhenanus, Grasshoff. The work first appeared in German in the *Dyas Chymica Tripartita* (Frankfurt, 1625) by Heerman Condeesyanus (a pseudonym of Grasshoff's).
8 Aristotle, *On the Art of Poetry*, Ch. 6 ('A Description of Tragedy'). Catharsis is from Greek καθαιρειν, to cleanse, or purify.
9 Erlich, Avi, *Hamlet's Absent Father* (Princeton, 1978). Cited TLS, 21 April 1978, p. 443.
10 'Neither wholly guilty, nor wholly innocent. She is entangled by her destiny, and by the anger of the gods, in an unlawful passion which appals her.' Racine, *Phèdre* (Preface), in *Théâtre Complet*, II, p. 197.
11 Sophocles, *King Oedipus* (Penguin edn, trans. E. F. Watling, pp. 28, 29.)
12 Shakespeare, *Hamlet*, I, v, 189–90. I am concentrating on the second of three 'types' of tragedy which E. M. W. Tillyard discerns in *Hamlet*: 'The second type of tragic feeling has to do with sacrificial purgation and it is rooted in religion. . . . The aim is to rid the social organism of a taint. The audience will be most moved as the victim is or represents one of themselves' (*Shakespeare's Problem Plays*, p. 20).
13 On 'individuation' – the harmonizing of rational conscious and chaotic unconscious – see JCW IX (i), *Conscious, Unconscious, and Individuation*; Fordham, Ch. 4 ('Religion and the Individuation Process').

14 'An inquiry into the separation and synthesis of psychic opposites in alchemy' is the sub-title of Jung's last book, *Mysterium Coniunctionis* (1955–6), JCW XIV.
15 *Rosarium Philosophorum*, cited (without page ref.) JCW XII 335 n. 46.
16 JCW XII 6.
17 Shakespeare, *King Lear*, IV, vi, 110: III, iv, 28.
18 Bradley, Lecture VII, p. 235.
19 Shakespeare, *King Lear*, III, ii, 70–1.

THE WHEEL OF FIRE

20 Shakespeare, *King Lear*, IV, vii, 45–8.
21 Ibid., II, ii, 173.
22 Ibid., II, iv, 71–3.
23 Shakespeare, *Henry V*, III, vi, 31–4.
24 Shakespeare, *King Lear*, III, iii, 27.
25 Ibid., V, iii, 174.
26 T. S. Eliot, 'East Coker', III (*Collected Poems and Plays*, p. 181).
27 Shakespeare, *King Lear*, I, i, 90.
28 Eliot, 'East Coker', V (*Collected Poems and Plays*, p. 183).
29 Shakespeare, *King Lear*, IV, vi, 161; IV, ii, 49–50. The latter is not in F.
30 Ibid., I, iv, 284–8; III, ii, 7–8.
31 Ibid., III, iv, 147.
32 Ibid., I, iv, 277–8.
33 Ibid., V, iii, 320; IV, iv, 16–17; V, iii, 16–17.
34 Horatio warns Hamlet that the Ghost 'might deprive your soveraigntie of reason' (*Hamlet*, I, iv, 73).
35 Shakespeare, *King Lear*, II, iv, 57–8, 121.
36 Ibid., I, v, 8–9.
37 Ibid., IV, vi, 120–1.
38 Ibid., I, i, 144–5.
39 Ibid., IV, vi, 126–30.
40 Bacon, R., pp. 20–1.
41 Sherwood Taylor, *Alchemists*, p. 112, associates circulation with the 'Lullian' alchemy that took root in England in the fourteenth century. Yates, *Art of Memory*, p. 190, links this to the 'combinatory wheels' which are a feature of genuine Lullism.
42 'Hermes', *Tractatus Aureus* (1610 edn), pp. 262f (commentary by Dominicus Gnosius). Cited JCW XII 128.
43 MS fragments of Charnock's collected by Ashmole (MS 1445 pp. 37–9) describe chemical events after various numbers of circulations, e.g.: 'At the number off 561 there apperyd a lyttell star bright and shining whitter than the snowe.... At 600 and 10 there apperid a 5 or 6 small White pearles copped and round.' Notes like 'Circulacon 520 the seige off Troy' suggest that each one took a week (520 weeks equals 10 years, the length of the siege of Troy). See Taylor, 'Thomas Charnock', pp. 171–6.
44 Norton, Ch. V, TCB 82–4.
45 Bloomfield, TCB 315.
46 Ripley, *Compound*, I, sig. C4(v).
47 Ibid., II, sigs. D2(v)-D3(r).
48 Ibid. (recapitulation), sig. K3(v).
49 Khunrath, *Von Hylealischen Chaos*, p. 204. Cited JCW XII 124.

50 Dorn, *Philosophia Chemica ad Meditativam Comparata*, TC I 492. Cited JCW XII 381.
51 The arrow in the left hand of the wheel-turner is a version of the sign for Sulphur (⚧); the peg-leg of the other figure identifies him as Vulcan. These are by no means the only interpretations possible.
52 Bacon, R., p. 38.
53 Muir's note to IV, vii, 47 in AS *King Lear*, on the authority of H. W. Crundell.
54 Boehme, Jakob, *De Signatura Rerum* (trans. Ellistone), Ch. IV, 28. Cited JCW XII 166.

AN ALCHEMICAL RECONSTRUCTION

55 Shakespeare, *King Lear*, I, ii, 107, 112–13, 118.
56 No less than three quartos of *Eastward Hoe* were published in 1605, suggesting its popularity on stage; *Volpone* was described on its title-page in the Folio *Works* of 1616 as 'acted in the yeere 1605', though Herford and Simpson evoke the ambiguities of old style/new style dating, and suggest early 1606 as the date of the first performance.
57 Keats, letter to Richard Woodhouse, 27 October 1818.
58 I have shamelessly borrowed this device from T. Walter Herbert's *Oberon's Mazéd World*, in which 'a judicious young Elizabethan contemplates *A Midsummer Night's Dream*'.

7 The Transmutation of King Lear

THE KING

1 Shakespeare, *King Lear*, I, i, 1–5.
2 Ibid., I, i, 37–41.
3 Ibid., I, i, 62.
4 Ibid., I, i, 63.
5 Ibid., I, i, 135–7.
6 Sonnet CXIV, 2.
7 Shakespeare, *King Lear*, I, i, 110–20. 'Mysteries of Hecate' is F2's emendation of F's 'miseries of Hecatt'.
8 Ripley, *Cantilena*, vv. 3–5, 7, 10.
9 'Hermes', *Tractatus Aureus*, Ch. VII (Regardie, p. 40).
10 Sendivogius, p. 10.
11 Sendivogius – Waite's translation (MH II 99–100): J. F. (p. 10) has 'air' instead of 'ore', probably a printer's error.
12 Ibid., p. 10.
13 Valentine, *Twelve Keys*, p. 233.
14 Trismosin, *Splendor Solis* (Tractatus III of *Aureum Vellus*, 1598), trans. J. K. (1920), pp. 29f. Cited JCW XIV 331.
15 Maier, *Symbola*, p. 380.
16 See Shirley, p. 64. The author was Petrus Bonus of Ferrara; the work was written *c.* 1330, and first published by Janus Lacinius (Venice, 1546).
17 Bonus, *Petiosa Margarita Novella*, sig. VIII (v). My translation.

THE DRAGON

18 Shakespeare, *King Lear*, I, i, 124, 233–4, 262–4.
19 Ibid., I, i, 177–8.
20 Ibid., I, i, 127–32.
21 Ibid., I, i, 163–4, 180–1.
22 Ibid., I, iv, 117–19.
23 Ibid., I, i, 145–9.
24 Ibid., I, i, 117–21.
25 Ripley, *Verses belonging to Ripley's Scrowle*, TCB 378; 'Hermes', *Tractatus Aureus*, Ch. III (Regardie, 37); Valentine, *Twelve Keys*, p. 238; see also the anonymous verses (TCB 354) about 'A dragon lying in his deepe denne. . . . Leprouse huge and terrible in sight'.
26 Ripley, *Compound*, III, sig. E(v).
27 *Aureliae Occultae Philosophorum Duo Partes*, TC IV 501f. Cited JCW XIII 218.
28 Andrewes, TCB 279.
29 See reproductions, Rola, plate 20 (*Rosarium*) and p. 105 (*Phil. Ref.*).
30 See reproduction, Taylor, *Alchemists*, pl. 1; another early Greek version of Ourobouros is in Rola, pl. 1; contemporary versions are in JCW XIII pl. 7 (from Horapollo, *Selecta Hieroglyphica*, 1597) and pl. 13 (Reusner, *Pandora*, 1588); and in Maier, *Atalanta*, Emblema XIV ('Hic est Draco caudam suam devorans'), p. 65.
31 JCW XIII 293.
32 Lambspringk, MH II 286–7.
33 Shakespeare, *King Lear*, III, i, 7–9: 'his white hair, Which the impetuous blasts . . . . Catch in their fury, and make nothing of.' AS, following the 1608 Quarto: the passage is not in F.
34 From the poem 'Verus Hermes', in *Prodromus Rhodostauroticus, Parergi Philosophici* (1620). Cited JCW XIII 228.

THE DAUGHTER

35 Shakespeare, *King Lear*, I, i, 69. I have preserved the reading in the 1608 Quarto, rather than F's 'I am made of that selfe-mettle as my Sister', which AS follows. The latter is rhythmically superior, but the former serves my chemical reading better. 'Mettle' was anyway no more than a variant spelling of 'metal', though we have preserved it as a separate word.
36 Ibid., I, i, 123–4.
37 Ibid., I, i, 213–15. AS preserves 'best' from the Quarto: it is omitted in F.
38 Ibid., II, iv, 135–6, 161–2.
39 Ibid., I, iv, 271, 317.
40 Ibid., I, iv, 291–8.
41 Ibid., II, iv, 173–5.
42 Ibid., II, iv, 224.
43 Valentine, *Twelve Keys*, Waite's trans (MH I 352) rather than the verse rendition in the 1670 trans (p. 288).
44 Paracelsus, *Opus Paramirum*, lib. IV (*De Matrice*), Huser I 78.
45 Shakespeare, *King Lear*, IV, vii, 28–9.
46 ibid., I, i, 196–7.
47 Ibid., I, i, 250–3.

48 'Hermes', *Tractatus Aureus*, Ch. III (Regardie, p. 36).
49 Lambspringk, MH I 274–5.
50 Grasshoff, MH I 13.
51 Ripley, *Compound* (Preface), sig. B4(r).
52 Valentine, 'The Manual Operations', Ch. 3 ('Of the Spirit of Mercury), in *Last Will and Testament*, p. 486.
53 Philalethes, *Fons*, MH II 262.
54 Philalethes, *Ripley Reviv'd* (Rola, p. 25).
55 JCW XII 23.
56 Boehme, *De Signatura Rerum* (trans. Ellistone), Ch. IV, 27. Cited JCW XII 166.
57 *Tractatus Aristotelis (ad Alexandrum Magnum)*, TC V 884. Cited JCW XII 252.
58 Valentine, *Twelve Keys*, p. 233.
59 Cited Read, *Prelude*, p. 201.
60 Ibid., loc. cit.
61 Sendivogius, p. 74. 'He [sc. Mercury] is a Beast, and yet hath the Wings of a Bird; he is Poison, yet cureth the Leprosie.'
62 Valentine, *Twelve Keys*, p. 233.

THE MORTIFICATION

63 Shakespeare, *King Lear*, I, i, 288–300.
64 Ibid., I, i, 307–8.
65 Ibid., I, iii, 20–1 (AS, following Quarto: not in F).
66 Ibid., I, iv, 82–4.
67 Ibid., I, iv, 146–56 (AS, following Quarto: not in F).
68 Ibid., I, iv, 199–202.
69 Ibid., I, iv, 238–9.
70 Ibid., I, iv, 255–7.
71 Ibid., II, iv, 203–6.
72 Ibid., II, iv, 239–41, 248-50.
73 Ibid., II, iv, 262–5.
74 Philalethes, *Ripley Reviv'd* (Rola, p. 25).
75 Ripley, *Compound*, II, sig. D2(v).
76 Tymme, sig. H3(v).
77 Shakespeare, *King Lear*, I, iv, 279–81.
78 Philalethes, *Introitus*, Ch. XXV, MH II 191.
79 Shakespeare, *King Lear*, III, i, 14; III, iv, 111.
80 Ibid., III, iv, 109.
81 On the 'Vitriol acrostic', see Ch. 1, n. 16.

THE FOOL

82 Shakespeare, *King Lear*, I, iv, 107–9, 168–9, 179–80.
83 Ibid., I, iv, 136–9.
84 Ibid., I, iv, 223–5.
85 Ibid., IV, vi, 193, 184–5.
86 Steward, Julian, *The Ceremonial Buffoon of the American Indian*, pp. 189f. Cited Willeford, p. 83. Enid Welsford, *The Fool* (London, 1935), p. 253, says: 'Lear's tragedy is the investing of the King with motley: it is also the crowning and apotheosis of the Fool.'
87 Shakespeare, *King Lear*, I, iv, 209, 323–4.

88 Ibid., I, iv, 77–8.
89 Cited in *King Lear through the Ages* on BBC Radio 3.
90 Tate's adaptation, complete with love-affair between Edgar and Cordelia and a happy ending, was first performed in 1681.
91 Shakespeare, *King Lear*, V, iii, 305.
92 Willeford, Pl. 26. He compares this to the bisexual emblem of alchemical *coniunctio*: 'The hermaphroditism of the fool, like that of Mercurius, is an expression of the fact that complementary and opposing qualities of many kinds coexist in him' (p. 181).
93 Dorn, *De Tenebris Contra Naturam*, TC I 470; Aegidius de Vadis, *Dialogus inter Naturam et Filium Philosophiae*, TC II 105. Both cited JCW XIII 217.
94 See the drawings from Cantara's *Cabala Mineralis* (BM Add. MS 5245), JCW XII, pl. 121; and *Speculum Veritatis* (Vatican MS, Cod. Lat. 7286) in Rola, pl. 3.
95 Philalethes, *Ripley Reviv'd*, p. 100. Cited JCW XII 146.
96 Shakespeare, *King Lear*, II, iv, 71–2.
97 JCW XII 66. The term '*servus fugitivus*' is glossed by Ruland: 'Servus fugitivus, i.e. Mercurius; Hermes propter humiditatem fugitivam sic nominat'.
98 Sendivogius, p. 64, 67, 71.
99 Shakespeare, *King Lear*, I, iv, 120, 116, 188.
100 Sendivogius, p. 64.
101 Shakespeare, *King Lear*, I, iv, 154 (AS following Quarto: not in F).
102 Sendivogius, p. 67.
103 Shakespeare, *King Lear*, I, v, 23.
104 Ibid., I, iv, 158–61. (AS following Quarto: not in F).
105 Jonson, *Eastward Hoe*, I, ii, 83–5.
106 Shakespeare, *Winter's Tale*, IV, iii, 25.
107 Shakespeare, *King Lear*, IV, vii, 38–9: 'Wast thou fain, poor father, To hovel thee with swine and rogues forlorn?'
108 Virgil, *Aeneid*, IV 356.
109 Jonson, *Timber: or Discoveries*, 1883–4 (BJ VIII 621); *Epigrams*, CXXXIII ('On the Famous Voyage'), 99 (BJ VIII 86).
110 Maier, *Symbola*, p. 600. The words are spoken by the Erythraean Sibyl, Herophyle, during the course of Maier's allegorical journey (each of the four continents representing an element). A condensed version of this *peregrinatio* is published as the *Allegoria super Secreta Chymiae*, MH II 199f.
111 Shakespeare, *King Lear*, V, iii, 16–17.
112 *Larousse Encyclopaedia of Mythology*, p. 207.
113 Ibid., p. 123.
114 *The Dialogue of Cleopatra and the Philosophers*. Cited, from Berthelot's *Collections des Anciens Alchimistes Grecs*, pp. 289f, by Sherwood, *Alchemists*, p. 56.
115 JCW XII 335–6.
116 Rilke, *Orpheus. Eurydike. Hermes* (1904), trans. J. B. Leishman (in *Selected Poems*, pp. 40–1).

THE FIRE

117 Shakespeare, *King Lear*, II, iv, 147–9.
118 Ibid., II, iv, 198–9.

119 Ibid., II, iv, 280–8.
120 Ibid., III, ii, 6.
121 Ibid., III, ii, 1–24.
122 Ibid., III, i, 7–9 (AS, following Quarto: not in F).
123 Ibid., III, ii, 48–9; III, iv, 2–3.
124 Ibid., III, iv, 28–36.
125 Ibid., II, iv, 210–11.
126 Ibid., III, iv, 105–12.
127 Ibid., III, iv, 6–7, 12–13.
128 Ibid., IV, vi, 86–93.
129 JCW XII 25.
130 Sendivogius, p. 31; 'Hermes', *Tractarus Aureus*, Ch. IV (Regardie, p. 38).
131 Zosimos, *Visions* (JCW XIII 60).
132 Norton, Ch. VII, TCB 105.
133 Ripley, *Compound*, III, sig E(v).
134 Norton, Ch. V, TCB 61.
135 Sendivogius, p. 119.
136 See Ch. I, n. 92.
137 Philalethes, *Ripley Reviv'd*, (Rola, pp. 25–6).
138 Ibid., (Rola, pp. 26, 28).
139 Shakespeare, *King Lear*, IV, vii, 41–2.
140 Trismosin, *Splendor Solis*, trans. J.K., p. 29f. Cited JCW XIV 331.

THE BLACKNESS

141 Shakespeare, *King Lear*, III, ii, 67–73.
142 Ibid., IV, i, 5–6.
143 Ibid., IV, i, 18–21.
144 Ibid., III, iv, 39–40.
145 Ibid., II, iii, 7–9.
146 Ibid., III, iv, 161, 176, 184.
147 Ibid., III, vi, 6–7.
148 Philalethes, *Ripley Reviv'd* (Rola, p. 29), citing Bernard Trevisan.
149 Shakespeare, *King Lear*, II, iii, 20–1.
150 Ripley, *Compound*, II, sig. D2(r).
151 Bacon, R., p. 32.
152 Ripley, *Compound*, V, sig. E4(r).
153 Ibid., loc. cit.
154 Valentine, *Twelve Keys*, pp. 256–7.
155 Sendivogius, p. 108.
156 Tymme, sigs G3(v)–G4(r).
157 Sendivogius, p. 58.
158 Donne, Elegie VIII, JD I 91.
159 Vaughan (Regardie, p. 166).
160 Paracelsus, *Philosophia ad Athenienses*, I i, HAWP II 249.
161 Ripley, *Cantilena*, vv. 12–16.
162 Sendivogius, p. 58.
163 Philalethes, *Metallorum Metamorphosis*, MH II 243.
164 Shakespeare, *King Lear*, I, iii, 20–1 (AS, following Quarto: not in F).
165 Ibid., I, ii, 73–4.
166 Ibid., I, iv, 179–81.
167 Ibid., II, iv, 56–8.
168 Jordan, Edward, *A Brief Discourse of a Disease called Suffocation of the*

NOTES TO PAGES 199–212 273

    Mother (1605), p. 5. Cited Muir, note to II, iv, 56–7 in AS *Lear*.
169 Forester, p. 47.
170 JCW XII 71.
171 Ibid., 150–1.
172 Maier, *Symbola*, p. 380. Jung is referring both to this and to the immersion of Thabritius in *Visio Arislei* when he glosses the 'depths of the sea', JCW XII 329.

THE DEW

173 Shakespeare, *King Lear*, III, vi, 86–7, SD (Capell's: not in F) after IV, vi, 80; IV, vi, 117, 130.
174 Ibid., IV, vi, 195, 136.
175 Ibid., IV, vi, 153–61, 166–70.
176 Ripley, *Compound*, V, sigs F(r)–F(v).
177 Shakespeare, *King Lear*, II, ii, 169–70.
178 Ibid., III, i, 46–7.
179 Ibid., IV, iv, 8–9, 15–18; IV, vii, 14–17, 26–9.
180 Ibid., IV, vi, 206–8.
181 Tymme, sig. S4(r).
182 Paracelsus, *Opus Paramirum*, lib. IV (*De Matrice*), Huser I 78. Philalethes, *Fons*, MH II 269. *Rosarium Philosophorum*, cited JCW XII 79, from *Artis Auriferae* II 223. 'Hermes', *Tractatus Aureus*, Ch. II (Regardie, p. 36).
183 Tymme, sig. B4(v).
184 Sendivogius, pp. 70–1.
185 Siebmacher, MH I 80.
186 Grasshoff, MH I 48.
187 Philalethes, *Ripley Reviv'd* (Rola, p. 25).
188 Philalethes, *Fons*, MH II 269.
189 Lambspringk, MH I 302–4.
190 Valentine, 'The Manual Operations', Ch. 3, in *Last Will and Testament*, p. 486.
191 Paracelsus, *Paragranum* III (HAWP II 151).
192 Norton, Ch. V, TCB 77.
193 Ripley, *Cantilena*, v. 28.
194 Ibid., vv. 30–3, 36–7.

THE STONE

195 Shakespeare, *King Lear*, IV, vii, 71–2, 77, 83–4.
196 Ibid., V, iii, 8–19.
197 Donne, 'The Good-morrow', JD I 7.
198 Donne, 'Valediction: forbidding mourning', JD I 50. 'Dull sublunary lovers love (whose soule is sense)'.
199 Shakespeare, *King Lear*, V, iii, 20–5.
200 Ibid., V, iii, 297–300.
201 Ibid., III, iv, 74–5.
202 The numbers (3 and 4) and shapes (triangle and square inside the circle of the globe) suggest the old conundrum of 'squaring the circle'. Maier's *De Circulo Physico Quadrato* (1616) discussed various alchemical ramifications of this: 'Gold is the image of the absolute circle written upon Nature. . . . Within the golden circle is a quadrature of four equal

parts (the four elements)', etc. See Craven, *Count Michael Maier*, pp. 50–1; JCW XII 124f; and Maier's 21st Emblem in *Atalanta Fugiens*, bearing the motto: 'Fac ex mare & foemina circulum, inde quadrangulum, hinc triangulum, fac circulum & habebis lap. Philosophorum.'
203 'Hermes', *Tractatus Aureus*, Ch. III (Regardie, pp. 36–7).
204 Grasshoff, MH I 49.
205 Shakespeare, *King Lear*, IV, iv, 12–15.
206 Grasshoff, MH I 48.
207 Shakespeare, *King Lear*, IV, i, 36–7; V, iii, 170–1; IV, vii, 14–15.
208 Valentine, *Twelve Keys*, p. 277; 'Testament of Arnold de Villanova' in Sendivogius, MH II 83 (omitted in J. F.'s trans); Geberus, *Investigatio Perfectionis*, trans. Richard Russell (1678), cited Holmyard, p. 132; Bringer (ed.), *Azoth* (1613), rubric accompanying emblem of *nigredo*, cited (apparently from French edition of 1659) Read, *Prelude*, pp. 270–1; Khunrath, *Von Hylealischen Chaos*, p. 204, cited JCW XII 124.
209 Ripley, *Compound*, II, sig. D3(r).
210 Shakespeare, *King Lear*, V, iii, 290.
211 Notes to *King Lear* from Johnson's edition of 1765, in Wimsatt, W.K. (ed.), *Dr Johnson on Shakespeare*, p. 126.
212 Shakespeare, *King Lear*, V, iii, 271–3, 257–63.
213 HAWP I 289. See Ch. 1, n. 11.
214 Donne, 'The Sunne Rising', JD I 11.
215 Shakespeare, *King Lear*, V, iii, 262–3, 265–7.
216 HAWP I 289. See Ch. 1, n.11.
217 Shakespeare, *King Lear*, V, iii, 263.
218 Ibid., V, iii, 305–11.

CODA: THOMAS TYMME'S CHYMICALL PATHOLOGY

219 Pomeranz, Herman, 'Medicine in the Shakespearean Plays and Era' (*Medical Life*, XXXXI, October 1934, p. 522). Cited Edgar, p. 245.
220 Pagel, pp. 149–50.
221 Shakespeare, *King Lear*, II, iv, 56–8.
222 Paracelsus, *Vom Fallendt: Causa*, Huser I 543.
223 Tymme, sigs X3(v)–X4(v). Tymme also notes that these inner 'meteors' of madness derive from 'continuall vapours and exhalations which are lifted up from the lower belly (which we fitly compare with the earth) into the aire, that is to say into the uppermost region of the body, the braine' (sig. X2(v)). This ties in again with Lear's conception of madness 'climing' from the 'element below'.
224 Ibid., sig. X4(v); Shakespeare, *King Lear*, III, i, 10–11 (AS, following Quarto: not in F).
225 Paracelsus, *Liber de Caducis*, Huser I 593. Cited Pagel, p. 167.
226 Shakespeare, *King Lear*, II, iv, 199–200, 286–7.
227 Paracelsus, *Liber de Caducis*, Huser I 593. Cited Pagel, p. 167.
228 Tymme, sigs S2(v)–S3(v).
229 Paracelsus, *De Virtutibus Rerum*, Bk. I. Cited Pagel, p. 146.
230 Shakespeare, *King Lear*, IV, iv, 12–15.
231 Paracelsus, *Paragranum* I, Huser I 210.
232 Tymme, sigs S4(v)–T(v).
233 Tymme, sig. G4(v). The definition of transmutation, quoted at the head of the chapter, is sig. Aa4(r), from 'The Conclusion of this Treatise'.
234 Thomas Creede flourished as printer and bookseller between 1593 and

1617: over 70 titles are listed to him in the Stationers' Register. The *Mirror* would have been printed at the Sign of the Catherine Wheel, Thames Street, but the *Chymicall Physicke* at the Eagle and Child, Old Exchange (whither Creede moved business in 1600). Shakespeare would have recognized his device from other title-pages, for Creede printed quartos of *Richard III* (1598 and 1602), *Romeo and Juliet* (1599), *Henry V* (1602) and the *Merry Wives of Windsor* (the appalling First Quarto of 1602). See McKerrow, pp. 80–1; *The Library*, April 1906, pp. 155–7.
235 The verb is correctly '*virescere*' (to grow green).
236 Shakespeare, *King Lear*, I, iv, 117–18.

8 Shakespeare's Chemical Theatre

THE SECRET CAVE

1 Derek Traversi, 'The Last Plays of Shakespeare', in Ford (ed.), *The Pelican Guide to English Literature*, II, p. 257.
2 Shakespeare, *Cymbeline*, V, v, 447.
3 Shakespeare, Pericles, III, ii, 103–4 (Quarto of 1608: not in F).
4 Shakespeare, *Cymbeline*, V, v, 351–2.
5 Shakespeare, *Winter's Tale*, IV, iv, 95–7.
6 Ibid., V, ii, 89.
7 Ibid., V, iii, 37–9.
8 Shakespeare, *The Tempest*, I, ii, 399–404.
9 The scene 'features' six characters, though only five appear. Posthumus is the absent presence.
10 Shakespeare, *Cymbeline*, III, vi, 3, 7.
11 Ibid., III, iii, 2–9.
12 Ibid., III, iii, 79.
13 'The scene is incomparable in its technical virtuosity. Wendell enumerates twenty four distinct dénouements' (Nosworthy, notes to last scene – V, v – in AS *Cymbeline*).
14 Shakespeare, *Winter's Tale*, III, ii, 134–6.
15 'Hermes', *Tractatus Aureus*, III (Regardie, p. 37).
16 Forman does not specify where and when he saw *Cymbeline*, but the other Shakespeare plays he recorded in his 'Bocke of Plaies' were seen at the Globe in 1611 (*Macbeth*, 20 April, and *Winter's Tale*, 15 May). See Schoenbaum, pp. 214–15; and pl. 176, for a facsimile of Forman's account of *Cymbeline* (from MS Ashmole 208, f. 206).
17 The date is unlikely to be authentic, since it refers to the supposed discovery of Christian Rosencreutz's tomb in 1604. This emblem appears in the *Geheime Figuren der Rosenkreuzer* (1785). It is, nevertheless, absolutely typical of this genre of alchemical-Rosicrucian illustration, and may be based on a seventeenth-century original.
18 Shakespeare, *Cymbeline*, III, iii, 69–70.
19 Ibid., III, iii, 83–5.
20 MH I 3. Preface to 1678 edn.
21 Shakespeare, *Cymbeline*, III, vii, 27–8.
22 Virgil, *Aeneid*, VI 258f.
23 'ΜΥΣΤΗΡΙΟΝ profecto vere DIVINUM et quod spectatores omnes praesertim penitiores in admirationem et amorem illius merito rapit.' The

cave is in fact described as the '*porta*' (door, entrance) which leads into the '*amphitheatrum*'.

24 Shakespeare, *Cymbeline*, I, vi, 40–2. There is certainly some alchemical play on this liquor variously believed to be a 'thing most precious' and a 'drug of damn'd nature', a cordial 'which hath the king Five times redeem'd from death' and a poison 'murd'rous to the senses'.

25 Ibid., III, vi, 138–9.

26 Ibid., IV, ii, 197–8.

27 This SD first appears in the Globe edition (ed. Clark and Wright, 1884), but no doubt articulates the obvious gestural conclusion of Imogen's speech. When Lucius discovers her, he asks her: 'Who is this thou mak'st thy bloody pillow?' (IV, ii, 362–3).

28 Ripley, *Compound*, X, sig. I3(v).

29 Ibid., V, sig. F(r).

30 *Visio Arislei* (version in *Rosarium Philosophorum: Artis Auriferae* II 246f). Cited JCW XII 337. I have altered Gabricus to Thabritius for continuity. See Ch. 6, n. 4.

31 Shakespeare, *Cymbeline*, V, iv, 138–40.

32 The lion belongs to the same nucleus of alchemical symbols as the dragon, suggesting transformation as a symbiotic (self-) devouring and (self-) generating process: 'his venom becomes the great medicine.' Shakespeare's 'Lyon's whelpe' might be glossed as 'matter undergoing transformation'. Cf. Andrewes, TCB 279: 'Our Lyon ... soone can overtake the Sun, and suddainely can hym devoure.... Whan he hath the Sun up eate, he bringeth hym to more perfection.... This Lyon maketh the Sun sith soone to be joyned to his sister the Moone: By way of wedding a wondrous thing, thys Lyon should cause hem to begett a King.' Thus the Lion 'oversees' the mortification of Raw Stuff (Sun = *aurum vulgi*) and the generation of the Stone (King): the pivotal point is once again the 'incestuous' chemical wedding (between Sun and Moon). The 'Lyons whelpe' can be seen as a precise and intentional alchemical image (as is 'embraced by tender air', to suggest the ministrations of *anima mercurii*).

33 Maier, *Themis Aurea* (Preface), sigs. A3(v)-A4(r). This anonymous translation, published in 1656, was dedicated to Elias Ashmole. The work, subtitled 'The Laws of the Fraternity of the Rosie Crosse', is an apologia for the Rosicrucians. The theme of the chemical wedding is prominent in Rosicrucian alchemy, cf. Andreae's *Chymische Hochzeit*, published two years earlier than *Themis*.

34 Maier, *Themis Aurea* (Preface), sig A3(v); Shakespeare, *Cymbeline* V, v, 443.

35 Arthos, J., 'Pericles, Prince of Tyre: A Study in the Dramatic Use of Romantic Narrative', *Modern Language Review*, lii, (October 1957), p. 583. Cited Introduction to AS *Pericles*, p. lxxvii.

36 The King's Men acquired the lease of the Blackfriars theatre on 9 August 1608 and began playing there slightly over a year later. It accommodated a seated audience of 700 (as opposed to the milling 3,000 at the Globe); admission was sixpence (a penny for the groundlings at the Globe). 'At Blackfriars, in a glittering and hushed atmosphere, the King's men catered for a sophisticated clientele.... Theatre historians have seen the transfer as profoundly influencing the stagecraft of Shakespeare's last plays' (Schoenbaum, pp. 213–4). See Bentley, G. E., *Shakespeare and his Theatre* (Lincoln, Nebraska, 1964).

37 Shakespeare, *Henry V*, Prologue, 11–14.
38 Shakespeare, *The Tempest*, IV, i, 118–20.
39 Ibid., IV, i, 148–50.
40 Shakespeare, *A Midsummer Night's Dream*, Epilogue (spoken by Puck), V, i, 422–5.
41 Shakespeare, *Hamlet*, II, ii, 524–30. I have retained F's 'warm'd' (527), though most modern editions have 'wann'd', following the Quartos.
42 Ripley, *Compound*, II, sig. D3(r).
43 Shakespeare, *The Tempest*, IV, i, 156–7.

Bibliography

1 Anthologies and Collected Editions

ALCHEMICAL

MH   *Musaeum Hermeticum Reformatum et Amplificatum ... Continens Tractatus Chimicos XXI praestantissimos* (2 vols, Frankfurt, 1678; expanded from original edition, Frankfurt, 1625). Trans. A. E. Waite, *The Hermetic Museum Restored and Enlarged* (2 vols, London, 1893).

TC   *Theatrum Chemicum, praecipuos selectorum auctorum Tractatus de Chemiae et Lapidis Philosophici antiquitate, vertitate, jure, praestantia, & operationibus, continens,* ed. Lazarus Zetzner (3 vols, Ober-Ursel, 1602; and vol. IV, Strasbourg, 1613; vol. V, Strasbourg, 1622; vol. VI, Strasbourg, 1661).

TCB   *Theatrum Chemicum Britannicum. Containing Severall Poeticall Pieces of our Famous English Philosophers,* ed. Elias Ashmole (London, 1652).

Huser   *Aureoli Philippi Theophrasti Bombasts von Hohenheim Paracelsi ... Opera, Bücher und Schrifften,* ed. Johann Huser (2 vols, Strasbourg, 1603; condensed from original edition, Basle, 1589–90).

HAWP   *The Hermetic and Alchemical Writings of Paracelsus,* trans. A. E. Waite, based on the Geneva folio of 1658 (2 vols, London, 1894).

JCW   *The Collected Works of C. G. Jung,* ed. Herbert Read, Michael Fordham, Gerhard Adler, trans. R. F. C. Hull (19 vols, London, 1953–). Especially vols:
      XII   *Psychology and Alchemy* (revised edition, 1968); trans. from *Psychologie und Alchemie* (1944; revised edition, 1952).
      XIII   *Alchemical Studies* (1967); trans. from various works including *Paracelsus als Geistige Erscheinung* (1942); *Der Geist Mercurius* (1948); *Die Visionen des Zosimos* (1954); *Der Philosophische Baum* (1954).
      XIV   *Mysterium Coniunctionis: An Inquiry into the Separation and Synthesis of Psychic Opposites in Alchemy* (second edition, 1970); trans. from *Mysterium Coniunctionis* (1955–6).
      XV   *The Spirit in Man, Art and Literature* (1966); trans. from various works including *Paracelsus* (1929) and *Paracelsus als Artz* (1941).

LITERARY

F   *Mr. William Shakespeares Comedies, Histories, & Tragedies. Published according to the True Originall Copies*, eds John Heminges, Henry Condell (London, 1623). Except where stated in the Notes, all Shakespeare quotations are taken from this edition. (F = first Folio).
AS   *The Arden Shakespeare*; Individual plays, ed. various (London, 1899–). Except where stated in the Notes, all Shakespeare line-references are taken from this edition.
BJ   *Ben Jonson*, ed. C. H. Herford, Percy and Evelyn Simpson (11 vols, Oxford, 1925–52).
JD   *The Poems of John Donne*, ed. Herbert J. C. Grierson (2 vols, Oxford, 1912).
TN   *The Works of Thomas Nashe*, ed. Ronald B. McKerrow; reprinted with corrections and supplementary notes, ed. F. P. Wilson, (5 vols, Oxford, 1958).

2 Individual Texts

Agrippa, Heinrich Cornelius, *De Occulta Philosophia* (Lyons, 1531); trans. J. F. (John French?), *Three Books of Occult Philosophy* (London, 1651).
Andreae, Johann Valentin, *Chymische Hochzeit Christiani Rosencreutz* (Strasbourg, 1616); trans. Ezekiel Foxcroft, *The Hermetic Romance, or The Chymical Wedding* (London, 1690).
Andrewes, Abraham, *The Hunting of the Greene Lyon* (late 15th century?), published in TCB, 278f).
Anon., The 'Rosicrucian manifestos': *Fama Fraternitatis, des Loblichen Ordens des Rosenkreutzes* (Cassel, 1614) and *Secretioris Philosophiae Confessio Brevis . . . cum Confessione Fraternitatis R.C.* (Cassel, 1615); trans. Thomas Vaughan, *The Fame and Confession of the Fraternity of R.C.* (London, 1652); in Yates, *Rosicrucian Enlightenment*, pp. 238f.
Anthonie, Francis, *The Apologie of Aurum Potabile* (London, 1616).
Arber, Edward, *A Transcript of the Registers of the Company of Stationers of London* (5 vols, London, 1875–94).
Aubrey, John, *Brief Lives* (late 17th century), ed. Oliver Lawson Dick (London, 1949).
Backhouse, William, *The Magistery* (1633, published in TCB, 342).
Bacon, Francis, *Of the Proficience and Advancement of Learning, divine and humane* (London, 1605).
Bacon, Roger (Attrib.), *The Mirror of Alchimy* (London, 1597).
Bald, R. C., *John Donne: A Life* (Oxford, 1970).
Bloomfield, William, *Bloomefields Blossoms, or The Campe of Philosophy* (1557, published in TCB, 305f).
Boas, Marie, *The Scientific Renaissance, 1450–1630* (London, 1962).
Bradley, A. C., *Shakespearean Tragedy* (London, 1904); paperback 1965.
Burckhardt, Titus, *Alchemie* (Olten, 1960); trans. William Stoddart, *Alchemy: Science of the Cosmos, Science of the Soul* (London, 1967).
Charnock, Thomas, *The Breviary of Naturall Philosophy* (1557, published in TCB, 291f).
Chaucer, Geoffrey, *The Canterbury Tales* (c. 1386–1400), ed. A. C. Cawley (London, 1958).

Craven, J. B., *Doctor Robert Fludd* (Kirkwall, 1902).
Craven, J. B., *Count Michael Maier* (Kirkwall, 1910).
Davies, Sir John, *Nosce Teipsum (and) Hymnes of Astraea in Acrosticke Verse (and) Orchestra* (London, 1622).
Debus, Allen G., *The English Paracelsians* (London, 1965).
Debus, Allen G., 'Renaissance Chemistry and the Work of Robert Fludd', in Debus and Robert P. Multhauf, *Alchemy and Chemistry in the Seventeenth Century* (Los Angeles, 1968).
Debus, Allen G., *The Chemical Dream of the Renaissance* (Cambridge, 1968).
Debus, Allen G., 'The Paracelsian Compromise in Elizabethan England', *Ambix*, vol. VIII (1960), pp. 71f.
Debus, Allen G., 'Mathematics and Nature in the Chemical Texts of the Renaissance', *Ambix*, vol. XV (1968), pp. 1f.
Dee, John, *Monas Hieroglyphica* (Antwerp, 1564); trans. (with facsmile of original edition), C. H. Josten, 'A Translation of John Dee's "Monas Hieroglyphica" ', *Ambix*, vol. XII (1964), pp. 8f.
Descartes, René, *Principia Philosophiae* (Amsterdam, 1644); trans. Elizabeth S. Haldane and G. R. T. Ross, *The Principles of Philosophy*, in *The Philosophical Works of Descartes* (2 vols, Cambridge, 1911).
Dorn, Gerhard, *Dictionarium Paracelsi* (Frankfurt, 1583); trans. J. F. (John French?), *A Chymical Dictionary*; in Sendivogius, pp. 305f.
Duncan, Edgar Hill, 'Donne's Alchemical Figures', *English Literary History*, vol. IX (1942), pp. 257f.
Duncan, Edgar Hill, 'The Alchemy in Jonson's "Mercury Vindicated" ', *Studies in Philology*, vol. XXXIX (1942), pp. 625f.
Duncan, Edgar Hill, 'Jonson's Use of Arnald of Villa Nova's "Rosarium" ', *Philological Quarterly*, vol. XXI, iv (October 1942), pp. 435f.
Duncan, Edgar Hill, 'Jonson's "Alchemist" and the Literature of Alchemy', *Publications of the Modern Language Association of America*, vol. LXI (1946), pp. 699f.
Duncan, Edgar Hill, 'Chaucer and "Arnold of the Newe Toun" ', *Modern Language Notes*, vol. LVII (1942), pp. 31f.
Duncan, Edgar Hill, 'The Literature of Alchemy and Chaucer's "Canon's Yeoman's Tale" ', *Speculum*, vol. XLIII (October 1968), pp. 633f.
Eccles, Mark, *Christopher Marlowe in London* (London, 1934).
Edgar, Irving, *Shakespeare, Medicine and Psychiatry* (New York, 1971).
Evans, R. J. W., *Rudolf II and His World* (Oxford, 1973).
Everard, John, *The Divine Pymander of Hermes Mercurius Trismegistus in XVII Books* (London, 1650).
Everard, John, *Hermes Trismegistus His Second Book called Asclepius* (London, 1657).
Ferguson, John, *Bibliotheca Chemica* (2 vols, Glasgow, 1906).
Ferguson, John, 'Some English Alchemical Books', *The Journal of the Alchemical Society*, vol. II (1913), pp. 1f.
Fioravanti, Leonardo, *A Ioyfull Iewell*; trans. John Hester (London, 1579).
Fioravanti, Leonardo, *A Compendium of the Rationall Secretes of . . . L. Phioravante*; trans. John Hester (London, 1582).
Fordham, Frieda, *An Introduction to Jung's Psychology* (London, 1953).
Forester, James, *The Pearle of Practise* (London, 1594).
French, Peter, *John Dee: The World of an Elizabethan Magus* (London, 1972).
Geoghegan, D., 'Gabriel Plattes' Caveat for Alchymists', *Ambix*, vol. X (1962), pp. 97f.
Godwin, Joscelyn, *Robert Fludd* (London, 1979).

BIBLIOGRAPHY 281

Grasshoff, Johann (attrib.), *Tractatus Aureus de Lapide Philosophico* (Frankfurt, 1625); trans. A. E. Waite, *The Golden Tract concerning the Stone of the Philosophers* (MH, I, 7f).

Halliwell, James Orchard (ed.), *The Private Diary of Dr. John Dee, and the Catalogue of his Library of Manuscripts* (London, 1842).

Hart, W. H., 'Observations on Some Documents relating to Magic in the Reign of Queen Elizabeth', *Archaeologia*, vol. XL (1866), pp. 389f.

Harvey, Gabriel, *Pierces Supererogation, or A New Prayse of the Old Asse* (London, 1593). In Grosart, Alexander B. (ed.), *The Works of Gabriel Harvey* (London, 1884), vol. II.

Herbert, T. Walter, *Oberon's Mazéd World* (Baton Rouge, 1977).

Hermann, Philip, *An Excellent Treatise teaching howe to cure the French-Pockes*; trans. John Hester (London, 1590).

'Hermes Trimegistus', *Poemandres* and *Asclepius*, see Everard.

'Hermes Trismegistus', *The Emerald Table* (Latin MSS – 'Tabula Smaragdina' – from c. 1200); trans. Anon., *The Smaragdine Table of Hermes Trismegistus of Alchimy*; in Bacon R., pp. 16f.

'Hermes Trismegistus', *Tractatus Vere Aureus de Lapidis Philosophici Secreto*, ed. Dominicus Gnosius (Leipzig, 1610); trans. Anon., (Atwood?), *A Suggestive Inquiry into the Hermetic Mystery* (London, 1850); in Regardie, p. 30f.

Hester, John, *The Key of Philosophie* (London, 1596).

Hester, John, see Fioravanti, Hermann, Holland, Penotus, pseudo-Paracelsus, Quercetanus.

Holland, Isaac, *Fragmentum ex Theoriis* (Latin edition, 1647); trans. John Hester, *A Fragment out of the Theorickes of Is. Isaacus Hollandus*; in pseudo-Paracelsus, sig. F2(v)f.

Holmyard, E. J., *Alchemy* (London, 1957).

Holt, David, 'Jung and Marx: Alchemy, Christianity, and the Work against Nature', lecture at the Royal Society of Medicine, London, 21 November 1974.

Hooykaas, R., *Religion and the Rise of Modern Science* (Edinburgh, 1972); paperback 1973.

Hume, David, *History of England* (6 vols, London, 1805).

Josten, C. H., 'William Backhouse of Swallowfield', *Ambix*, vol. IV (1949), pp. 1f.

Kearney, Hugh, *Science and Change 1500–1700* (London, 1971).

Knights, L. C., *Drama and Society in the Age of Jonson* (London, 1937); paperback 1962.

Lacey, Robert, *Sir Walter Raleigh* (London, 1973); paperback 1975.

Lambspringk, *De Lapide Philosophico*, ed. Nicolas Barnaud (Prague, 1599); trans. A. E. Waite, *Concerning the Philosophical Stone*, MH I 271f.

Levi, Eliphas (Alphonse Louis Constant), *Dogme et Rituel de la Haute Magie* (2 vols, Paris, 1855–6); trans. A. E. Waite, *Transcendental Magic* (London, 1896); paperback 1968.

Lodge, Thomas, *A Fig for Momus* (London, 1595); Epistle 7 (sig. I(v)f) is *The Anatomie of Alchymie*.

Lyly, John, *Gallathea* (London, 1592).

Maier, Michael, *Symbola Aureae Mensae Duodecim Nationum* (Frankfurt, 1617).

Maier, Michael, *Atalanta Fugiens, hoc est Emblemata Nova de Secretis Naturae Chymicae* (Oppenheim, 1618).

Maier, Michael, *Themis Aurea, hoc est de Legibus Fraternitatis R.C. Tractatus* (Frankfurt, 1618); trans. N.L., T.S., and H.S., *Themis Aurea, The Laws of*

the Fraternity of the Rosie Cross (London, 1656).
McKerrow, Ronald B. (ed.), *A Dictionary of Printers and Booksellers in England . . . 1557–1640* (London, 1910).
Metlitzki, Dorothee L., *The Matter of Araby in Mediaeval England* (Yale, 1977).
Montaigne, Michel de, *Les Essais* (Paris, 1588); trans. John Florio, *The Essayes of Lo: Michaell de Montaigne* (London, 1603).
Mylius, Johann Daniel, *Philosophia Reformata* (Frankfurt, 1622).
Norton, Thomas, *The Ordinall of Alchemy* (1477; published in TCB, 1f).
Pagel, Walter, *Paracelsus: An Introduction to Philosophical Medicine in the Renaissance* (New York, 1958).
(pseudo-) Paracelsus, *Centum Quindecim Curationes Experimentaque* (Lyons, 1582); trans. John Hester, *A Hundred and Foureteene Experiments and Cures of . . . Paracelsus* (London, 1596).
Penotus, Bernardus Georgius, *An Apologeticall Preface*; trans. John Hester, in pseudo-Paracelsus, sig. A3(r)f.
Philalethes, Eirenaeus, *Brevis Manuductio ad Rubinum Coelestum* (Amsterdam, 1668); trans. A. E. Waite, *A Brief Guide to the Celestial Ruby* (MH II 246f).
Philalethes, Eirenaeus, *Metallorum Metamorphosis* (Amsterdam, 1668); trans. A. E. Waite, *The Metamorphosis of Metals* (MH II 227f).
Philalethes, Eirenaeus, *Fons Chemicae Veritatis* (Frankfurt, 1678, in MH), trans. A. E. Waite, *The Fount of Chemical Truth* (MH II 261f).
Philalethes, Eirenaus, *Introitus Apertus ad Occlusum Regis Palatium* (Amsterdam, 1667); trans. A. E. Waite, *An Open Entrance to the Closed Palace of the King* (MH II 159f).
Philalethes, Eirenaus, *Ripley Reviv'd* (London, 1678); in Rola, p. 23f.
Plat, Hugh, *The Jewell House of Art and Nature* (London, 1594; revised edition, 1653).
Porta, Giovanni Battista della, *Magia Naturalis sive de Miraculis rerum naturalium* (Naples, 1558); trans. Anon., *Natural Magick . . . in Twenty Books* (London, 1658).
Puttenham, George, *The Arte of English Poesy* (London, 1589).
Quercetanus (Joseph Duchesne), *De Ortu et Causis Metallorum* (Lyons, 1576); trans. John Hester, *A Breefe Aunswere of Iosephus Quercetanus . . . concerning the Original and Causes of Mettalles* (London, 1591).
Quercetanus (Joseph Duchesne), *Sclopetarius, sive de curandis vulneribus* (Lyons, 1576); trans. John Hester, *The Sclopotarie of Iosephus Quercetanus* (London, 1590).
Raleigh, Sir Walter, *The History of the World* (London, 1614).
Rattansi, P. M., 'Alchemy and Natural Magic in Raleigh's "History of the World" ', *Ambix*, vol. XIII (1965), pp. 122f.
Read, John, *Prelude to Chemistry* (London, 1936).
Read, John, *The Alchemist in Life, Literature and Art* (Edinburgh, 1947).
Read, John, *Humour and Humanism in Chemistry* (London, 1947).
Reidy, J., 'Thomas Norton and the "Ordinall of Alchimy" ', *Ambix*, vol. VI (1957), pp. 59f.
Regardie, Israel, *The Philosopher's Stone* (St Paul, Minnesota, 1970).
Ripley, George, *The Compound of Alchymy* (London, 1591).
Ripley, George, *Cantilena* (c. 1450–70); trans. Anon., 'George Ripley's Song' (MS, before 1581, MS Ashmole 1445, VIII, pp. 2–12); *Ambix*, vol. II (1946), pp. 177f.
Rola, Stanislas Klossowski de, *Alchemy: The Secret Art* (London, 1973).

Rowse, A. L., *The England of Elizabeth* (London, 1950); paperback 1973.
Rowse, A. L., *Shakespeare the Man* (London, 1973); paperback 1976.
Rowse, A. L., *The Case Books of Simon Forman* (London, 1974); paperback 1976.
Ruland, Martin, *Lexicon Alchemiae, sive Dictionarium Alchemisticum* (Frankfurt, 1612).
Sadler, Lynn Veach, 'Relations between Alchemy and Poetics in the Renaissance and Seventeenth Century', *Ambix*, vol. XXIV (1977), pp. 69f.
Salinger, Leo, 'Comic Form in Ben Jonson: Volpone and the Philosopher's Stone', in Marie Axton and Raymond Williams (eds), *English Drama: Forms and Development* (Cambridge, 1977), pp. 48f.
Sargent, Ralph, *At the Court of Queen Elizabeth: The Life and Lyrics of Sir Edward Dyer* (London, 1935).
Schoenbaum, S., *William Shakespeare: A Documentary Life* (Oxford, 1975).
Schuler, Robert M., 'William Blomfild, Elizabethan Alchemist', *Ambix*, vol. XX (1973), pp. 75f.
Scot, Reginald, *The Discoverie of Witchcraft* (London, 1584).
Sendivogius, Michael, *Novum Lumen Chemicum* (Prague, 1604); trans. J. F. (John French), *A New Light of Alchymy* (London, 1650; second edition, 1674).
Shirley, John William, 'The Scientific Experiments of Sir Walter Raleigh, the Wizard Earl, and the Three Magi in the Tower, 1603-17', *Ambix*, vol. IV (1948), pp. 52f.
Sidney, Sir Philip, *The Defence of Poesie* (London, 1595).
Siebmacher, Johann Ambrosius, *Wasserstein der Weysen* (Frankfurt, 1619); trans. A. E. Waite, *The Sophic Hydrolith, or Water Stone of the Wise* (MH I 69).
Stillman, John Maxson, *The Story of Early Chemistry* (New York, 1924); paperback (*The Story of Alchemy and Early Chemistry*) 1960.
Stolcius, Daniel, *Viridarium Chymicum figuris ... adornatum* (Frankfurt, 1624).
Taylor, F. Sherwood, *The Alchemists* (London, 1952); paperback 1976.
Taylor, F. Sherwood, 'Thomas Charnock', *Ambix*, vol. II (1946), pp. 148f.
Thorndike, Lynn, *History of Magic and Experimental Science* (6 vols, New York, 1923-41).
Tillyard, E. M. W., *The Elizabethan World Picture* (London, 1943); paperback 1972.
Tillyard, E. M. W., *Shakespeare's Problem Plays* (London, 1950); paperback 1970.
Tymme, Thomas, *The Practise of Chymicall and Hermeticall Physicke, for the Preservation of Health* (London, 1605).
Urdang, Georg, 'How Chemicals Entered the Official Pharmacopoeias', *Archives Internationales d'Histoires des Sciences*, vol. VII (1954), pp. 303f.
Valentine, Basil, 'Zwölff Schlüssel', in *Von dem Grossen Stein der Uralten* (Eisleben, 1599); trans. Anon., 'A Practick Treatise together with the XII Keys', in *The Last Will and Testament of Basil Valentine* (London, 1670).
Valentine, Basil, *Triumphwagen Antimonii* (Leipzig, 1604), trans. I. H. (Hughes?), *The Triumphant Chariot of Antimony* (London, 1660).
Vaughan, Thomas, *Coelum Terrae, or The Magician's Heavenly Chaos* (London, 1650); in Regardie, p. 163f.
Waite, Arthur Edward, *The Real History of the Rosicrucians* (London, 1887); paperback 1977.
Waite, Arthur Edward, *The Lives of the Alchemystical Philosophers* (London,

1888); paperback (*Alchemists through the Ages*) 1970.
Ward, Robert, 'What is Forced by Fire: Concerning some Influences of Chemical Thought and Practice upon English Poetry', *Ambix*, vol. XXIII (1976), pp. 80f.
Willeford, William, *The Fool and His Sceptre* (London, 1969).
Yates, Frances A., *Giordano Bruno and the Hermetic Tradition* (London, 1964).
Yates, Frances A., *The Art of Memory* (London, 1966); paperback 1969.
Yates, Frances A., *Theatre of the World* (London, 1969).
Yates, Frances A., *The Rosicrucian Enlightenment* (London, 1972).
Yates, Frances A., *Shakespeare's Last Plays: A New Approach* (London, 1975).
Yates, Frances A., *The Occult Philosophy in the Elizabethan Age* (London, 1979).

Index

Note Anonymous texts are indexed by title; all other texts by author.

acetum philosophorum, 94, 96
Aegidius de Vadis, 180
Aesop, 51
Agrippa, Heinrich Cornelius, 8, 44, 47, 50
albedo, see colours, alchemical
Albertus Magnus, 32
alchemy: Arabic, 25, 32, 138, 143; allegorized, *see* allegories, alchemical; Bohemian, 19–22, 45, 81, 85, 88; and capitalism, 113–19; as chemical theatre, 6, 22, 47–8, 51, 52, 54, 91–2, 93, 106, 127, 142–4, 158–60, 180–4, 206, 214–15, 225–39; and Christianity, 5, 52–3; and classical mythology, 98–9, 179, 183, 233–5; continuity of alchemical tradition, 25, 41, 43, 65, 141; and the Creation, 5–6, 35–6; definitions and synopses of, 25, 26, 61, 215–16; in Donne's poetry, 70, 103, 119–35, 223; elemental and metallurgical theories of, 25–6, 27, 31, 56, 103; Elizabeth I and, 17–19; fraudulent, 7–14, 19–22, 69–71, 112–17; Greek, 25, 49, 95, 136–8; and the Hermetic tradition, 48–53, 69, 81, 85; illegality and heterodoxy of, 6, 12, 13, 76; in Jonson's plays and masques, 6–7, 10–11, 27, 71–2, 97–102, 107–19, 182–3; Jungian interpretation of, 5, 6, 143, 184, 200–1; language of, 9, 23, 26, 54, 90–8, 102–6, 118–19, 135, 182, 200–1, 222–3, *see also* symbols, alchemical; literature of, 6–7, 23–54, 81–92, 136–41; medicinal aspects of, 16, 26–7, 55–80, 81, 84–5, 86, 88, 106, 123, 193, 198, 204, 207, 218–22; medieval English, 32, 34, 41, 81, 83, 87–8, 90, 138–9; microcosmology of, 4, 5–6, 35–6, 47–8, 92–3, 124, 184; modern image of, 1–3, 6–7, 152; in Nashe's writings, 70–1, 77, 78–9, 105, 106; and Nature, 3–4, 7, 27, 58, 61–2, 86, 101–2, 227–30; psychic and spiritual meanings of, 4–5, 6, 50–4, 69, 81, 83–4, 86–7, 122–

35, 141–4, 154–239; renaissance of, 41–4, 50–4, 81–90, 140–1, 152; and Renaissance magic, 41–4, 46–8, 50, 54, 212; scepticism towards, 2, 7–10, 13–14, 22, 52, 70–2, 77, 78–80, 96–8, 102, 112–17; in Shakespeare's plays and poetry, 6–7, 27, 72–80, 104, 143–239; and tragedy, 136–44, 149, 152, 158, 160; in various Elizabethan and Jacobean writings, 7, 8, 13, 17, 19–20, 69–70, 97, 102–6
Alchymia, Lady, 14, 17, 43, 67, fig. 1
alembic, 72, 75, 103, 130, 134, 148, 196, 200
allegories, alchemical, 89, 91–2, 136–41, 157–60, 197–8, 207–9, 214–15, 233–5
Alnwick Castle, 17
amber, white, 67, 199
ambergris, 16
Andreae, Johann Valentin, 88–9, 152
Andrewes, Abraham, 163
Anhalt, Christian of, 65
anima (alchemical), 3, 5, 30, 32, 38, 85, 92, 116, 126–30, 133, 135, 168, 170, 178, 179, 182, 183, 204, 233
anima (Jungian), 200
anima mercurii, see Mercury, Our
animals, alchemical, *see* symbols, alchemical
Anrach, Georgius, fig. 31
Anthonie, Francis, 4, 6, 15, 67–8, 153
antimony, 5, 16, 60, 66, 72, 84–5, 96, 171
Apocrypha, the, 151
Apollo, 146, 194
aqua ardens, 94, 164, 171
aqua fortis, 11, 94, 171
aqua regia, 94
aqua vitae, 52–3, 70, 118, 207
Aquinas, Thomas, 124
arcana, 57, 59–61, 62, 65, 70, 73, 74, 75, 80, 123, 170, 193, 204, 207, 222; *see also* Paracelsism
archetypes, 5, 50, 200–1, 236; *see also*

symbols, alchemical
archeus, 61, 62, 86
argent-vive, *see* mercury, common
Arnald of Villanova, 27, 32, 43, 86, 87, 98, 102
Aristotle, 25, 43, 132, 141
arsenic, 10–11, 60, 73
Artephius, 123
Asclepius, see Hermes Trismegistus
Ashmole, Elias, 7, 21, 22, 23, 36
assaying, 10, 182
astrology, 5, 13, 20, 43, 45, 56, 59, 62, 179
astromancy, 12
astronomy, 13, 43, 48
Athanor, *see* furnace
Aubrey, John, 15, 16, 43
Aurelia Occulta, 93, 163
aurum nostrum, see gold, philosophical
aurum potabile, see gold, cordial of
aurum vulgi, see gold, common
Avicenna, 43, 57
Azoth, 53, 93, 96, 97

Backhouse, William, 15, 28–9, 225
Bacon, Francis, 1–3, 6, 7, 96, 152–3
Bacon, Roger, 25, 32, 43, 89; *The Mirror of Alchimy*, 23–32, 34, 41, 46, 48, 51, 54, 90, 99, 102, 115, 223, fig. 2
Baker, George, 65
Baker, Thomas, 44
balm, *see* balsam
balneum, 79, 94, 98, 128, 171, 194, 230
balsam, 12–13, 66, 68, 71, 76, 83, 103, 124–5, 128, 134, 167, 168, 182, 198, 203, 204, 221
Barnaud, Nicolas, 21, 88; *see also* Lambspringk
Barrow, Henry, 76
Bartholomew the Englishman, 32
Beckett, Samuel, 121–2
Bedford, Countess of, *see* Harrington, Lucy
Bedo, William, 12
Bible, the, 46
Blackfriars, 76, 236
blackness, the, *see* colours, alchemical
Bloomfield, Miles, 15
Bloomfield, William, 3–4, 6, 18, 96, 149
Bodenstein, Adam von, 62, 64
Boehme, Jakob, 151, 171
Bonus, Petrus, 32, 127, 160, fig. 12
Bostocke, Robert, 65–6, 67, 68
Boyle, Robert, 1, 16
Bradley, A. C., 143–4
British Museum, the, 20, 75
Bruno, Giordano, 44
Buckingham, Duke of, *see* Villiers, George

Buckley, John, 12
Burghley, Lord, *see* Cecil, William
Butler, Samuel, 57

Cabbalism, 20, 43–5, 50, 51, 52, 53, 56, 88, 98
caduceus, 143, 183
calcination, 35, 36–7, 98
Calvin, John, 45
Camden, William, 19
Campanella, Tommaso, 44
Carey, Elizabeth, 77
Carey, George, second Lord Hunsdon, 76–9
Carey, Henry, first Lord Hunsdon, 77
Cecil, William, first Lord Burghley, 17, 19, 21, 114
Celsus, Aulus Cornelius, 56
Chamberlain's Men, the, 77
Chaos, Our, *see* Raw Stuff
Chapman, George, 11, 77, 107
Charles I, 65
Charles II, 16
Charnock, Thomas, 15, 17, 139, 148
Chaucer, Geoffrey, 8, 10, 102
Christian II of Saxony, 85
chrysosperm, 97
chymicall physick, *see* Paracelsism
Chymicall Wedding, the, 38–9, 88–9, 91, 138, 140, 212, 232–6, figs. 9, 29, 30, 31
Chymische Hochzeit Christiani Rosencreutz, see Andreae, Johann
chymists, 6, 16, 65–9, 70–2, 73, 75–6, 81, 106, 109, 131, 182, 222–3, fig. 6
cibation, 35, 40
circulation, 118, 147–51, 164
citrine, *see* colours, alchemical
clay of wisdom, *see lutum sapientiae*
coagulation, *see* congelation
coining, 10, 11-13, 111, 113, 182
colcothar, 60
colours, alchemical: blackness (*nigredo*), 39, 41, 88, 97, 98, 128, 136, 138, 139, 140, 143, 150, 158, 164, 190, 191, 194, 197–8, 201, 203; greenness (*viriditas*), 38, 93–4, 96, 97, 99, 163–4, 209; peacock's tail (*cauda pavonis*), 99–100; redness (*rubedo*), 39–40, 96, 97, 110, 141, 148, 150, 213–14, 216, 235; whiteness (*albedo*), 39–40, 89, 98, 99, 138, 140, 148, 150, 203, 206, 215, 235; yellowness (*citrinitas*), 99
Combach, Ludwig, 21
congelation, 28–9, 32, 35, 41, 54, 98, 127, 149, 195
coniunctio, see conjunction
conjunction, 5, 35, 38–9, 41, 126, 138–

INDEX

40, 198, 212–14, 232–6, figs. 9, 23, 24, 29, 30, 31, *see also* Chymicall Wedding, the
contraria contrariis, 65, 69–70; *see also* Galenism
Cook, Roger, 17
Copernicus, Nicolas, 43
copper, *see* Venus
coral, red, 16
corpus mundi, 133
corpus subtile, 29, 87, 136, 138–9, 207, 211
Creede, Thomas, 223, fig. 25
Cremer, Abbot, 90
creta vitrioli, 66
Croll, Oswald, 21, 64
cucurbite, 148

Davies, Sir John, 17, 104–5
De Alchemia, 25
De Alchimia, 207, 209, fig. 22
Debus, Allen G., 65
Dee, Arthur, 15, 68
Dee, John, 9, 15, 18, 20–1, 43–4, 50, 68, 69, 85, 100, 153; *Mathematicall Preface*, 43–4; *Monas Hieroglyphica*, 18, 43, 44–8, 50–2, 54, 81, 84, 87–8, 93, 205, fig. 4; 'Prefatory verses' to Ripley's *Compound*, 54; *Propadeumata Aphoristica*, 52; *Testamentum*, 43
Democritus, 65
Descartes, René, 3
Derby, Earl of, *see* Stanley, Ferdinando
Deucalion, 233–6
Devereux, Robert, second Earl of Essex, 19
dissolution, 4, 29–30, 32, 35, 37, 41, 54, 96, 98, 121, 127, 131–3, 136, 149, 151, 158, 163, 168, 175, 194–5, 197, 198, 199
distillation, 4, 16–17, 19, 43, 59, 65, 68, 69, 70, 73–7, 79–80, 105, 106, 134, 148, 149, 193, fig. 6
Donne, John, 19, 23, 106, 153; alchemical motifs in the *Songs and Sonnets*, 119–28, 130–5, 152; in the *Divine Poems*, 133; 'An Anatomie of the World', 125, 128–30, 132, 133; 'The Anniversarie', 122; 'The Dissolution', 131–3, 223; 'The Crosse', 70; 'Ecclogue', 103; *Elegie* VIII, 'The Comparison', 103, 196; *Elegie* XVIII, 'Loves Progress', 112; 'Elegie on the Lady Marckham', 130, 133–4; 'The Extasie', 122–3, 124–8, 130, 133; 'The Good-morrow', 121, 122, 123, 211; 'The Indifferent', 119; 'Loves Alchymie', 1–2; 'Loves Growth', 123;

'Nocturnall upon S. Lucies Day', 124, 133–4; 'Song' ('Goe, and catch a falling starre'), 119; 'The Sunne Rising', 122, 217; 'A Valediction: forbidding mourning', 123, 211; verse-letter to 'E of D', 105; verse-letter to Lucy, Countess of Bedford, 124–5; verse letter to Henry Wootton, 70; 'Womans Constancy', 121
Dorn, Gerhard, 5, 6, 11, 53–4, 64, 69, 85, 87, 88, 94, 151, 180, 216
Drake, Sir Francis, 66, 115
Drury, Elizabeth, 128–9
Duchesne, Joseph, *see* Quercetanus
Dudley, Robert, Earl of Leicester, 15, 43
Duncan, Edgar Hill, 98
dung, 111–12, 180
Dürer, Albrecht, 56
Dyer, Sir Edward, 15–16, 18, 21, 43

Ebreo, Leone, 124
Edward IV, 33–4
Einstein, Albert, 37
El Dorado, 114
elements, first (or simple), 131, 133, 151, 223
elements, the four, 3–4, 25–6, 30, 37–8, 131–3, 151
Eliot, T. S., 145–6
Elixir, the, 2, 12–13, 22, 26–7, 32, 72, 75, 96, 97, 130, 134, 207; *see also* Stone, Philosophers'
Elizabeth I, 17–19, 21, 43, 114, 189
Emerald Table, the, *see* Hermes Trismegistus
epilepsy, 60, 220
Erasmus, Desiderius, 8, 57
Erastus, Thomas, 64
Euclid, 43
Euridyce, 217
Evans, Robert, 20, 21
Evelyn, John, 16
exaltation, 35, 40, 110, 139

Febure, *see* Lefevre
fermentation, 35, 40, 100
Ficino, Marsilio, 44, 46–7, 49
Field, Richard, 76
filius macrocosmi, 39, 52
Fioravanti, Leonardo, 67, 72
fire, 12, 30, 38, 40, 46, 51, 61, 64, 79, 83–4, 86, 94, 95, 100, 103, 105, 131–3, 151, 163–4, 168, 171, 172, 185, 189–91, 194, 195–6, 198, 199, 205, 215, fig. 17; *see also* furnace
Flamel, Nicolas, 87, 98–9, 102
Fletcher, John, 19–20
Florio, John, 121

INDEX

Fludd, Robert, 15, 68, 90, 153, fig. 19
Forester, James, 15, 76–80, 153, 199
Forman, Simon, 9, 15, 67, 77, 78, 79, 85, 228, 232
Frederick II of Denmark, 64
French, Peter, 44
Frobenius, Johannes, 57
furnace, 7, 8, 9, 17, 19, 30, 32, 53, 57, 64, 72, 76, 97–8, 103, 132, 151, 189–91, 194, 230, figs. 1, 6, 17, 26

Galen, 57, 65, 74, 78
Galenism, 56, 60, 69, 71, 123
Galid, *see* Yazid, Khalid ibn
gamaaea, 47, 51
Garland, John, *see* Hortulanus
Garland, Edward, 21
Garland, Francis, 21
Garter, Order of the, 78–9, 89
Geber, 7, 32, 43, 86, 95
Geheime Figuren der Rosenkreuzer, fig. 26
gematria, 12
geomantics, 62
Gesner, Conrad, figs. 1, 6
Gilbert, Adrian, 15
Glass, Our, *see* vessel
Globe theatre, the, 153, 180, 228, 236
gnosticism, 49, 164
gold, common, 1, 4, 7–9, 10, 11, 12, 17, 21–2, 26–8, 31, 40, 45, 51, 60, 62, 95, 96, 103–4, 107–10, 112–17, 124, 150, 157, 158, 163, 194, 207
gold, cordial of, 26–7, 60, 68, 117
gold fulminate, 84
gold, light, 11, 182
gold, philosophical, 26–8, 31, 32, 53–4, 83, 87, 93, 102, 105, 110, 115–17, 123, 157
Goldsmiths' Company, the, 110
Goodridge, John, 77
Grasshoff, Johann, 110, 140–1, 143, 148, 169, 206, 214–15
Greenwood, John, 76
Grierson, Herbert, 124
guaiac, 57
Gunpowder Plot, the, 17, 152
Gwynne, John, 43

Hajek, Tadeas, 21
Hariot, Thomas, 9, 16
harlequin, 178
Harrington, Lucy, Countess of Bedford, 124
Harsnett, Samuel, 9
Harvey, Gabriel, 19, 75, 103, 105
Hecate, 156
heliocentricity, 43

Henri IV of France, 64
Henry, Prince of Wales, 16
Henslowe, Philip, 228
Herbert, Henry, second Earl of Pembroke, 66
Herbert, Mary, Countess of Pembroke, 15, 66
Herbert, William, third Earl of Pembroke, 66
Hermann, Philip, 67
Hermes Psychopompos, 183–4
Hermes Trismegistus, 23, 43, 48–50, 65, 86, 92, 97; *Asclepius*, 49–50; *Emerald Table*, 23, 25, 48–9, 116, 148; *Pimander*, 49–50; *Tractatus Aureus*, 49, 111, 148, 157, 163, 169, 205, 212–14, 228
Hermeticism, 43, 48–53, 56, 69, 81, 85
Hester, John, 15, 35, 36, 64, 66–7, 68, 69, 72, 75–6, 79, 81, 85
Hippocrates, 57, 65
Hirschvogel, Augustin, fig. 5
Hoefer, Ferdinand, 2
Hoghelande, Ewald van, 22, 85
Hollandus, Isaac, 67, 100
homoeopathy, 60, 65, 69–70, 123, 221
Hortulanus, 23, 25, 30
humours, comedy of, 107–8
humours, the four, 60, 108
Huser, Johann, 64
hydrochloric acid, 84
hysterica passio, 67, 199, 218, 219

ignis innaturalis, *see* fire
Iliaster, 61
incest, 138–41, 197–200, 212–14, 232–3
individuation process, 143
iron, *see* Mars
iron pyrites, *see* marcasite

James I, 16, 65, 89, 152, 158
Jonson, Ben, 22, 43, 106, 129, 134, 153, 182–3; *The Alchemist*, 7, 9, 10, 14, 19, 27, 43, 71, 78, 97–100, 109, 111–15, 129, 182; *The Divell is an Asse*, 118–19; *Eastward Hoe*, 6, 10–11, 107–12, 115, 119, 152, 182–3, 223; *Mercurie Vindicated*, 100–2; *Volpone*, 57, 71–2, 113–17, 119, 152
Johnson, Samuel, 216
Josten, C. H., 44
Jung, Carl Gustav, 5, 6, 143, 164, 180, 184, 188, 200–1
Jupiter, 146, 179, 233

Keats, John, 153
Kelley, Edward, 15–16, 19–22, 26, 43, 100, 114

INDEX

Keynes, John Maynard, 114
Khunrath, Heinrich, 52–3, 69, 74, 151, 230–1, figs. 7, 28
Kisley, Thomas, 68

Lambe, John, 9
Lambspringk, 88–9, 94–5, 153, 164–5, 168, 169, 206–7, figs. 11, 13
Lannoy, Cornelius de, 18
lapis philosophorum, see Stone, Philosophers'
Laski, Albert, Palatine of Sieradz, 20
laudanum precipitatum, 67
lead, *see* Saturn
lead acetate, see *saccharum plumbi*
Lefevre, Nicolas, 16
Leicester, Earl of, *see* Dudley, Robert
Leonicinus, 56
Libavius, Andreas, 22
limbeck, *see* alembic
Lodge, Thomas, 8, 13, 97
Lully, Raymond, 2, 19, 25, 32, 43, 72, 86, 87, 96, 100, 103, 130
Luna, 9, 10, 45, 51, 97, 98, 105, 114, 139, 175, 207, 230
Luther, Martin, 56
lutum sapientiae, 49, 97
Lyly, John, 7, 102–3

Macready, William Charles, 177–8
magic, 1, 2, 6, 13, 56, 182, 236–8; angel, 20, 43–4, 50, 77; Egyptian, 49–50; shady, 8–9; Renaissance, 20, 44–8, 53, 62
magnesia, 52, 97
magnum opus, the, 28, 34–41, 45, 47–8, 51, 52, 54, 83, 91, 111, 133, 136–41, 142, 149, 158, 168, 215–16, 223
Maier, Michael, 9, 65, 90, 111, 158, 171, 183, 200, 233–6, figs. 14, 15, 17, 29
Manardus, 56
manikin, pissing, 180
marcasite, 60, 97
Marlowe, Christopher, 12–13, 74, 76, 105
Mars, 179
Marston, John, 11, 107
massa confusa, see Raw Stuff
mathematics, 20, 43, 44–5, 50, 212
mathesis, 44, 50
Maximilian II, Emperor, 45
Mayerne, Theodore Turquet de, 65, 66, 68
Medici, Cosimo de', 49
Medicine, the Universal, *see* Stone, Philosophers'
Melton, John, 9
menstruum, 37, 96, 97

Mercurius, the god, 47, 49, 179, 182, 183–4
mercury, common, 8, 10, 21, 26, 31–2, 60, 66, 95, 99, 105, 107–11, 114, 170, 171, 179–83, 223
Mercury, Our (or philosophical), 16, 31–2, 34, 36–7, 39, 45–8, 49, 81, 88, 92–6, 97, 99, 100-2, 110, 118–19, 125, 129, 130, 132, 138, 139, 164, 168–72, 175, 179–83, 198–9, 205–9, 221, 223, 225, 233, 236, 239, figs. 14, 20, 21, 22, 25
Michelspacher, Stefan, 230, fig. 27
Middleton, Thomas, 105
Moffett, Thomas, 15, 66–7, 73
Montaigne, Michel de, 119, 121
moon, *see* Luna
Morienus, 43
mortification, 126, 130, 171, 172–6, 194
Mosan, James, 68
Moses, 5, 49, 132
mother, rising of the, see *hysterica passio*
Musaeum Hermeticum, 31, 230
musk, 16
Mylius, Johann Daniel, 35, 164, fig. 30

Nashe, Thomas, 9, 70–1, 77, 78–9, 105, 106
neo-Platonism, 49, 124, 127–30
Nero, 193–4
Newgate prison, 12, 76
Newton, Sir Isaac, 2
nigredo, see colours, alchemical
Northumberland, Earl of, *see* Percy, Henry
Norton, Samuel, 15, 18
Norton, Thomas, 8, 28, 30, 90, 123–4, 148, 189, 205, 207
numerology, 43

Odington, Walter of, 34
Oedipus, 51, 142
Oporinus, Johannes, 57
Orpheus, 184–5, 217
Otto, Count Palatine, 64
Ourobouros, 164, fig. 13

Paracelsians, 62–9, 73, 75–8, 132; caricatured, 70–2
Paracelsism, 58–62, 65–73, 75, 78–80, 81, 84–5, 86, 88, 123, 193, 198, 204, 207, 218–22; astral diagnosis and treatment, 59–60, 67; chemical remedies, 59, 60, 62, 64, 66–9, 71–2, 73, 76, 78–80, 84–5, 199; homoeopathic principles, 60, 65, 69–70, 123, 221; Paracelsist Nature, 58, 61–2, 72–3, 222; *see also arcana*;

chymists; virtues; *etc.*
Paracelsus, 3, 6, 15, 19, 54, 55–62, 64, 65, 68, 72–4, 83, 85, 88, 100, 103, 123, 132, 193, 217, 218, 220–2, fig. 5; *Archidoxis*, 62; *Bücher und Schrifften* (Collected Works), 62–3; *De Caducis*, 220; *De Compositione Metallorum*, 108; *Elixir Solis*, 88; *Von der Frantzosischen Kranckheit*, 57; *Grosse Wundarznei*, 56, 58, 62; *Imitatio*, 57; *Labyrinthus Medicorum Errantium*, 58, 61, 62; *De Natura Rerum*, 11, 58, 62; *Paragranum*, 56–7, 59–60, 207; *Paramirum*, 56, 58, 60, 168, 205; *Philosophia ad Athenienses*, 197; *Philosophia Sagax*, 46, 58, 62; *De Vita Longa*, 62
Parsons, Fr Robert, 16
Pascal, Blaise, 3
peacock's tail, *see* colours, alchemical
pearl, 16, 18, 60, 66, 93, 227
Peirithous, 184
pelican, 148, 212
Pembroke, Earls and Countess of, *see* Herbert, Henry, William and Mary
Penotus, Bernardus, 67–9
Percy, Henry, ninth Earl of Northumberland, 16, 17, 77, 85, 153, 160
Pernety, Dom A. J., 26
petasus, 183
Peterson, John, 18
Petrarca, Francesco, 8
pharmacopoeias, 62, 65, 66, 222
Philalethes, Eirenaeus, 4, 27, 29–30, 94, 170, 175, 180, 191, 194, 196, 198, 205, 206
Philips, Judith, 15
Pico della Mirandola, Giovanni, 44, 50
piger Henricus, 98
Pimander, *see* Hermes Trismegistus
plaintain, water of, 67
Plat, Sir Hugh, 9–10, 13
Plato, 43, 49
Platonism, 46, 49, 50, 128–30, 135
Plattes, Gabriel, 9
Plotinus, 124
Pluto, 91
poison, 16, 60, 73, 84–5, 93, 94–5, 113, 139, 163, 165
Poole, John, 12, 76
Porta, Giovanni Battista della, 12
prima materia, 25, 35, 37, 112, 151, 164, 191, 196–7
projection (alchemical), 9, 10, 21, 30, 35, 40, 118
projection (Jungian), 5, 200
putrefaction, 35, 39, 41, 98, 125, 128, 130, 138, 139, 140, 150, 158, 190, 194–8, 233
Puttenham, George, 69–70
Pyrrha, 233–6
Pythagoras, 43, 65

Quercetanus, 35, 36, 37, 39, 40, 44, 64, 67, 85, 90, 152, 220
quicksilver, *see* mercury, common
Quiller-Couch, Sir Arthur, 178
quinine, 16
quintessence, 3–4, 5, 18, 30, 32, 36, 38, 47, 58, 62, 74, 77, 80, 92, 105, 111, 116, 123, 125, 128, 132, 134, 135, 168, 170, 193, 204, 207, 209

Rabbards, Ralph, 34
Rabelais, François, 56
Racine, Jean, 142
Raleigh, Sir Walter, 15, 16–17, 49, 68, 73, 76, 153
Ramus, Peter, 44
Raw Stuff, 5, 28, 35–6, 45, 47, 95–6, 111, 113, 114, 134, 149, 150, 157–8, 163–5, 170, 175, 181, 189, 194–5, 198, 205, 239
Read, John, 171
realgar, 10–11
rebis, 31, 212, fig. 24
Red Stone, the, *see* colours, alchemical
Rilke, Rainer Maria, 55, 184–5
Ripley, Sir George, 34, 43, 88, 89, 100, 153, 163; *Cantilena*, 34, 139, 140, 157, 197–8, 199, 207, 209; *The Compound of Alchymy*, 32–41, 51, 54, 81, 83, 84, 87, 88, 94, 99–100, 102, 110, 126, 130, 144, 149–50, 151, 163, 170, 175, 189, 194–5, 203, 223, 232–3, 239, fig. 3; *Medulla Alchemiae*, 34; *The Mystery of Alchymists*, 34; the 'Scrowle', 34; *Vision*, 34, 94
ros caeli, *see* symbols, alchemical (dew)
Rosarium Philosophorum, 26, 94, 115, 136, 143, 164, 205, 212, figs. 8, 23
Rosenberg, Count, *see* Vok, Vilem
Rosencreutz, Christian, 87, 88–9, fig. 26
rosewater, 73
Rosicrucians, 15, 83, 87–90, 228; manifestos, 87–8, 90
Rowse, A. L., 76
Royal College of Physicians, 66–7
rubedo, *see* colours, alchemical
Rudolf II, Emperor, 19–21, 45, 65, 85, 88
Ruland, Martin, 125

saccharum plumbi, 66, 84
St Albans *Gesta Abbatum*, 33–4
Sartre, Jean-Paul, 121–2

Saturn, 26, 83, 179, 194
Scheutz, Michael, *see* Toxites
Scot, Reginald, 8, 9, 102
seed, metallic, 100, 157, 191, 195–6
Sendivogius, Michael, 28, 31, 85–6, 90, 92, 95, 100–2, 124, 152, 153, 157, 172, 180–1, 183, 189, 195–6, 198, 205
separation, 5, 35, 37–8, 41, 126, 136, 149, 191
Seton, Alexander, 15, 85
Severinus, Peter, 64
Shakespeare, William, 6, 8, 14, 15, 19, 23, 35, 70, 75–8, 106, 143, fig. 32; *All's Well That Ends Well*, 74–5; *Antony and Cleopatra*, 104; *Coriolanus*, 143; *Cymbeline*, 225, 227–36; *Hamlet*, 121–2, 142, 236, 238–9; *Henry V*, 74, 145, 236–8; *Henry VIII*, 90; *Julius Caesar*, 104; *King John*, 104; *King Lear*, 6, 41, 67, 143–224, 225, 227, 228, 232; *Macbeth*, 70, 75; *The Merry Wives of Windsor*, 78–80; *A Midsummer Night's Dream*, 22, 238; *Othello*, 143; *Pericles*, 75, 90, 225, 235; *The Rape of Lucrece*, 76; *Richard II*, 18; *Romeo and Juliet*, 72–3, 75, 78; *The Sonnets*, 73–4, 78, 104, 105, 156; *The Tempest*, 14, 115, 182, 225, 236, 238–9; *Timon of Athens*, 106, 143; *The Winter's Tale*, 27, 183, 225; *Venus and Adonis*, 74, 76, 78, 125
Sidney, Sir Philip, 15, 43, 66, 105
Siebmacher, Johann, 26, 126, 134, 205–6
silver, *see* Luna
similia similibus, *see* homoeopathy
skrying, *see* magic, angel
Soerensson, Peter, *see* Severinus
Sol, 9, 10, 45, 51, 53–4, 97, 105, 108, 116, 134, 150, 163, 175, 190, 214, 230
solution, *see* dissolution
solve et coagula, 28–9, 32, 37, 39, 93, 109, 127, 128, 138, 143, 147, 149, 168, 195, 225
Southwell, Robert, 103
spagyrites, 62, 68
Speculum Veritatis, 151, fig. 10
spirit, 3–4, 5, 6, 13, 29–32, 37, 46–9, 51, 52, 59–60, 66, 68, 74, 77, 78, 81, 92–6, 98, 100, 111, 128–9, 136, 148, 168, 170–1, 204–5, 207, 211, 217, 222, 238–9
spiritus mundi, 46–7
Sphinx, the, 51
Stanley, Ferdinando, fourth Earl of Derby, 17, 77
Stationers' Register, 76
Stillman, John Maxson, 2

Stolcius, Daniel, 171, 212, figs. 9, 16, 18, 24
Stone, Philosophers', 2, 5, 10, 15, 21, 26, 27–8, 29–31, 32, 34, 35, 39, 40, 41, 45, 48, 52, 54, 75, 81, 83, 84, 85, 86, 95–6, 97, 98, 106, 111, 112, 114, 123–4, 127, 128, 136, 141, 150, 157–8, 164, 169–70, 176, 195, 197, 211–17, 227–8, 230
sublimation, 4, 5, 10–11, 28–30, 35, 40, 104, 109–11, 148, 149, 182
sulphur, 8, 31–2, 39, 46, 52, 60, 96, 97, 99, 100, 138, 140, 151, 189, 220, 233, fig. 10
sulphuric acid, 4, 38, 94
sun, *see* Sol
symbols, alchemical: basilisk, 96, 164; bird, 28, 92, 172, 232; blood, 91–2, 97; castle, 4, 34–5, 41, 83, 84, 88, 91, 126, 176; cave, 225, 227–31, figs. 26, 27, 28; cocatrice, 96; coitus, 39, 91, 138–40, 197–8, 232–6, figs. 8, 29, 30, 31; crow, 39, 97, 99, 228, fig. 18; dew, 46, 53, 87, 93, 95, 139, 140, 205–6, 209, 225; dove, 88; dragon, 36, 38, 91–2, 93–5, 96, 97, 99, 113, 151, 160, 163–5, 166, 168, 171–2, 182, 190, 198, 199, 207, 209, 212–14, 215, 228, figs. 13, 22; eagle, 28, 30, 51, 88, 91, 92, 96, 97, 228; egg, 34, 45, 51, 239; garden, 91, 140; golden fleece, 27–8, 98–9; grapes, juice of, 94; grave, 91, 232–3; hart, flying, 97; hen, 228; honey, 93; king, 4, 89, 91, 94, 96, 98, 110, 138–41, 157–60, 165, 168, 171–2, 175, 181, 182, 189, 197–9, 205, 206–7, 212–16, 223, 228, 230, figs. 9, 11, 12, 15, 16, 17, 23, 24; lion, 38, 89, 93–4, 96, 97, 99, 140, 163, 209, 228, 233–5; marriage–bed, 91, 197, 233; milk of the virgin, 93, 94, 97, 209, 213–14; mother, 196–200, 204, 213–14, 215; oven, 91, fig. 17; palace, *see* castle; panther, 97, 99–100; peacock, 99–100; phoenix, 28, 83–4; prison, 138, 140; queen, 89, 91, 140, 143, 170, 203, 223–4, 230, figs. 14, 20, 21, 22; rain, silver, 46, 93, 95, 205–7; raven, 88; red man, 41, 97, 141, 150, 158, 212; rose, 89, 91–2; salamander, 28, 96, 163; scarab beetle, 51; serpent, 29, 30, 36, 93, 163, 171–2, 175, 182, 199, 205; stag, fugitive, 180; swan, 91, 92; toad, 36, 94, 96, 97, 163; tree, 97; white woman, 41, 97, 141, 203; wheel, 41, 100, 101, 144, 147–51, 152, 164, 223, fig. 10; wolf, 91, 171–2, 175, fig. 15

syphilis, 57, 77–8

Tantalus, 217
tartar, 60, 84
Tate, Nahum, 177
Taylor, F. Sherwood, 35
Themis, 233
Theseus, 184
Thölde, Johann, 83, 87–8
Thorndike, Sybil, 178
Thurnheisser, Leonhardt, 64, 207, fig. 20
Tincture, see Stone, Philosophers'
touchstone, 107, 182
Tower of London, 12, 16–17, 153
Toxites, Michael, 62
Tractatus Aristotelis, 171
Tractatus Aureus, see Hermes Trismegistus
tragedy, 141–7, 149, 151, 166, 217–18
transmutation, 1, 4, 7, 11, 21–2, 26, 27, 28, 41, 43, 51, 54, 62, 69, 85, 95–6, 102, 105, 112, 114, 122, 129–30, 134, 136, 148, 149, 153, 154, 158–9, 164–5, 171, 205, 215–16, 217, 222, 233
Traversi, Derek, 225
Trevisan, Bernard, 98
Trismosin, Salamon, 28, 158
Trithemius, Johannes, 44, 50, 56
True Chronicle History of King Leir, 152
Turba Philosophorum, 138, 205, 207, fig. 21
Turner, Charles, fig. 32
Turner, William, 65
turris circulatorius, 97–8
tutty, 97
Tymme, Thomas, 15, 44, 64, 153; translation of *Monas Hieroglyphica*, 42, 44, 45–6, 90, 92, fig. 4; *The Practise of Chymicall Physicke*, 5, 6, 44, 85, 90, 109–10, 124, 125, 131, 132, 152, 154, 175, 196, 205, 218–23, fig. 25

urine, 11, 180

Valentine, Basil, 6, 81–3, 87–8, 92–3, 152, 170, 207; *Triumphwagen*

Antimonii, 4–5, 83, 84–5, 88, 90; *Zwölff Schlüssel*, 83–4, 87–8, 89, 91, 92, 93, 96, 129, 153, 158, 163, 168, 171, 195, 236
Vallensis, Robert, 99
vapour, *see* spirit
Vaughan, Thomas, 196–7
Venus, 9, 10–11, 26, 114, 115
Verus Hermes, 165
vessel, the Hermetic, 4, 5, 6, 30, 32, 34, 35, 45, 48, 49, 53, 91, 106, 136, 138, 200, 206, 239
Villiers, George, first Duke of Buckingham, 9
Virgil, 183, 230
virtues, 57–8, 59, 60, 62, 64, 69, 73, 129, 165, 204, 205, 222
Visio Arislei, 138–40, 198, 233
vitriol, 4, 38, 60, 84, 94, 163, 164, 171, 175, 220; the 'vitriol acrostic', 4, 175–6, 228
Vok, Vilem, Count Rosenberg, 19, 21
volatility, *see* spirit
Vulcan, 61, 62, 69, 86, 92, 100, 102, 132, 151, 207, 215, fig. 10

Waldkirch, Conrad, 64
Walsingham, Sir Francis, 66
Webster, John, 78
West, Alice, 9
West, John, 9
whiteness, the, *see* colours, alchemical
Whitgift, John, 21
Wilton House, 15, 66
Wittelsbach, Ernst of, 58
Wootton, Henry, 70
Wriothesley, Henry, third Earl of Southampton, 68
Württemburg, Friedrich, Duke of, 85, 99

Yates, Frances A., 44, 87–8, 89, 90
Yazid, Khalid ibn, 23, 25

Zetzner, Lazarus, 86–7, 89
Zosimos of Panoplis, 49, 136, 138